MATERIALS SCIENCE AND TECHNOLOGIES

INJECTION MOLDING: PROCESS, DESIGN AND APPLICATIONS

MATERIALS SCIENCE AND TECHNOLOGIES

Additional books in this series can be found on Nova's website under the Series tab.

Additional E-books in this series can be found on Nova's website under the E-book tab.

MATERIALS SCIENCE AND TECHNOLOGIES

INJECTION MOLDING: PROCESS, DESIGN AND APPLICATIONS

PHOEBE H. KAUFFER
EDITOR

Nova Science Publishers, Inc.
New York

CONTENTS

For permission to use material from this book please contact us:
Telephone 631-231-7269; Fax 631-231-8175
Web Site: http://www.novapublishers.com

Additional color graphics may be available in the e-book version of this book.

LIBRARY OF CONGRESS CATALOGING-IN-PUBLICATION DATA
Hardcover ISBN: 978-1-61761-307-4
eBook ISBN: 978-1-61761-420-0

Injection molding : process, design, and applications / editor, Phoebe H. Kauffer.
p. cm.
Includes bibliographical references and index.
ISBN 978-1-61761-307-4 (hardcover)
1. Injection molding of plastics. I. Kauffer, Phoebe H.
TP1150.I554 2010
668.4'12--dc22

2010026942

Published by Nova Science Publishers, Inc. † New York

PREFACE

Injection molding is one of the most versatile and important manufacturing processes, capable of mass-producing complicated plastic parts in a variety of complex shapes with high dimensional precision. It is a major processing technique for converting thermoplastic and thermosetting materials with the aid of heat and pressure into complicated parts, consuming worldwide approximately 32% of all plastics. This book presents current research data in the study of injection molding from across the globe, including an overview of injection molding as a manufacturing technique for pharmaceutical applications; melt/solid weldline in over injection molding; metal injection molding of Co for biomedical applications; and the application of ultrasonic technology in the injection molding process.

Chapter 1- Injection molding (IM) originated from and has been widely used in the plastic processing industry for numerous applications, but is a relatively new technique to the pharmaceutical industry and academia as drug delivery technology. Lately, the interest in IM has grown increasingly for biomedical and pharmaceutical applications, since IM offers several advantages over traditional pharmaceutical processing techniques. This is reflected in the increasing number of publications and patents found in the scientific literature. Moreover, as via the advent of high through-put screening in the drug discovery process more than 40% of new chemical entities found exhibit poor solubility characteristics, it is a major challenge for formulation scientists to increase the solubility of such compounds. IM and hot-melt extrusion offer the chance to circumvent this problem by producing solid dispersions or solid solutions resulting in improved drug bioavailability. In addition, IM has been applied to provide controlled or modified drug delivery via the homogeneous embedding of drug particles in release-controlling polymers. This book chapter reviews materials and processing aids used by this technology and discusses various medical applications such as the production of medical devices, (biodegradable) medical implants and stents, bone-analogue composites, tissue-engineered scaffolds and vaginal rings. This review focuses on different pharmaceutical drug delivery systems including controlled release delivery devices, sustained release matrices and IM capsules with respect to their physicochemical characterization. The possibility of a continuous operation system with few processing steps, the potential of automation and reduction in labor cost, the availability of an increasing number of active pharmaceutical ingredients and polymers demonstrates that IM is a versatile production technology for the pharmaceutical industry, offering exciting prospects for the future.

Chapter 2 - Sequential injection molding, also called over-injection molding (OIM), which offers in situ process to join thermopolastics has demonstrated considerable importance in the field of polymer processing due to short manufacturing times and low cost. In comparison with the conventional injection molding, sequential overinjection molding includes multi-stage operation: the first part of desired figure is molded by one-stage injection, and then the melt for the second part is subsequently injected over the surface of the first part molded in the previous process. So the bonding characteristics of the interface between the first and second part are very important for the molded products. Some works have been reported on the studies of the relationships between interfacial adhesion and processing parameters used in sequential injection molding.

Combined with the special thermomechanical environments in the vicinity of the interface formed during injection molding, the complex crystalline structures can be obtained at semicrystalline polymer interfaces due to the fountain flow along injection direction. Therefore, it is important to study the interfacial crystalline structure and its effect on the final adhesion of overinjection molded semicrystalline polymer interfaces.

In this part, the thermomechanical environments, such as melt temperature, mold temperature, injection pressure and injection speed will be changed to investigate the diverse interfacial crystalline structures and their morphology development of PP/PP overinjection molded interface. Based on the experimental results, a formation mechanism of diverse interfacial crystalline structures will be proposed. In addition, the relationship between interfacial crystalline structures and interfacial bonding strength will be discussed.

Chapter 3 - Smoothed Particle Hydrodynamics (SPH) is a meshless and fully Lagrangian method. It has been widely applied in simulating fluid flows with complex moving boundaries, such as free surfaces, and dynamic response of materials involving large deformation, fracture and fragmentation. In the simulation of flow problems this method usually solves the conservation equations explicitly and the stable results can be obtained only for momentum dominated flows with low viscosity and pressure. However, in polymer processing, the fluid is non-Newtonian with the viscosity being as high as $O(10^3)$ to $O(10^4)$ *Pa-S*, and under pressure which can be as high as $O(10^6)$ to $O(10^{10})$*Pa*. The standard SPH algorithm is infeasible in this case.

In this chapter, the authors give a concise review of SPH with emphasis on the SPH formulation, stability problems and enforcing boundary conditions; and describe how to apply SPH to the simulation of injection molding flows. The viscosity of the fluid is modeled by a power-law function. The fluid is considered compressible under high pressure. The pressure of fluid is calculated in terms of realistic state equations for polymer melts, such as the Tait equation, Sun-Song-Yang's equation or a weakly compressible equation. To relieve the stability constraints on the time step, the authors propose a semi implicit SPH formulation for high viscous non-Newtonian flow, so that a reasonable time step can be adopted in simulations. An iterative algorithm is used to solve the resulting system equations of the SPH formulation. This method is completely matrix-free and robust.

It has been known that inconsistency and tensile instability are the intrinsic weakness of standard SPH. For a highly viscous flow under high pressure, the SPH system becomes highly unstable. To make the system more stable, we use an artificial pressure to stabilize the SPH system thus avoiding the instability.

The present method is firstly tested in simulating simple flows: the transient slit flow and simple shear flow. The simulated flow profiles are in good agreement with the analytical solutions. Secondarily, the fountain flow is simulated. The simulated free surfaces for the Newtonian and shear-thinning fluids are in agreement with those of finite element simulations. Finally we apply the method to the simulation of the moulding flows of Newtonian and non-Newtonian fluids. The authors obtain stable results for the power-law moulding flow with the highest zero shear rate viscosity being $1.22\times10^4 Pa\cdot s$ and the lowest power-law exponent 0.294, Reynolds number is 3×10^{-4} to 6×10^{-5}, and the highest pressure is in the range $O(10^8)$ to $O(10^{10})Pa$.

Chapter 4 - As one of the most widely employed methods, injection molding can be characterized by high productivity, high degree of automation and good dimensional stability. However, molding complicated parts, multigated mold cavities and cavities containing inserts may generate serious difficulties in terms of mold filling and final production especially. In fact, molding of such parts usually produces weldline once the melt fronts have joined either by impingement flow or around an insert [1, 2]. Weldlines also form when jetting occurs along the flow path where the flow is suddenly accelerated (e.g. near the gate) [3].

Weldlines can be categorized as cold and hot weldline [4, 5]. In previous one, two melt fronts meet head to head and additional flow could not take place after combination. This type leads lowest strength. Hot weldlines (stream lines or melt lines) form when two or more melt streams recombine to each other such as breaking up and rejoining of flow around pins. An additional flow after recombination occurs.

Chapter 5 - The cobalt alloys are used for surgical implant applications and their properties are influenced by the carbon content. They can be produced with different carbon contents. The increase in the carbon content results in an increase in hardness but also in a corresponding decrease in ductility. It is then important to find a proper combination of the carbon content and the processing route (sintering plus any post sintering treatments) to optimize mechanical properties. The strain hardening behavior of the material was underlined and justified by strain induced transformation of austenite into martensite. In addition, the role of carbides on both wear and corrosion resistance was unknown.

Chapter 6 - In the past four decades, the ultrasonic technology has been introduced for polymers, composites and polymer-metal joint welding. Since the ultrasonic energy can vibrate the polymer melts without additional heating, this clean, reliable and efficient technology was fast spread and developed in a variety of polymer related areas, such as melting, mixing and welding. With the boom of sensor, actuator and transducer since 1980's, the ultrasonic is also involved in this technological development trend. The special features of ultrasonic in refection, diffraction and interference accomplished its applications in displacement metering and crack detection. This chapter demonstrates an overview on the state of the art in application of ultrasonic technology, focusing on polymer processing fields, particularly injection molding process. High ultrasonic frequency can produce heat and oscillation in materials from the result of high frequency stresses, which sparks the initiation of applying it for injection molding process to homogenize and increase the dispersion of the molten material, either in the liquid stage or in the solidifying stage. The special designed ultrasonic transducer provides a valid method to monitor the polymer melts filling during injection molding process. Therefore, the focus aspects of this chapter are placed on describing the main application principles of ultrasonic technology in equipment innovation, parts properties improvement and process on line monitoring of injection molding process.

Chapter 7 - The injection mold is a high precision tool responsible for the production of most plastic parts used everywhere. Its design is considered critically important for the quality of the product and efficient processing, as well as determinant for the economics of the entire injection molding process. However, typically, no formal engineering analysis is carried out during the mold design stage. In fact, traditionally, designers rely on their skills and intuition, following a set of general guidelines. This does not ensure that the final mold design is acceptable or the best option. At the same time, mold makers are now highly pressured to shorten both leading times and cost, as well as to accomplish higher levels of mold performance. For these reasons, it is imperative to adopt new methods and tools that allow for faster and higher integrated mold design. To that end, a new global approach, based on the integration of well-known quantitative techniques, such as Design for Six Sigma (DFSS), Structural Equation Modeling (SEM), Axiomatic Design (AD) and Multidisciplinary Design Optimization (MDO) is presented. Although some of these methods have been largely explored, individually or in combination with other methodologies, a quantitative integration of all aspects of design, in such a way that the whole process becomes logical and comprehensible, has not yet been considered. To that end, the DFSS methodology, through its IDOV roadmap, was adopted. It is based on the ICOV Yang and El-Haik proposal, establishing four stages for the design process: *Identify*, which aims to define customers' requirements/expectations; *Design*, where the creation of a product concept, and its system-level design, is performed; *Optimization*, in which all the detailed design, through product optimization, is handled; and finally, *Validation*, where all product design decisions are validated, in order to verify if the new designed entity indeed meets customer and other requirements. As a result, this approach tackles the design of an injection mold in a global and quantitative approach, starting with a full understanding of customer requirements and converting them into optimal mold solutions. In order to validate it, an integrated platform was developed, where all different analysis modules were inserted and optimized through an overseeing code system. The results attained highlight the great potential of the proposed framework to achieve mold design improvements, with consequent reduction of rework and time savings for the entire mold design process.

Chapter 8 - In the present work the authors have studied the interfacial adhesion characteristics of bi-layer structures obtained by over molding. The main injection molding parameters, such as injection temperature, injection and holding pressure and injection speed have been modified to the extent permitted by both the process and materials. Several surface treatments were applied to the base substrate to evaluate the effect of the surface roughness on the adhesion level obtained with the over molding process. In addition a coupling agent (PP-g-MA) was added to the polymer to improve the adhesion level. The results showed that the adhesive fracture toughness increased as temperature and surface roughness were raised. Depending on the selected parameters, different types of failure, ranging from adhesive to cohesive, were noticed. In metal-polymer systems, maximum levels of adhesion were obtained when the coupling agent was used and over molding was performed on previously torch-heated metal plates. The creation of chemical bonds between polymer and metal leads to maximum adhesion strength and cohesive type failure. For the case of polymer-fabrics an experiment design was carried out and statistical analysis was performed. Results showed that increased adhesion was achieved at lower injection temperatures.

In: Injection Molding: Process, Design, and Applications ISBN: 978-1-61761-307-4
Editor: Phoebe H. Kauffer

Chapter 1

OVERVIEW OF INJECTION MOLDING AS A MANUFACTURING TECHNIQUE FOR PHARMACEUTICAL APPLICATIONS

T. Quinten, T. De Beer, J.P. Remon and C. Vervaet
Laboratory of Pharmaceutical Technology,
University of Ghent, Ghent Belgium

ABSTRACT

Injection molding (IM) originated from and has been widely used in the plastic processing industry for numerous applications, but is a relatively new technique to the pharmaceutical industry and academia as drug delivery technology. Lately, the interest in IM has grown increasingly for biomedical and pharmaceutical applications, since IM offers several advantages over traditional pharmaceutical processing techniques. This is reflected in the increasing number of publications and patents found in the scientific literature. Moreover, as via the advent of high through-put screening in the drug discovery process more than 40% of new chemical entities found exhibit poor solubility characteristics, it is a major challenge for formulation scientists to increase the solubility of such compounds. IM and hot-melt extrusion offer the chance to circumvent this problem by producing solid dispersions or solid solutions resulting in improved drug bioavailability. In addition, IM has been applied to provide controlled or modified drug delivery via the homogeneous embedding of drug particles in release-controlling polymers. This book chapter reviews materials and processing aids used by this technology and discusses various medical applications such as the production of medical devices, (biodegradable) medical implants and stents, bone-analogue composites, tissue-engineered scaffolds and vaginal rings. This review focuses on different pharmaceutical drug delivery systems including controlled release delivery devices, sustained release matrices and IM capsules with respect to their physicochemical characterization. The possibility of a continuous operation system with few processing steps, the potential of automation and reduction in labor cost, the availability of an increasing number of active pharmaceutical ingredients and polymers demonstrates that IM is a versatile production technology for the pharmaceutical industry, offering exciting prospects for the future.

1. Introduction

Injection molding is one of the most versatile and important manufacturing processes capable of mass-producing complicated plastic parts in a variety of complex shapes with high dimensional precision. It is a major processing technique for converting thermoplastic and thermosetting materials with the aid of heat and pressure into complicated parts, consuming worldwide approximately 32 %w of all plastics. Basically, IM is considered as a repetitive, cyclical process in which polymers are heated and gradually molten in the injection unit of an injection molding machine, which resembles in principle an extrusion process. This molten material is then transferred by means of an injection step into a closed and shape-specific mold cavity, basically duplicating the cavity of the mold. After solidification, the article is recovered by opening the mold to release the product. The process is closely related to hot-melt extrusion (HME), a process in which materials are molten and then passed through a die to form an extrudate upon cooling. The injection molding cycle is however more complex than hot-melt extrusion, since it involves moving and stopping the melt into the mold rather than having a continuous flow of melt without interruptions. And in contrast to HME where the material is shaped as it is pushed through a die, during IM the final product is obtained in a cooled mold made of two parts that alternatively open and close.

The process was patented in 1872 by John and Isaiah Hyatt, who build the first injection molding machine (IMM) to mold camphor-plasticized cellulose nitrate. The development and optimization of this technique was strongly accelerated due to the huge demand for inexpensive, mass-produced parts during World War II [1]. In 1946, James Watson Hendry and Beck (1952) developed the first screw injection molding machines, still accounting for the vast majority of all injection molding machines, as this design allowed a more precise control over the speed of injection and the quality of articles produced [2]. In addition, a screw-IMM allows materials to be compounded, so that e.g. active pharmaceutical ingredients (API), functional excipients and processing aids can be mixed thoroughly with thermoplastic polymeric carriers before being injected into parts. A further development involved the processing of thermosetting materials. In the 1970s, gas-assisted injection molding technology was introduced, which permitted the production of complex, hollow articles that cooled quickly. This advancement greatly improved design flexibility as well as the strength and surface finish of manufactured parts while reducing production time, cost, weight and waste. Modern IMM enable completely automated processing while production parameters such as temperature, pressure and time allocated for each stage are all precisely controlled [3].

Although injection molding found its origin in the plastic processing industry, the pharmaceutical industry has grown more and more interested in IM as processing technology for medical applications due to the tight competition in the fast growing market of the health care industry all over the world. Increasing research and development costs, expiry of patents, the lack of new drug products in the pipeline, product life cycle management, the recent global financial crisis combined with economical recession and budget pressure on public health systems have all arisen the need for innovative pharmaceutical production technologies today. Innovation will become more and more important for sustaining the productivity and profitability of the pharmaceutical industry. With respect to these issues, IM as novel processing technology permits the production of high-quality novel drug delivery systems in a cost-effective way which fulfill the strict regulatory requirements and increasing demands set by the EMEA, FDA and other regulatory authorities [4]. Regarding product life cycle management, novel drug delivery systems can be prepared via IM for already marketed drug products in order to increase the market value, competitiveness and patent life. Additionally, these dosage forms became extremely popular due to their relatively low development cost and time required to introduce the product to the market as compared to new chemical entities [5].

Injection molding offers numerous advantages: a) a simple, continuous set-up with few production steps that can easily be scaled-up, since mixing, melting, homogenizing and shaping of the product occur in a single step, b) a reduction in labor forces due to the extensive automation of the process c) elimination of solvent residuals or a time-consuming drying step as solvents or water are not used, d) large-scale industrial production at a high production rate by implementing multi-cavity molds, e) design flexibility allowing the manufacturing of various shapes, and f) minimal material loss make IM a viable, competitive and worthwhile alternative for a variety of pharmaceutical and biomedical applications as compared to conventional production technologies commonly applied in the pharmaceutical industry. Additional benefits include that low-dosed compounds may be dissolved in the molten carrier ensuring content uniformity of the product, that drug particle shape and size exert a minimal effect on the production of e.g. tablets, and that the compressibility of the drug can be neglected [6].

The disadvantages of IM are mostly cost-related such as the investment in equipment, facilities, high startup and running costs and the need for a specific mold design.

The increasing number of thermo-stable active pharmaceutical ingredients and the variety of thermoplastic polymers introduced to the market allow the processing of a wide range of materials, offering a promising future for injection molding as pharmaceutical production technology. In order to successfully design IM drug delivery systems, the selection of the pharmaceutical grade polymer is critical as it often dictates the processing conditions and drug release characteristics. All components must be thermally stable at the processing temperature during the whole injection molding cycle and the polymer needs to be processed at relative low temperatures to ensure the stability of the API in the final dosage form, limiting the formulation scientist sometimes in his options. The physicochemical properties of the thermoplastic carrier materials, as well as drug substances and other excipients (e.g. formulation additives, processing aids, release modifying agents, etc.) must be carefully analyzed and characterized since they may have a profound impact on the processability, product stability and performance (e.g. drug release) [7,8]. The physical and chemical stability of the IM dosage form is influenced by the nature of the polymer and excipients, the physical state of the drug in the final dosage form, and the storage and packaging conditions [9].

Currently, injection molding has been applied for a plethora of pharmaceutical/medical applications varying from the development of oral drug delivery dosage forms to the design of complicated stents or implants [10,11]. In terms of drug delivery, the oral drug administration still remains the route of choice for the majority of clinical applications. During the past 30 years, extensive research has been performed on the development of modified and controlled-release drug delivery systems, offering many advantages over traditional dosage forms such as more desirable drug concentration-plasma levels resulting in a better therapeutic efficiency. One approach that has received considerable attention as a means of prolonging drug release has been the incorporation of drug in solid polymers. These polymers are molten during processing, and act as thermal binders and drug release retardants upon cooling and solidification, allowing API release over longer periods in a highly controlled manner [12,13].

Some drugs possess ideal characteristics for absorption throughout the gastrointestinal tract, whereas others present difficulties such as low solubility or limited permeability. Since the advent of high through-put screening and combinatorial chemistry in the drug discovery process, more than 40% of the new chemical entities exhibit poor solubility characteristics. It still remains a major challenge for formulation scientists to increase the solubility of such compounds, however, IM and hot-melt extrusion offer the chance to circumvent this problem by producing solid dispersions or solid solutions resulting in improved drug bioavailability [6].

The high precision of the IM process, the ability for mass production due to short production time, and the possibility to accurately control the three-dimensional structure of the product, favour IM as an ideal candidate for the production of a variety of controlled-release drug delivery systems, medical devices [14], (biodegradable) medical implants and stents [10,11], bone-analogue composites [15], tissue-engineered scaffolds [16] and vaginal rings [17]. The diversity of these applications in the pharmaceutical and (bio)medical field demonstrates the potential of injection molding as drug delivery technology, which increasingly gained importance as evidenced from the scientific output and patents issued during the last years.

2. Materials Used in Injection Molding

Innovation in pharmaceutics is mainly presented by the discovery and development of new APIs. However, formulation development is an integral part of drug discovery and development. These formulations are prepared for drug compounds at both discovery stage (discovery lead) and preclinical stages (preclinical lead), and are administered via various routes (e.g. oral and intravenous) to animals after extensive in-vitro testing. They aim at evaluating these new compounds on a broad range of pharmaceutical interests such as pharmacology, pharmacokinetics and toxicology [18]. The last step in the drug development process is the clinical trial step (on humans), if successful followed by the introduction of the drug or device on the market. Studies have estimated the cost per new drug to vary between 500 million dollar and more than 2 billion dollar, depending on the therapy or the developing firm [19]. This is an issue that should be addressed by identifying new and better ways to improve efficacy and effectiveness of drug discovery and clinical trials. Besides, the development of new drug delivery systems as well as the exploitation of innovative drug delivery routes (e.g. transdermal, pulmonary, nasal, etc.) and/or processing technologies may provide methods to improve the efficacy and/or safety of the APIs [20].

These drug delivery systems are often complex mixtures of active pharmaceutical ingredients and functional excipients, such as meltable carriers, plasticizers and other processing aids in case of IM. The meltable carrier used in injection molding is usually a thermoplastic polymer. Polymers are a wide class of natural or synthetic substances composed of very large molecules, having a wide range of mechanical, physical and chemical properties and which can be easily manufactured into complicated shapes. Thermoplastics become soft when they are exposed to heat and harden when they are cooled regardless of how often this process is repeated.

In spite of this diversity, all materials must have a certain degree of thermal, physical and chemical stability, and must meet high levels of purity and safety, meeting the strict regulatory requirements, e.g Generally Recognised As Safe status (GRAS), Good Manufacturing Practice (GMP) and Environmental, Health and Safety (EHS) guidelines. Furthermore, these materials should easily melt inside the extruder and solidify quickly in the mold, resulting in short injection molding cycles. Therefore, polymeric materials suitable for thermal processing require either a low glass transition temperature, or a low melting point in case of semi-crystalline polymers. The availability of a broad selection of various polymers and functional excipients for injection molding applications, offers an interesting prospective for continuous research, formulation and drug development.

2.1. Carriers

The selection of the carrier material for pharmaceutical drug delivery systems strongly depends upon the application. As mentioned in the introduction, the choice of the pharmaceutically approved polymer is critical in the formulation process as its properties not only dictate the processing conditions, but also govern the dissolution characteristics of the dosage form (drug release kinetics and release mechanism). For example, after oral administration of a sustained release device, the entire drug content should preferentially be released within 24h. In contrast, medical implants, stents and vaginal rings should slowly deliver the API over a period of weeks or even months, illustrating the need for a wide range of different polymers with specific characteristics capable of tailoring drug release according to the particular application of the specific drug delivery device. Moreover, these carriers should be pharmacologically inactive and nontoxic.

In general, a number of design options are available to control or modulate drug release from a drug delivery system. The majority of oral controlled-release delivery systems manufactured via IM fall in the category of matrix or reservoir systems. From a technological point of view, a matrix (or monolithic) drug delivery system can be defined as a system that controls the drug release from a polymeric or matrix agent, wherein the drug is uniformly dissolved or dispersed. These systems include both swellable and non-swellable matrices and the solid polymer can be either lipophilic or hydrophilic.

In matrix systems, the drug is homogeneously embedded in a polymeric matrix and is released via (Fickian) diffusion, swelling (in case of hydrophilic polymers) and/or erosion (process of material loss from the polymer bulk) [21]. Any or all of these mechanisms may occur in a given release system depending on the matrix properties, generating slow drug release to the biological environment in a highly reproducible way.

In contrast, reservoir drug delivery devices consist of a core containing solid drug particles, which is totally or partially enclosed by an inert membrane. This membrane surrounding the drug core controls drug release. Drug release from matrix or reservoir systems can be affected by physiological factors like pH, fed/non-fed condition which affects gastric emptying [22,23].

Biodegradable materials have been considered for various uses, including agricultural, packaging and medical applications. In particular biomedical (e.g. bone screws, bone plates, contraceptive reservoirs, staple structures, nano- or micro-sized drug delivery vehicles) and tissue-engineering (e.g. plain membranes for guided tissue regeneration and multifilament meshes or porous structures for tissue engineering) applications often use these biodegradable materials as they are easily processed via HME or IM. Bioerodible polymers are based on their ability to degrade in an aqueous environment by hydrolytically- or enzymatically-induced polymer chain scission by which polymers are cleaved into oligomers and monomers. Therefore, medical devices produced with these materials do not require surgical removal after a period of time [24].

Mathematical modeling of the drug release can yield important information about the mass transport processes that are involved in the release of therapeutic agents from all these drug delivery systems. It can provide insight in the release mechanisms involved in the drug release process and can be helpful for the optimization of existing drug delivery systems or the development of new systems [25,26].

The processing conditions of a thermoplastic carrier system are determined by its chemical stability and physical properties such as glass transition temperature, molecular weight and melting point (in case of a semi-crystalline polymer). During processing, polymers are subjected to thermal stress due to the relatively high production temperature used to soften the material and to mechanical shear stress created by the action of the screws. This material is even subjected to more stress during the injection step, as high shear rates are generated while the material moves in the (cold) runner towards the mold. In order to limit temperature increases from viscous heating and to facilitate mold filling, low viscosity grade polymers are usually preferred. All these stresses can compromise the stability of the polymers, resulting in polymer chain scission, chemical depolymerization, unzipping of substitute groups, thermal and/or oxidative degradation; hence limiting its use. Crowley et al. (2002) studied in detail the chemical/thermal stability of polyethylene oxide in function of storage, molecular weight, screw speed and process temperature for sustained-release tablets prepared by melt extrusion, identifying that all these factors affected polymer stability (figure 1). In addition, the type of plasticizer and antioxidant also influenced both PEO stability and drug release [27].

The compatibility between the active substance and carrier is also a major parameter, as it not only influences the drug release profile but also the stability of the system during and after processing [28]. This is of particular importance for drugs in the amorphous state, that are physically unstable as they are susceptible to recrystallisation. Forster et al. (2001) described the formation of glass solutions of the poorly soluble drugs indomethacin, lacidipine, nifedipine and tolbutamide with polyvinylpyrrolidone (PVP) and polyvinylpyrrolidone-co-vinyl acetate by melt extrusion. All these drugs exhibited H-bonding with PVP between the carbonyl group of the pyrrole ring of the polymer and a H-donor group of the drug and it was speculated that differences in the physical stability of drug/polymer extrudates were due to differences in H-bonding between the components [29].

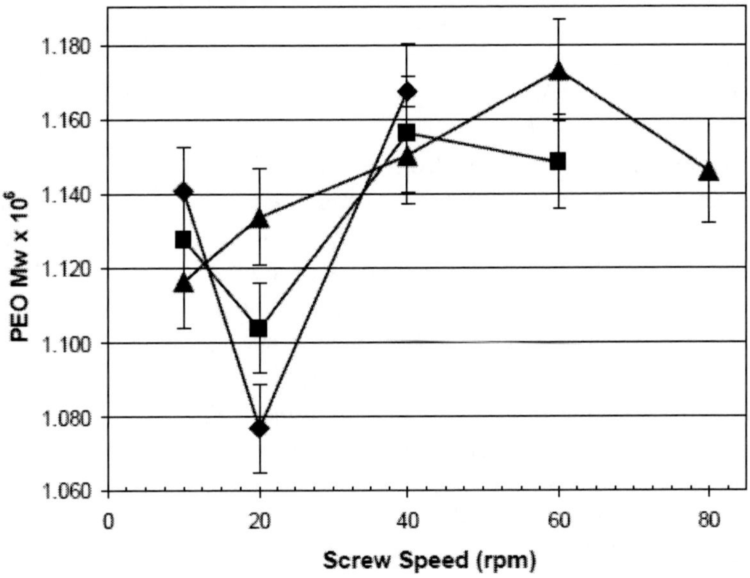

Figure 1. Influence of extrusion temperature and screw speed on the weight average molecular weight of PEO 1M, following extrusion of a formulation containing 80% PEO 1M and 20% chlorpheniramine maleate (CMP). Temperature of the different zones of the extrusion barrel (from feed zone to die: (◆) 70, 80, 100 and 105°C; (■) 80, 90, 110 and 120°C; (▲) 85, 100, 120 and 140°C. Each point represents the mean ± standard deviations, n=6; (obtained from Ref. 27)

Depending on the ultimate application of the medical device, chemical resistance, water absorption, oxygen transmission, elasticity, impact strength and heat deflection temperature may be among the important properties to be considered when selecting the appropriate plastic material with respect to its relative cost [30]. Moreover, medical devices which come into contact with the body or bodily fluids must always be sterilized before use. The sterilisation of medical devices from thermoplastic polymers can sometimes be problematic since common sterilisation techniques by means of steam, hot water or heat are not applicable, as they can cause plastic deformation/melting and extensive hydrolytic or oxidative degradation, all having a detrimental effect on the mechanical properties [31]. Therefore, medical devices are commonly sterilised by high energy ionizing radiation (gamma and electron beam) or ethylene oxide gas, causing polymer chain scission or the possible formation of by-products and/or residuals, respectively.

The properties of the polymeric carrier can have a profound impact on drug release. Omelzuck et al. (1992) illustrated that the release rate of theophylline, chosen as model drug, was influenced by the molecular weight of poly lactic acid (PLA) in melt extruded matrix tablets. The release of theophylline slowed progressively down as the molecular weight of PLA increased, however, above a critical chain length similar release profiles were obtained. This was attributed to a reduction in tablet swelling at higher molecular weight of the incorporated PLA grade, resulting in slower drug diffusion [32]. An effect of the polymer viscosity grade was also reported for injection molded matrix tablets composed of ethylcellulose, the highest EC viscosity grades provided faster drug release rates, attributed to a higher level of water-uptake, tablet swelling and erosion during drug dissolution [33].

Many injection molding applications make use of water-soluble, water-insoluble and/or biodegradable thermoplastics such as polylactide (PLA), polyvinyl alcohol (PVA), polyethylene vinyl acetate (EVA), polyglycolide acid (PGA), polycaprolactone (PCL), polyanhydrides, polylactide-co-glycolide (PLGA), polyethylene glycol (PEG) or cellulose-derivates such as ethyl cellulose (EC), hydroxypropyl methylcellulose (HPMC),

Recently, there has been an increasing interest in the nonfood use of biopolymers such as starch and soy, because of their total biodegradability and worldwide availability at low cost. Unfortunately, native starch has poor mechanical properties, and cannot be processed by melt-based methods without being degraded. Native starch can be transformed into a thermoplastic material through thermo-mechanical treatment in the presence of suitable plasticizers such as water, glycerol or lecithin. Thermoplastic corn starch was injection molded into biodegradable composites, containing chitine as the reinforcing phase resulting in materials with higher modulus and decreased elongation at break [34].

Another approach to improve the mechanical properties of starch involves blending with synthetic polymers such as PLA, PCL, EVA, This has potential for many biomedical applications, including scaffolds for tissue engineering of bone and cartilage, materials for bone fixation and replacement, controlled-release carriers for drugs and bioactive agents and hydrogels [35].

In this regard, Tummala et al. (2006) investigated the influence of two different plasticizers, glycerol and D-sorbitol on the mechanical properties of bio-plastics, derived from mixtures of soy protein plastic with polyester amide. Soy proteins were converted to soy protein plastics via extrusion with a plasticizer or cross-linking agent and the mechanical properties were considerably improved by mixing with a biodegradable polyester as its processing window matches that of soy protein plastic [36].

2.2. Plasticizers

Plasticizers are functional excipients primarily included in the formulation of hot-melt extruded drug delivery systems in order to improve the manufacturing conditions and the physical and mechanical properties of the final product, however, they are less commonly applied for injection molded applications. They are usually low molecular weight compounds that bind to the polymer by van der Waals forces and allow the polymer chain segments to have a greater free volume and higher mobility due to decreased intermolecular interactions. This results in a reduction of tensile strength, elastic modulus, polymer melt viscosity and glass transition temperature of the polymeric material. During hot-melt processes, even for short periods only, materials are subjected to high temperatures which may cause both drug and carrier degradation. Plasticizers act by decreasing the glass transition temperature (T_g) of the polymer, thus allowing lower production temperatures, which is desirable for temperature-sensitive materials and enhances the overall stability while processing. In certain cases, the processing of heat-labile substances such as peptides and/or heat sensitive drugs is possible. Plasticizers also reduce the shear forces necessary to extrude a polymer, improving the workability and flexibility [7,8].

Several prerequisites have to be fulfilled by plasticizers, namely good efficiency, polymer-plasticizer compatibility, thermostability, and volatility/permanence. The efficiency of a plasticizer refers to the amount of plasticizer necessary to lower the polymer T_g. Compatibility refers to the similarity in chemical structure between plasticizer and polymer, a higher similarity resulting in a better compatibility.

Permanence relates to the ability of the plasticizer not to evaporate from the system. During IM, plasticizers are subjected to elevated thermal conditions that could lead to physical destabilization through vaporization, thereby leading to undesirable effects. Low-molecular-weight plasticizers are generally volatile and tend to evaporate in their free state or from a plasticized product upon exposure to elevated thermal conditions. Moreover, for plasticizers with low boiling points, evaporation occurs continuously, even at low temperature, although it may be undetectable even by sensitive analytical techniques. Because of this plasticizer loss, the plasticized product will gradually lose its flexibility and distensibility [37]. Smaller plasticizer molecules have higher diffusion rates into the polymer matrix, which results in higher plasticizer efficiency. However, due to their size, they also present higher volatility and consequently reduced permanence. It is necessary to balance these two parameters when selecting which type of plasticizer is most suitable.

As plasticizers remain in the final product, they can alter the properties and performance of the dosage form. Several authors have demonstrated that including plasticizers in hot melt extruded dosage forms can affect the drug release properties. Thumma et al. (2008) reported that drug release from melt extruded PEO (PolyOx WSR N-80: PEO N-80) matrices was influenced both by the plasticizer type and concentration. A faster release of tetrahydrocannabinol-hemiglutarate resulted from the inclusion of water-soluble plasticizers (polyethylene glycol 8000, triacitin), and from higher plasticizer concentrations. However, slower release rates were observed for formulations containing increasing concentrations of water-insoluble plasticizers (vitamin E succinate, acetyltributyl citrate). Triethyl citrate did not affect drug release (figure 2) [38].

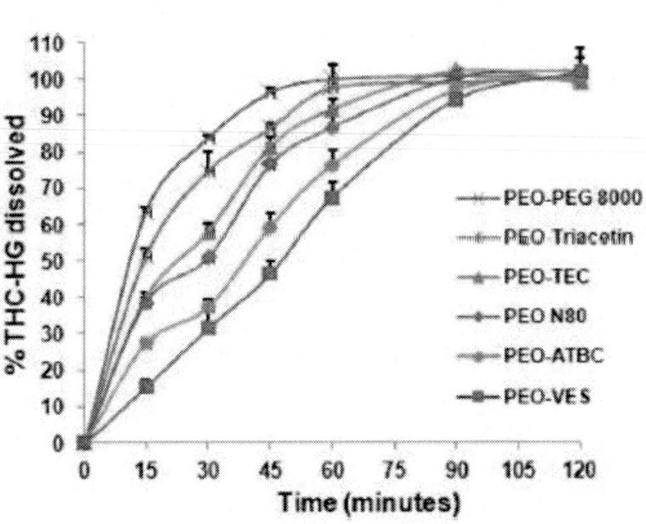

Figure 2. Effect of plasticizer type on the release of tetrahydrocannabinol-hemiglutarate (THC-HG) from PEO matrices (PEO N80: Polyox WSR N-80) (n=3). The films were fabricated at 110°C (Formulation: PEO/plasticizer/THC-HG 85/10/5, w/w). PEG 8000: polyethylene glycol 8000; TEC: triethyl citrate; ATBC: acetyltributyl citrate; VES: vitamin E succinate; (obtained from Ref. 38)

In contrast, some IM and HME applications use supercritical carbon dioxide (CO_2) as temporal plasticizer, which was effective in lowering the T_g but without being present in the final product. Carbon dioxide is volatile and escapes from the mold or die since it is gaseous upon expansion to atmospheric conditions. In this manner, biodegradable polylactide and water-soluble polyvinylalcohol were injection molded at low processing temperatures into foamed, highly porous and interconnected matrices for use as tissue-engineering scaffolds [39]. Verreck et al. (2006) studied the effect of pressurized CO_2 as temporal plasticizer on the hot melt extrusion of polyvinylpyrrolidone-co-vinyl acetate 64, Eudragit® E100 and ethylcellulose (20cps) obtaining foam-like structures which showed faster dissolution rates due to an increased specific surface area and porosity [40]. After characterization of the matrices Lyon et al. (2007) stated that the plasticization effect of CO_2 only occurred in the barrel of the extruder, and no viscosity reduction is observed after processing. These extrudates were thermally stable, and the use of CO_2 during the extrusion process resulted in faster drug release of carvedilol compared to unassisted HME, due to a higher internal surface area in the foam-like structure of the sample [41].

In addition, surfactants can be used as plasticizers as they can serve two purposes: plasticization of the polymer to aid polymer processing and solubilisation of the API to enhance bioavailability. Gebremeskel et al. (2007) demonstrated that various surfactants (Tween-80, docusate sodium, Myrj-52, Pluronic-F68 and sodium lauryl sulphate) showed a plasticizing effect on the processing of different hydrophilic polymers (PVP-K30, Plasdone-S630, HPMC-E5, HPMCAS and Eudragit L-100) with a poorly soluble API, thereby effectively lowering the T_g of the polymer-drug blends and reducing the melt viscosity during processing. Moreover, including surfactants in the formulation resulted in a considerable increase of the drug dissolution rate [42].

Moreover, many authors have reported that APIs included in a formulation and processed by a melt extrusion process exert a plasticizing effect on polymers, effectively lowering its glass transition temperature. For example, ibuprofen was a plasticizer for ethylcellulose as a solid solution was formed by means of HME, showing a single glass transition indicating completely miscibility between drug and polymer. The plasticizing efficiency was of the same magnitude as for the traditionally used plasticizers [43].

The effect of ibuprofen, chlorpheniramine maleate and metoprolol tartrate on the thermal, mechanical and diffusional properties of polyacrylate-based films was investigated and quantified [44]. The glass transition temperature of these films decreased significantly with increasing drug content, whereas the film flexibility and drug release rate increased, indicating that the three drugs were efficient plasticizers for Eudragit RS. Drug release was controlled by diffusion, for which an increase in the initial drug content resulted in faster drug diffusities and hence accelerated release rates.

Repka et al. (1999) investigated the effect of different plasticizers (polyethylene glycol 8000, acetyltributyl citrate, triethyl citrate and polyethylene glycol 400) and model drugs (hydrocortisone and chloropheniramine maleate) on the physico-mechanical properties and stability of hydroxypropylcellulose-based films. All drugs and plasticizers, except polyethylene glycol 400, exhibited excellent plasticizer qualities and showed an adequate drug stability in melt extruded films after 12 months of storage. The plasticizer type and production temperature influenced the drug stability and affected the physico-mechanical properties (tensile strength, percentage elongation, and Young's modulus) of the films over time [45].

Most pharmaceutical grade plasticizers are in a liquid state (e.g. organic esters (citrates and phthalates) and polyols (glycerol, low Mw PEG)), and a homogeneous blend of the plasticizer with a powder blend containing the active ingredient must be obtained before the solid mass is added to the extruder. An incomplete mixing of a polymer powder with a liquid additive has been shown to result in an unstable mass flow when feeding the mixture into the extruder [46]. In order to avoid this problem, the effect of a solid-state plasticizer (citric acid) on an acrylic polymer Eudragit® RS PO was investigated [47]. Citric acid monohydrate (CA MH) facilitated the extrusion of Eudragit RS PO, whereas the addition of anhydrous citric acid (CA) to the powder was less effective, attributed to the higher solubility of the monohydrate form in the acrylic polymer. Concentrations of up to 25% CA MH were soluble in extrudates, however the solubility depended upon the extrusion temperature. Extrudates with 25% CA MH required an extrusion temperature of at least 120°C to remain amorphous. Citric acid monohydrate increased the flexibility of Eudragit RS PO films, the decrease in tensile strength and elastic modulus and the increase in elongation of films was proportional to the amount of CA MH in the formulation, indicating plasticization of the polymer.

Solid-state plasticization is presumed to occur through distribution of the plasticizer in the molten polymer during extrusion, followed by the occupation of the active binding sites. This mechanism is similar to liquid plasticizers. Solid-state plasticization is beneficial as the mixing process before extrusion is more efficient, yielding a homogeneous powder formulation and resulting in a better uniformity of the final product. Additionally, improved flow properties assure a stable and continuous flow through the extruder [47].

2.3. Active Pharmaceutical Ingredients

As mentioned before, injection molding can only process thermostable pharmaceutical compounds, limiting the number of drug candidates that can be processed via this technique. This number can be increased by softening the process, thus protecting the API against harsh process conditions and favoring its stability. For example, including plasticizers and/or lubricants in the formulation can significantly reduce the processing temperature and shear rate while processing, improving the stability of the pharmaceutical compound. Zhu et al. (2004) illustrated that the inclusion of a plasticizer (triethylcitrate, TEC) and lubricant (glyceryl-monostearate, GMS) facilitated the thermal processing of chlorpheniramine maleate tablets. This resulted in a lower energy consumption and torque during processing, however, increasing GMS concentrations in the powder blend resulted in faster drug release rates. TEC lowered the T_g and melt viscosity, whereas GMS only reduced the melt viscosity while processing [48].

Injection molding and HME technology subject the drug and polymer to elevated temperatures, high pressure and extensive shear. The elevated temperature and the intensive mixing action generated by the screw can increase the solubility of the API in the polymeric carrier. Depending on the physico-chemical properties of all compounds (particularly polymer-drug miscibility and thermal properties of API and polymeric carrier), and on the processing conditions during the HME and/or IM process, crystalline drugs can melt or become solubilized in the polymer matrix. As a consequence, the drug may be present in the final product as undissolved crystalline particles embedded in the polymeric phase forming two discrete phases upon cooling (solid dispersion); or the drug and polymer are completely miscible and dispersed at the molecular level forming a single phase (solid solution); or as a mixture of both systems. In order to qualify as a solid solution, the drug/polymer mixture should show one single glass transition temperature and the drug should be present in the amorphous form. In general, dispersions in which no crystallinity can be detected are molecularly dispersed and the absence of crystallinity is used as a criterion to differentiate between solid solutions and solid dispersions. Since molecules in the amorphous state are thermodynamically metastable as compared to the crystalline state, the potential for crystallization during processing and storage is always present. Recrystallization and nucleation of drug molecules from the polymer melt is delayed during cooling of the extrudate due to reduced solute migration and the difficulty of nucleation in a highly viscous polymer medium. Furthermore, polymer viscosity increases dramatically when the temperature is lowered after extrusion and additives can be added to the formulation in order to prevent or reduce API recrystallization [6,7].

Prior to the HME or IM process, extensive knowledge of the drugs, polymers and other excipients that undergo processing is essential in designing a pharmaceutical dosage form. It is of extreme importance to characterize the state of the drug in detail, as it may have a profound impact on the processability, stability and drug release properties of the final dosage form. The availability of predictive tools (e.g. determining the cohesive energy by calculating solubility parameters to predict drug/polymer miscibility) and the characterisation of thermal and rheological properties can all help in selecting the appropriate materials and setting the process conditions, which can be both drug and time consuming [49].

Drug and polymer miscibility can be easily and directly visualised using hot stage microscopy. Moreover, theoretical approaches based on calculating the solubility parameters (δ) have increasingly received attention to assess polymer and drug miscibility which can help in predicting if glass solutions are likely to be formed [50]. The solubility parameter is a measure of the cohesive energy densities of materials [51]. The cohesive energy represents the total attractive force within a condensated state material and can be defined as the quantity of energy needed to separate atoms/molecules of a solid or liquid to a distance where the atoms or molecules possess no potential energy, so that no interactions occur between atoms or molecules. The difference between the solubility parameters of two materials gives an estimation of the likelihood that these components will be miscible. Components with similar solubility parameters are likely to be miscible ($\Delta\delta_t < 7MPa^{½}$) and components with $\Delta\delta_t >$ $10MPa^{½}$ are likely to be immiscible. This is because the energy of mixing released by interactions within the component is balanced by the energy released by interactions between the components. Greenhalgh et al. (1999) suggested that interactions between polar (δ_p) and hydrogen bounding (δ_h) significantly affect solubility and should be incorporated into the estimation of the total solubility parameter, which previously only accounted for dispersive forces (δ_d) [52]. As a consequence, most applications use the group contribution method described by Van Krevelen/Hoftyzer to determine the Hansen partial solubility parameter. In this way, based upon the chemical structure, intermolecular or Van der Waals forces (δ_d), intermolecular polar forces (δ_p), and intermolecular hydrogen bonding (δ_h) can be defined, which are all part of the total solubility parameter (δ_t). Consequently, by implementing these partial solubility parameters into the total solubility parameter, the miscibility of more polar molecules can be predicted [52].

A novel approach in this field comprises the use of molecular dynamics (MD) simulations to predict physicochemical properties and test models for polymers, excipients, drugs and mixtures, facilitating the preformulation process. Maus et al. (2008) have used MD to calculate solubility parameters, polymer-drug miscibility and glass transition temperature. The simulated data correlated well with experimental data obtained from differential scanning calorimetry of melt extruded drug products, indicating that the use of MD can substantially reduce the investment of time and money, by reducing the present trial and error principle in dosage form development [53].

In addition, differential scanning calorimetry (DSC) is indispensable to characterise the thermal properties of both drug and polymer prior and after processing, providing a means to describe drug and excipient miscibility in the molten state. The technique allows quantitative detections of transitions (melting point, glass transition temperature, recrystallisation, degree of crystallisation, degradation,..) by accurately measuring the energy that is required or liberated by the material during heating and/or cooling (endothermic and exothermic transitions of the system). The thermogram of a HME or IM product is compared to a physical mixture of the drug, polymeric carrier and other excipients. The lack of a melting peak in the thermogram of the final product is an indication that the drug is in the amorphous state rather than the crystalline form. The measurement of T_g in function of drug and polymer concentration provides information that can help in determining the process parameters (temperature, screw speed,..) since polymer processing is only possible above the T_g of the material. Additionally, DSC can be used to assess the miscibility of the API in the polymer, based on the shift in melting endotherm or glass transition temperature of the drug. According to the Gordon-Tayler equation, if the drug and polymer are miscible, the mixture will show a single T_g that ranges between the T_g of pure components and depends on the relative proportion of each component (weight fraction).

Different techniques are commonly used to characterize melt extruded and injection molded dosage forms. High-pressure liquid chromatography is the most commonly used technique to investigate drug stability, the presence of new peaks in the chromatogram can indicate drug degradation. Moreover, other thermo-analytical techniques such as hot-stage microscopy, differential scanning calorimetry, micro-calorimetry, dynamic mechanical thermal analysis (DMTA) and thermogravimetric analysis (TGA) are often applied to investigate the chemical stability, thermal behaviour and crystalline properties of actives and/or excipients in the final dosage form. An overview discussing the most important techniques and their domains of application in relation to critical steps in drug development is provided by Giron [54]. Non-thermal techniques include Fourier transform infrared spectroscopy (FT-IR), Raman spectroscopy, powder X-ray diffraction (XRD), solid-state nuclear magnetic resonance spectroscopy (NMR) and microscopic techniques (scanning electron and optical microscopy). In this regard, XRD is often used as it provides an individual fingerprint of all materials, and the absence or presence of peaks in the diffractogram can help in differentiating between solid solutions or dispersion.

Hydrolysis, solvolysis and oxidation are three primary mechanisms of drug degradation. One of the advantages of melt process techniques is that they are anhydrous, assuring that hydrolysis or solvolysis can not contribute to potential drug degradation. Since elevated temperatures can cause oxidative degradation of polymers and drugs, antioxidants can be included in the formulation in order to enhance the overall stability. Primary antioxidants (e.g. butylhydroxyanisol, butylhydroxytoluen) inhibit the oxidation reaction by capturing the formed peroxide radicals, reducing agents (e.g. ascorbic acid) are preferentially oxidized since they possess a lower reduction potential and synergistic antioxidants (e.g. disodium edetate, citric acid) complex metallic ions, which often induce oxidative reactions [7].

Although various drugs can be processed by IM and HME, new challenges have emerged as research is focusing on thermolabile therapeutic agents such as peptide and/or proteins. Some authors have investigated the possibility of processing unstable drugs via hot melt extrusion. Δ^9- tetrahydrocannabinol (THC) which is sensitive to heat, air and light was chosen as model drug and included in melt extruded films based on polyethylene oxide (PEO) or hydroxypropylcellulose (HPC) [55]. The effect of processing temperature (80 and 120°C for PEO matrices; 120, 160 and 200°C for HPC matrices), processing time (10 and 60 min) and storage conditions (storage temperatures: -18, 4, 25, 40°C) was investigated on the chemical stability of THC. In addition, 22 different formulation additives (anti-oxidants, lubricants, surfactants and plasticizers) were screened, from which 3 (polyethylene glycol-400, Capmul® PG-12 and vitamin E succinate) were selected based on the miscibility of polymer and additive and on successful fabrication of a film with sufficient flexibility. The inclusion of these aids positively affected the processing conditions. However, the drug was particularly unstable in vitamin E succinate films due to a chemical interaction between THC and vitamin E. Process temperature and residence time in the extruder both affected drug degradation, but processing temperatures of 110 and 140°C for PEO- and HPC-based systems, respectively, were suitable for THC incorporation in the polymer films, based on the thermal stability of the drug, polymer and additives. Contact with atmospheric oxygen resulted in oxidative drug degradation, which could be reduced by including antioxidants, ascorbic acid was the most effective for drug stabilisation in PEO-vitamin E succinate matrices. In addition, it was suggested that the degree of interactions (such as hydrogen bounds) between PEO and processing aids affected the permeability and hence diffusion of oxygen into the system, thus affecting drug oxidation. Also controlling the microenvironment pH via the inclusion of a buffer exhibited a significant stabilization effect on THC degradation [56].

Poorly compressible active pharmaceutical ingredients can be directly processed into tablets by means of IM, or via the cutting of HME produced extrudates into mini-tablets. In contrast to direct compression as manufacturing method, hot melt techniques require no compactibility of the pharmaceutical compound. This was illustrated by De Brabander et al., who described the incorporation of the poorly compressible drug ibuprofen into sustained release mini-matrices produced via HME [57]. Moreover, intensive mixing caused drug particles to de-aggregate in the polymer melt resulting in a more uniform distribution of fine particles. In this manner, HME offers a high throughput and low material loss, while preparing extrudates that possess excellent homogeneity [11].

The processing technique may have a profound impact on the drug release properties of the drug delivery system. Crowley et al. investigated drug release from tablets prepared by hot melt extrusion and direct compression, and reported a faster drug release for the latter type of matrices. Hot melt extrusion resulted in tablets with a lower overall porosity due to extensive densification of the molten mass during extrusion, compared to compaction of the powder during direct compression [58]. The effect of a post-processing thermal treatment was reported by Zhu et al. Tablets manufactured by direct compression, high-shear hot-melt granulation and hot melt extrusion were exposed to heat (60°C) for 0.25, 0.5, 1, 4, and 24h. The drug release rate for all tablet types decreased following post processing thermal treatment, and this reduction in drug release was attributed to an increase in intermolecular binding and entanglement between drug and polymer molecules that occurred during thermal processing. However, post-processing thermal treatment of the hot melt extrudate had a minimal effect on the release rate since the HME process already caused an extensive drug and polymer entanglement [59].

Functional excipients can be included in the formulation in order to modify drug release in drug delivery systems. These excipients are able to tailor drug release by altering the porosity or tortuosity of the dosage form. In this manner, HPMC and L-HPC have been successfully incorporated into EC-based injection molded matrix systems. HPMC as gelling agent and L-HPC as disintegrant were both chosen as matrix fillers to avoid the time-dependent drug release profiles characteristic for sustained release matrix systems. These polymers promoted drug release by opening the matrix structure from IM matrix tablets. Although these matrix systems offered a flexible system to tailor drug release, constant drug release rates were not obtained [33,60].

3. Injection Molding Applications

3.1 Solid Dispersions

The bioavailability of orally administrated drugs depends mainly on solubility and permeability. Oral absorption of drugs occurs in two sequential steps: dissolution followed by the permeation of the dissolved drug across the gastrointestinal membrane, whereby the rate limiting step determines the oral bioavailability of the drug. The advent of high throughput screening, automated synthesis and combinatorial chemistry in the drug discovery process has resulted in a vast number of drug candidates that are often highly lipophilic, of high molecular mass and poorly water-soluble, resulting in a low bioavailability. It is estimated that more than 40% of the marketed drugs are poorly water-soluble, and many promising lead components will never be marketed due to drug delivery issues.

Based on the biopharmaceutics classification system (BCS) active pharmaceutical ingredients are classified into four categories according to their solubility and permeability properties. For drug molecules exhibiting low solubility but reasonable membrane permeability (categorized as BCS Class II), the rate-limiting step in drug absorption is the drug dissolution step. A number of formulation strategies have been developed to improve the delivery of BCS II class drugs. They are based on techniques to increase the dissolution rate or to achieve sustained solubilisation of the drug. Although salt formation, solubilisation and particle size reduction have been commonly used to increase the dissolution rate, there are many practical limitations to these techniques [61,62]. For poorly soluble drugs with poor membrane permeability (BCS class IV drugs), formulation strategies can only improve absorption to a small extent due to the limited membrane permeability. These molecules often need chemical modification to obtain the appropriate physicochemical properties [63].

Consideration of the modified Noyes-Whitney equation provides some hints to improve the dissolution rate of poorly soluble compounds in order to minimize the limitations of a drug candidate for oral drug delivery.

$$\frac{dC}{dt} = \frac{A\,D\,(C_S - C)}{h}$$

which defines the dissolution rate *(dC/dt)* as a function of *A*, the surface area of drug exposed to the dissolution medium; *D*, the diffusion coefficient of the compound; C_s, the maximum solubility of the compound in the dissolution medium; *C*, the solubility of the compound in the dissolution medium at time *t* and *h*, the thickness of the diffusion boundary layer adjacent to the surface of the dissolving compound.

Different possibilities to improve the dissolution rate consist of: increasing the surface area available for dissolution by decreasing the particle size, optimizing the wetting characteristics of the compound surface, decreasing the boundary thickness, ensuring sink conditions for dissolution and very importantly improving the apparent solubility of the drug under physiologically relevant conditions [64, 65]. However, since the diffusion layer thickness is determined by the hydrodynamics of the system, the maintenance of sink conditions depend on the gastrointestinal permeability and the volume of luminal fluids, and the drug diffusion coefficient remains constant, the most attractive option to increase the dissolution rate is via enhancement of drug solubility through formulation approaches.

Various strategies have been attempted by formulation scientist to address these solubility issues. One of the possible strategies involves the formation of solid dispersions. This expression was first used by Sekiguchi and Obi to designate a new pharmaceutical approach to control drug release and enhance oral bioavailability of poorly water-soluble drugs by decreasing the particle size [66]. Because of such early promises in the bioavailability enhancement of poorly water-soluble drugs, solid dispersions have become one of the most attractive and active areas of research in the pharmaceutical field.

The term solid dispersions describes a family of dosage forms whereby one or more active ingredients are finely dispersed in a biologically inert solid state carrier or matrix. Solid dispersions are mostly prepared by melting (fusion method) or the addition of a solvent (solvent method). The fusion method consists of melting the carrier and drug, whereby the solid dispersion is formed upon cooling of the melt. With the solvent method, the drug and carrier are dissolved in a common organic solvent, followed by removal of the solvent via evaporation. Alternative methods to prepare solid dispersions include spray-drying, co-evaporation, co-precipitation, freeze-drying, roll-mixing and co-milling. Processing solid dispersions via the solvent method can cause environmental issues due to the use of organic solvents and health concerns because of possible residual solvents in the product. The solvent technique also requires explosion proof equipment, and the different operations involved in this technique are hard to scale-up. Therefore, the melting technique has become more widely utilized for the preparation of solid dispersions as processing can be accurately monitored and standardised via HME [67]. However, the melt extrusion approach for the fabrication of solid dispersions was only able to break through when many different carrier systems of pharmaceutically approved excipients became available, resulting in numerous publications in the scientific literature during the last ten years. Solid dispersions are classified into six major categories: (1) simple eutectic mixtures; (2) solid solutions; (3) amorphous (glassy) solid solutions; (4) amorphous precipitation of a drug in a carrier; (5) compound/complex formations and (6) glass suspension (Table 1) [67].

Table 1. Structure of solid dispersions (from Ref. 67)

	Glassy Solid Solution	Solid Solution	Eutectic	Glass Suspension	Amorphous Precipitation		Compound/ Complex Formation
					Crystalline Carrier	Amorphous Carrier	
Phases	1	1	2	2	2	2	1
Drug	Molecularly Dispersed	Molecularly Dispersed	Crystalline	Crystalline	Amorphous	Amorphous	Amorphous Or Crystalline
Carrier	Amorphous	Crystalline	Crystalline	Amorphous	Crystalline	Amorphous	Amorphous Or Crystalline

Solid dispersions aim to increase the bioavailability of the active pharmaceutical component, resulting in lower dosing of the dosage forms and reduced drug-plasma fluctuations after administration [64].

When the solid dispersion is exposed to aqueous media, the carrier dissolves and the drug is released as very fine, colloidal particles. Due to the enlarged surface area, the dissolution and hence bioavailability of poorly water-soluble drugs is enhanced. In solid solutions the drug is molecularly dispersed within the matrix, constituting a one-phase system, regardless of the number of components. Solid dispersions present several advantages to increase the drug dissolution rate: when the drug substance is present in the amorphous state, meaning that crystal lattice forces have already been overcome thus improving solubility. In case of molecularly dispersed systems, the drug particle size is maximally reduced, so that an isolated, dissolved molecule is obtained. Moreover, in case of solid dispersion containing a crystalline drug phase, selection of an appropriate carrier which exhibits a solubilising effect on the drug while dissolving can improve the dissolution rate. This carrier material can prevent drug agglomeration and increase the wettability and dispersibility of the drug in the dissolution media, thus generating a microenvironment conductive to dissolution [68]. An excellent review addressing the different mechanisms underlying the observed improvements in dissolution rate has been published by Craig [69].

Although the solid dispersion approach is very promising for solubility enhancement of poorly soluble drugs, very few products based on this technology are found on the market today. Problems limiting the commercial application of solid dispersions involve the method of preparation, the reproducibility of physicochemical properties, the formulation into dosage forms, the scale-up of the manufacturing process, and the chemical and physical stability of the drug and vehicle [70]. Also the stabilisation of solid solutions can sometimes provide issues, since molecules in the amorphous state are thermodynamically metastable compared to the crystalline state. In addition, the drug is often in the supersatured state when formulated as a solid solution. Due to these reasons, the risk for recrystallisation during processing and storage is always present. However, the use of recrystallisation inhibitors and the solubilising and stabilizing effect of the carrier system by means of specific intermolecular interactions with the drug can sometimes prevent this problem [71]. The critical factors affecting stability of amorphous systems include T_g, hygroscopicity, purity and storage conditions. Moisture exhibits a plasticizing effect, thereby lowering the T_g which increases the risk of conversion from the amorphous state to the crystalline state. Moreover, a careful selection of drug and polymer is required as specific interactions between drug and polymer can improve the stability of the system. Carriers with high T_g and high viscosity showed superior stability for amorphous drugs [72-74]. Choski et al. (2007) reported that amorphous indomethacin in solid solutions with Eudragit EPO, polyvinylpyrrolidone-vinyl acetate copolymer and polyvinylpyrrolidone K30 showed tendency to revert back to the crystalline form. However, the rate of reversion was dependent on the nature and concentration of the polymer, high concentrations of Eudragit EPO provided superior stabilisation of amorphous indomethacin [72].

Wacker et al. were the first to investigate injection molding as a process for manufacturing solid dispersions or solutions [75]. Solid dispersions based on polyethylene glycol 6000 and 5% w/w micronized hydrocortisone acetate were prepared by IM and compared to dispersions made by a hot melt method, which involved the melting of both drug and carrier at different temperatures (65-145°C) in a beaker, followed by stirring for 5 minutes. The resulting solutions or dispersions were poured into cooled molds to obtain discs. The IM tablets were only produced at 65°C, with an injection step and cooling step of 30 sec each. Discs prepared via the melt method at 85-125°C exhibited slower intrinsic dissolution rates compared to lower and higher temperatures (65 and 145°C, respectively), as these discs possessed a higher degree of crystallinity due to recrystallisation of the drug in the carrier, as seen with XRD. In contrast, injection molded tablets processed at 65°C showed a higher intrinsic dissolution rate compared to the melt technique, as intensive mixing of drug and carrier during the extrusion step prior to injection molding had a beneficial effect on the drug dissolution rate.

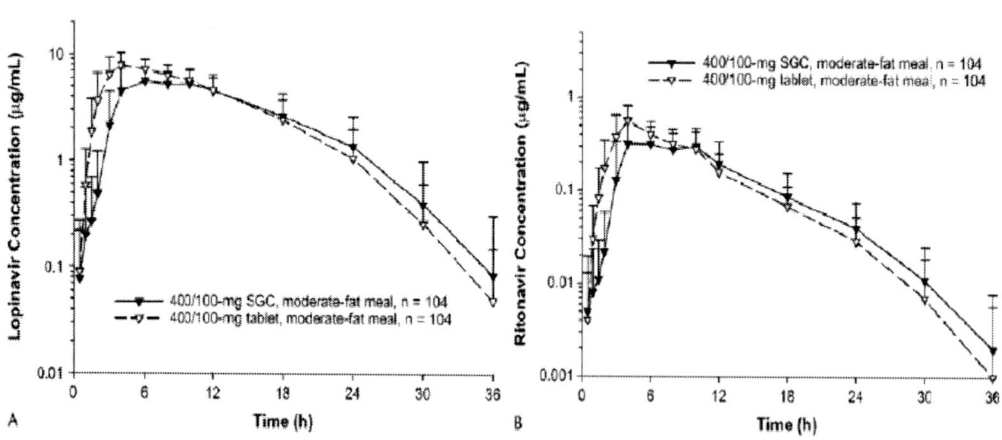

Figure 3. Plasma concentration after a single oral dose of lopinavir (400 mg) and ritonavir (100 mg), formulated in a soft gelatin capsule (▼) or hot melt extruded tablet (▽). The formulations were administered with a moderate-fat breakfast. A. Lopinavir, B. Rotinavir; obtained from ref. 78

The development of Kaletra® by Abott Laboraties is an interesting case study to illustrate the potential of HME to increase the solubility and bioavailibity of lopinavir and ritonavir (both HIV protease inhibitors, classified as BCS IV compounds) [76,77]. As conventional solid formulation methods (tablets and capsules) failed to produce adequate lopinavir and ritonavir plasma concentrations, the formulation of a soft gelatine capsule was required to obtain an efficient dosage form. The lopinavir/ritonavir soft gelatine capsule requires refrigerated storage and the intake of 6 capsules per day concomitant with food is required to maximize the bioavailability of lopinavir. However, via HME followed by in-line calendaring (shaping) both drugs could be processed in combination with polyvinlypyrrolidone polyvinylacetate into tablets, with both anti-HIV drugs completely dissolved at the molecular level in the matrix. After oral administration, these solid dispersion tablets were bioequivalent compared to the lopinavir/ritonavir soft gelatine capsules (figure 3) [78]. In addition, the drug plasma levels were considerably less variable using the HME-processed tablets and were less influenced by the diet (fasting, moderate-fat and high-fat meal). Other major advantages include a reduction of the daily tablet intake and the storage of the solid dispersion at room temperature.

3.2. Oral Controlled Release Drug Delivery Systems

Oral drug delivery is the most commonly employed route of drug administration, presenting several advantages, such as patient compliance, ease of ingestion, low cost and versatility. Despite their popularity, one of the major disadvantages of oral drug delivery systems is the fluctuation of the drug plasma concentration. Due to periodic administration of the dosage form, the drug plasma concentration may fluctuate widely, presenting peaks and valleys. Because of this fluctuation, the drug plasma concentration reaches toxic and/or sub-therapeutic levels, which can lead to the undesirable side effects or inefficient drug therapy, respectively. The drug plasma concentration is within the therapeutic window for only a fraction of the treatment period (Figure 4A).

Controlled release formulations are a promising approach in order to achieve and maintain therapeutically optimal drug plasma concentration for an extended period of time, as can be seen in Figure 4B. By controlling the delivery rate of the drug, the duration of the therapeutic action can be sustained. In general, the development of sustained release formulations offers many benefits over conventional dosage forms: controlled administration of a therapeutic dose at the desired delivery rate, less fluctuations of plasma concentration resulting in reduced toxicity and side effects and increased efficacy of the therapy, and reduced frequency of dosing improving patient compliance. Disadvantages include the longer time required to achieve therapeutic blood concentrations, possibly increased variation in bioavailability after oral administration, enhanced first-pass effect, risk on dose dumping (especially for reservoir-based systems), lack of dose flexibility and usually higher costs [13, 23, 79-80].

Not all drugs are suitable for controlled release: drugs possessing a very short or very long half-life, drugs having a significant first pass-metabolism, drugs with an absorption window in the gastrointestinal tract, drugs with a low water-solubility or drugs which plasma concentration is not related to the therapeutic effect.

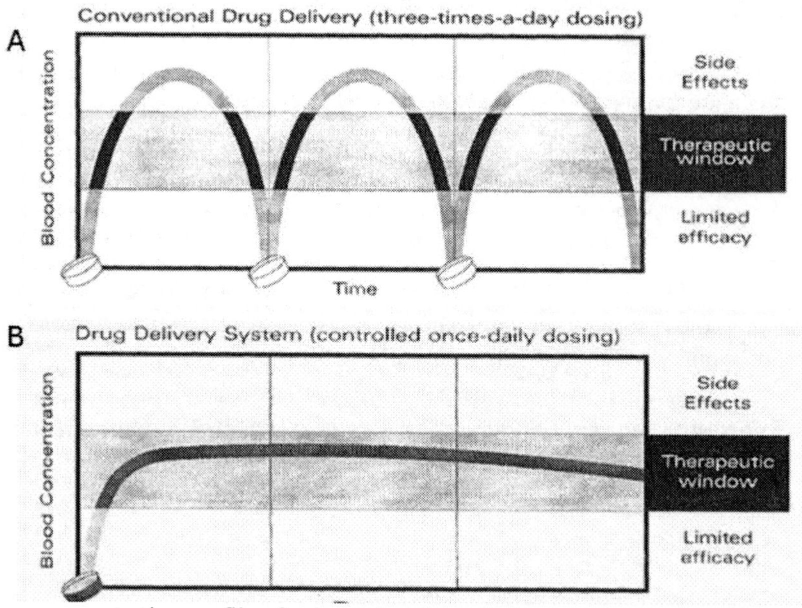

Figure 4. Plasma concentration profile of a drug administered via a conventional oral formulation (A) versus a controlled drug delivery system (B)

The first report regarding injection molding as manufacturing technique for sustained release preparations dates back to 1971. These authors conducted a preliminary study to investigate the sustained release properties of two types of thermoplastic polymers (epoxy resins and copolymers of vinyl acetate and β-methacrylic acid) as matrix carrier for injection molded discs. In this study, injection molded discs of both systems were able to sustain drug release in different buffer solutions (pH range of 1.2-2.5 and 6.7-7.3 for epoxy resins and copolymers of vinyl acetate and β-methacrylic acid, respectively) and ionic strengths (ranging between 0.1 and 0.4) over up to 45h [81].

An early review addressing different plastic processing techniques with emphasis on injection molding was provided by Hüttenrauch in 1974 for the preparation of sustained release preparations. This author discussed the design of injection molding machines (ram- and screw-based injection molding machines), provided information over the injection molding cycle, and discussed the properties of different thermoplastic materials [82].

In another study performed by the same group, the release of sodium chloride from polyethylene-based IM matrix tablets was studied using the square root of time model. The dimensions of the injection molded tablet affected drug release: sodium chloride release was linear proportional to the specific surface of of the polyethylene tablet. Drug release could also be enhanced by adding soluble salts (potassium nitrate) [83].

Cuff and Raouf selected mixtures of polyethylene glycol (PEG) 8000 and microcrystalline cellulose (MCC, used as thickening agent) as matrix carrier for injection molded matrix tablets, containing 14 different model drugs [84]. This study paid specific attention to the solubility of the model drug in the carrier, heat stability, drug particle size and shape and physical properties of the tablet. Two screening tools were used to identify compounds that are suitable for IM: the solubility and stability of a compound in molten PEG. Drugs with low solubility in the carrier yielded good tablets, however, two compounds with higher solubility in PEG (>5%) provided problems. Tafazolone tablets had a low erosion rate since the high solubility of the drug in PEG 8000 changed the properties of the PEG-MCC matrix. Fenoprofen calcium could not be processed as these tablets did not harden upon cooling. No influence of drug particle size and shape was found for compounds that are insoluble and stable in molten PEG.

Another application of injection molding that received considerable attention comprises the injection molded capsule made from starch as alternative to conventional hard gelatine capsules, produced via a dip-molding process [85]. Starch as natural hydrophilic polymer was selected as it is readily available and cheap. In case of starch, the amount of water affected the processing properties: injection molding of starch/water mixtures containing an equilibrium water content of 13-14% at ambient temperature and humidity resulted in materials that flowed satisfactorily without degradation in the temperature range between 140-190°C. By controlling the water content and the temperature changes along the barrel, the starch granules were broken down to form a melt. This starch melt containing 13-14% water formed an amorphous material at room temperature. Capsules caps and bodies were separately molded in multi-cavity molds, allowing a more precise control of the dimensions of the capsule-shaped drug delivery device (marketed as Capill®) compared to the dip technology and avoided the drying step used in the dip-molding process. The injection molded starch-based capsule is not only a step forward in capsule production, but also in polymer processing technology as it was the first use of elevated temperature to injection mold a native polymeric material containing a high water content into a stable final product. Moreover, the production of starch-based capsule via IM allowed a smaller closure area and thicker walls compared to capsules manufactured by dip-molding, thus improving capsule handling (filling and blistering) but without significantly affecting disintegration and drug release. This was illustrated when acetylsalicylic acid formulated in Capill® and conventional dip-molded hard gelatine capsules resulted in similar drug-plasma levels. These starch capsules allowed enteric coating and were stable at different storage conditions. The physicochemical properties of starch/water-based mixtures and its applications were described in several follow-up studies, discussing the rheological properties, melt viscosities, and thermal behaviour of thermoplastic potato starch-based capsules [86-89]. The development and characterisation of injection molded starch capsules is reviewed by Vilivalam et al. (2000), with respect to stability, enteric coating and in-vivo performance studied [90]. After blistering capsules were physically stable for at least 18 months. Furthermore, capsules coated with cellulose acetate phthalate or with mixtures of methacrylic acid copolymers remained intact in gastric medium, and reproducibly disintegrated in intestinal medium. Moreover, Capill® capsules coated with a mixture of Eudragit L100 and S100 (3/1-mixture which dissolved at pH 6.3) were successfully used for colon targeting, as scintigraphic images revealed that drug molecules were only released in the descending colon 8.5h after administration [91].

Thermoplastic starch was used for the production of sustained release tablets via extrusion granulation and injection molding. Since native starch is not thermoplastic, thermal processing is used to disrupt and transform the semi-crystalline structure of starch granules into a homogeneous amorphous material. This transformation usually makes use of different plasticizers. In this study, urea, glycerol and water were used as plasticizer, glycerol-monostearate and calcium stearate were selected as lubricant, and sodium benzoate was chosen as model drug. Raman spectroscopy and DSC confirmed that sodium benzoate was molecularly dispersed in the starch matrix, and the drug release from these IM tablets was successfully delayed [92-93].

Injection molding as a pharmaceutical technology to produce matrix tablets was investigated by Quinten et al. [33]. Mixtures of metoprolol tartrate (MPT) (30% w/w) and hydroxypropylmethylcellulose (HPMC) with ethylcellulose (EC) as matrix carrier were hot-melt extruded and subsequently injection molded into biconvex matrix tablets. Without plasticizer ethylcellulose required a process temperature of at least 140 °C (i.e. above EC's T_g of 133°C), however, these tablets were very brittle. Adding dibutylsebacate as plasticizer (20% w/p) reduced the T_g of EC (52.8°C) and improved the melt flow of the mixture, allowing to manufacture controlled-release tablets in a temperature range between 110 and 140°C. EC and MPT were stable under these process conditions. The incorporation of HPMC as hydrophilic filler in the formulation was essential as tablets without HMPC released less than 50% drug after 24h. Including 35, 40 and 50% HMPC in the EC matrices resulted in 60, 80 and 100% drug release after 24h, respectively. Drug release from these matrices was diffusion and swelling controlled. Higher EC viscosity grades (45 and 100 mPa.s) resulted in complete drug release, whereas lower viscosity grades showed incomplete release. A similar effect of the molecular weight of HPMC was observed, as higher viscosity grades disrupted the matrix structure more due to a higher degree of polymer swelling, resulting in faster and complete MPT release. However, these formulations also showed a significant burst release. Thermo-analytical characterisation (XRD, DSC, Raman spectroscopy) revealed that a two-phase solid dispersion was formed, as crystalline MPT was partially solubilised in EC.

In a next study, the effect of low-substituted hydroxypropyl cellulose (L-HPC) on drug release from the EC-based IM tablets was investigated. L-HPC was chosen to promote drug release from ethylcellulose-based IM tablets since it functions as disintegrant. The effect of processing temperature, matrix composition and EC viscosity grade on drug release was investigated. Drug release was significantly influenced by the EC and L-HPC concentration in the injection molded formulation. Processing tablets at higher temperatures, 140°C instead of 120°C, resulted in harder tablets and slower drug release rates. These differences were not attributed due to differences in porosity, as the tablet porosity was overall very low (<5%), and independent of the formulation and process conditions. In addition, higher EC viscosity grades resulted in a higher tablet hardness, however, as these tablets were more prone to erosion after being introduced to the dissolution medium their release rates were faster. Dynamic X-ray tomography revealed an anisotropic pore structure throughout the tablet, and the tablets possessed a skin-core structure caused by a molecular orientation of the molten polymers entering the mold after being injected [60].

Most oral drug delivery systems that have been discussed in this paper controlled drug release via diffusion from a matrix system. As the solvent in the gastro-intestinal tract is water, these systems are only capable of regulating drug release of good water-soluble drugs, but water-insoluble drugs present many challenges. Therefore, the Egalet® technology was proposed as a drug delivery system able to overcome the limitations of poorly water soluble drugs as it enables the controlled release of drugs through gradual erosion of a tablet [94,95]. The Egalet® systems consist of an impermeable shell enclosing a drug-containing core, the top and bottom of the cylinidrical-shaped core are in direct contact with the dissolution medium (Figure 5).

The outer shell is made of cetostearyl alcohol and ethylcellulose, while the matrix core comprises a mixture of polyethylene glycol monostearates and polyethylene oxides. The drug is distributed throughout the matrix, which is eroded slowly by gut movements and gastrointestinal fluids as it passes through the gut. The drug release mechanism is based on surface erosion, effected through water diffusion, polymer hydration, disentanglement, and dissolution. Drug release can be controlled by modifying the length and composition of the plugs. In addition, this design allows to built a lag time prior to drug release by implementing drug-free plugs in the outer layer. One of the major advantages of this system comprises the ability to obtain zero order release rates, by a constant erosion of the matrix [96].

Mixtures used for the shell, active core and plug were combined into the Egalet sysem during a reciprocating injection molding process that allowed a sequential molding of the shell and core contents within the die. This design provided an efficient manufacturing process and allowed an accurate dimensional control and fill. Marvola et al. (2004) proved that the in-vitro release rate of caffeine from an Egalet® system followed zero-order rates. However, a scintigraphic study in humans demonstrated that the in-vivo release depended on the place in the gastro-intestinal tract: the release rate was fastest in the small intestine and slowest when the device resided in the descending colon, resulting in deviations from the zero-order release kinetics observed in in-vitro tests [97].

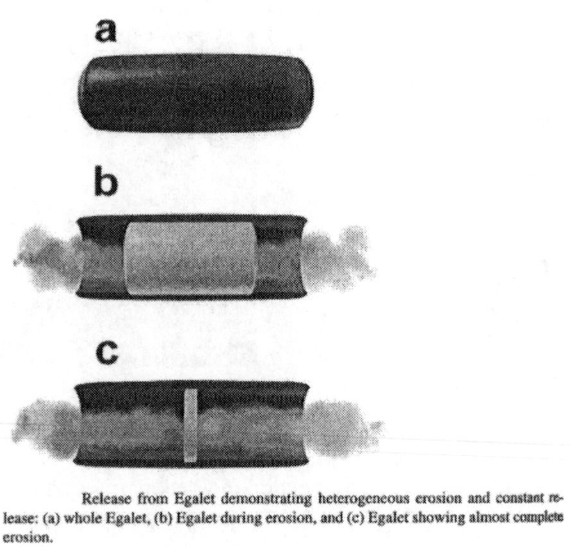

Figure 5. Illustration of the Egalet® technology (From Ref. 94)

Vaz et al. developed a double-layer co-injection molded soy protein-based drug delivery device in order to overcome time-dependent drug release rates frequently seen with monolithic matrix devices [98]. Matrices exhibit continuously diminishing delivery rates, and one of the most common approaches to overcome this problem is the modification of the drug delivery device by restricting the surface area exposed to the dissolution medium. Therefore, these authors assessed the feasibility of co-injection molding, that allowed the simultaneous molding of two polymers into parts with high dimensional accuracy and adjustable skin-core morphology. The injection molder was operated in mono-sandwish mode and the injection molding sequence (figure 6.A) include the independent plastication of the two materials in separate extruders; II) the extrusion of a limited amount of material A to the front of the extrusion barrel of material B while the valve towards the IM is closed, and III) the simultaneous injection of both materials due to the action of the horizontal unit. As material A is injected first in the mold, it will form the skin of the molded part due to the contact with cooled cavity walls, whereas material B flows within this skin, cooling down under mild thermo-mechanical conditions and forming the core of the material.

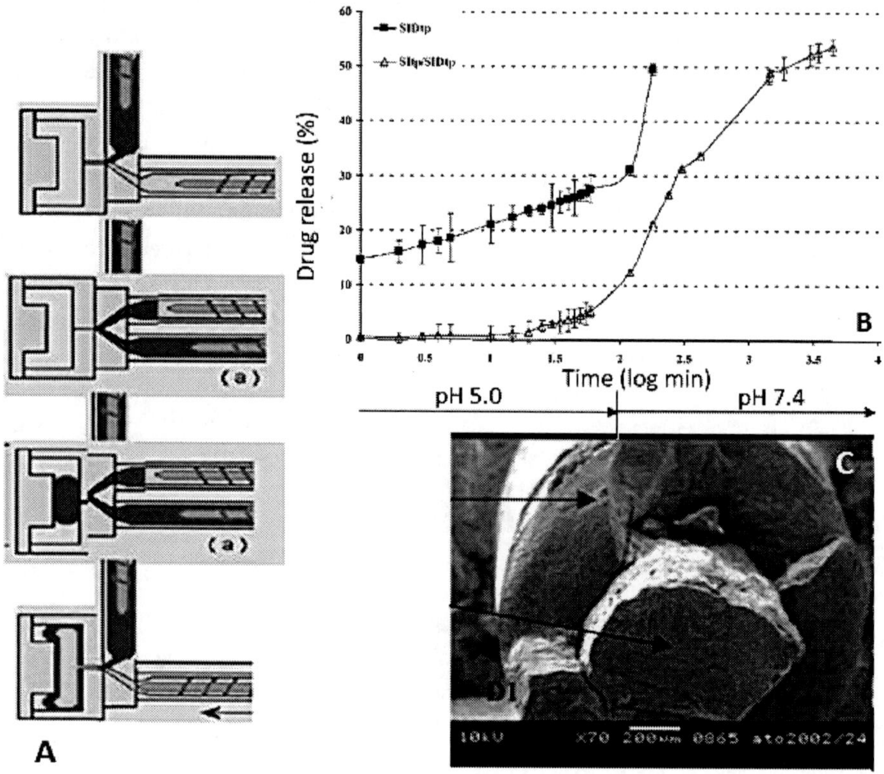

Figure 6. A: Injection molding sequence for manufacturing a skin-core drug delivery device; B: SEM picture illustrating the skin-core morphology comprising a drug-free outer layer and a drug-containing core; C: Dissolution profile of a bi-layered co-injection molded matrix system in different release media. Obtained from reference 98 and 100

This results in a system comprising a functional core material within a skin layer, the latter acting as an effective barrier for early drug diffusion (figure 6.B). Core material containing mixtures of soy protein isolate with glycerol as plasticizer (and in particular cases glyoxal as crosslinker) and theophylline as drug, pre-compounded via HME. The skin material was composed of similar mixtures but without drug, and some of the formulations were reinforced with hydroxyapatite.

The viscosity rate between core and skin layer played an important role in the formation of the interfacial shape, namely the thickness and uniformity of skin and core. The adequate selection of materials and optimization of process conditions during co-injection molding resulted in an accurate control of the thickness of the layers of the device. However, complex dissolution profiles were obtained due to the complex intricacy of these devices (Figure 6.C). The skin thickness and degree of crosslinking of the core both significantly affected drug release, whereas drug release was very sensitive to pH changes due to the amphoteric nature of soy protein [99,100]. At pH 5.0, the soy protein exhibited low solubility and drug release is controlled via drug diffusion, whereas at higher pH (7.4) the polymer matrix was more soluble. In the latter case, drug release is controlled via swelling, drug diffusion and polymer dissolution, resulting in faster release rates.

3.3. Medical Implants

Rothen et al. compared injection molding versus ram extrusion as manufacturing technique for polylactic acid-based implants containing the peptide vapreotide as model drug [11]. Both manufacturing techniques decreased the molecular weight of the polymer which was more pronounced for injection molded implants, as this process involved higher temperatures and shear forces. Both techniques did not affect the crystalline properties of PLA. Regarding drug stability, injection molded implants showed slightly higher peptide degradation (peptide purity: 92.0-97.0%) compared to melt extruded implants (peptide purity: 98.5-99.5%), probably due to the higher temperatures and greater shear forces employed during the IM manufacturing process. The most important difference between these two production methods was the implant degradation studied by scanning electron microscopy (figure 7). After 144h immersion in water/ methanol (50:50) the extruded sample was extremely porous and extensively fragmented, whereas the injection molded sample showed numerous cracks throughout the matrix, but without porosity. This resulted in a higher in-vitro release rate of vapreotide for the extruded implant. These differences were also assigned to the higher matrix density of injection molded implants and to the crystalline/amorphous ratio, which is supposed to be increased in the injection-molded implants, due to the specifity of the manufacturing process. Namely, the authors suggested that the IM process resulted in polymer chain orientation due to the elevated temperatures of the injection molding process, which may have affected the internal structure of the matrix, the mechanical properties and possibly the degree of crystallinity. As can be seen in figure 7, injection molded samples showed less cracks, were less porous and had significantly slower release rates during dissolution (±55% after 140h) compared to HME implants (±85% after 140h). Disadvantages of IM processing included higher processing temperatures (i.e. more peptide degradation), larger amounts of material needed for trial and more material loss.

Deng et al. evaluated the process conditions, in-vitro performance, morphology and microstructures of injection molded beads (Septacin®) containing gentamicin sulfate for the treatment of osteomyelitis [101]. A biodegradable copolymer of fatty acid dimer (FAD) and sebacic acid (SA) (ratio 1:1) was used as matrix carrier. Increasing the melt viscosity, decreasing the melt temperature and rapid solidifying of this matrix by cooling the mold, resulted in a flow-induced structural reinforcement of the implants. This resulted in implants that possessed a characteristic skin-core structure with an oriented skin layer and a non-oriented core. This micro-structure was generated by the shear and extensional chain orientation of the material during the flow into the mold, and the relative higher mechanical strength of the outer layer prevented the beads from cracking in water. These implants showed considerably less lateral cracks after 72h immersion in water, and had a lower drug release rate compared to beads lacking this outer skin layer (figure 8). In another study the impact of the production process was evaluated: physical mixtures were pre-compounded via twin-screw extrusion into pellets which were subsequently fed to an injection molder to produce Septacin® beads [102]. The next step involved the up-scaling of the manufacturing process, allowing a large scale fabrication of Septacin® implants by means of a production-scale injection molding machine. Cycle reproducibility was ensured as a high shot-to-shot weight precision was achieved by the injection molding process, which did not significantly affect the molecular weight of the copolymer [103].

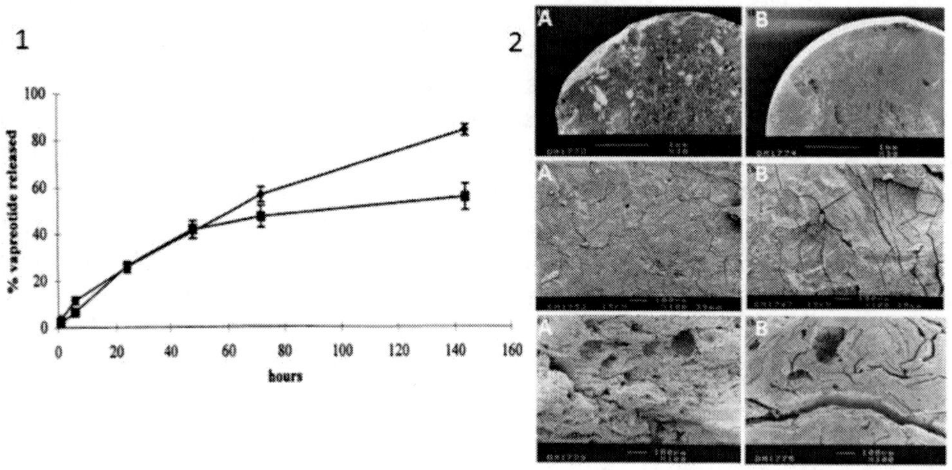

Figure 7. 1: In vitro drug release from polylactic acid implants prepared by extrusion (●) and by injection molding (■). 2: Scanning electron micrographs of the fracture surface perpendicular to the longitudinal axis of the cylinder for extruded (A) and injection molded (B) PLA samples, (I) original samples, (II) after 1h immersion, (III) after 144h immersion in water/methanol (50:50). Modified from reference 10

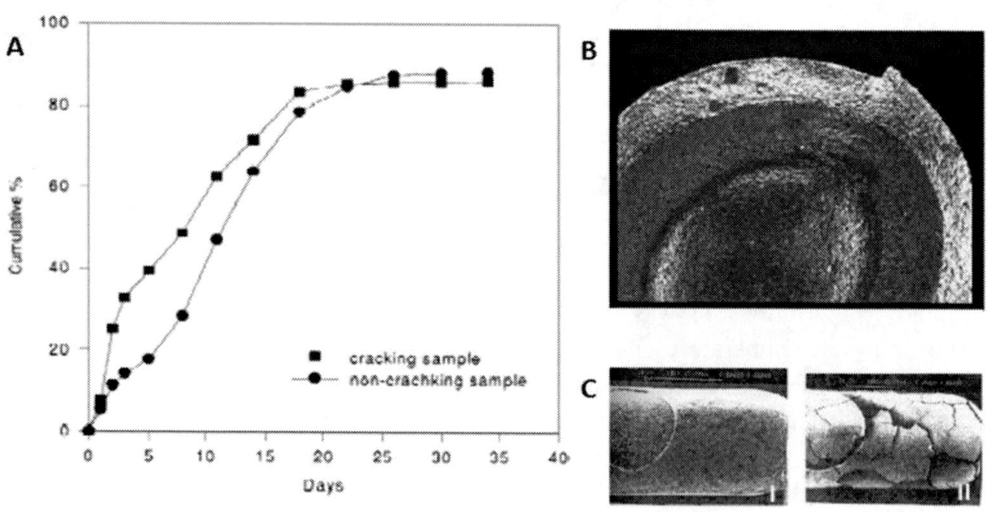

Figure 8. A: In vitro release profile of gentamycine from cracking (■) and non-cracking (●) Septacin beads. B: Skin-core structure of a non-cracking Septacin bead. C: Physical integrity of Septacin beads with (I) and without (II) the skin-core structure after 3-day immersion in water. Obtained from reference 105

Sterilization via γ-radiation of Septacin® beads (70-80 kGy) increased the crystallinity, polydispersity and weight-average molecular weight of the copolymer as a result of polymer chain extension in the copolymer backbone during radiation. In contrast, at higher radiation doses (94-103 kGy), the copolymer weight-average molecular weight decreased significantly [104]. Futhermore, it was demonstrated that Septacin® beads required a storage temperature below 0°C to avoid copolymer degradation. Storage at 25°C reduced the gentamicin release rate from the implant and it was suggested that the depolymerisation of the polyanhydride carrier might render the matrix more resilient to cracking. Water penetration into the implant increased the internal osmotic pressure and resulted in subsequent cracking of implant, which is an important factor in controlling the drug release. A heat treatment (2 weeks, 60°C) in the presence of nitrogen also significantly reduced the copolymer molecular weight by promoting polymer degradation and causing a decrease in stiffness of the implants. This improved the resistance to stress-related cracking in water as evidenced by the more intact implant structure of the beads after 25 days of dissolution. Hence, a post-molding heat treatment furthermore sustained drug release from Septacin® beads. In-vivo evaluation in humans demonstrated that these implants were capable of achieving high local drug concentrations at the implantation site with limited systemic exposure to the drug [105-106].

Another example of an injection molded implantable rod for the treatment of fungal osteomyelitis is provided by Soriano et al. These implants, made of poly-lactide (PLA) and poly(l-lactic) acid (L-PLA) with fluconazole as model drug, were studied in-vitro and in-vivo. Gamma-irradiation of these implants promoted polymeric degradation, however, the loading efficiency, chemical stability and solid state characteristics of drug and carrier were not affected. In-vivo studies showed that approximately 85 and 80% of the drug was released following almost zero-order kinetics within 12 and 4 weeks from PLA- and L-PLA-based rods, respectively [107].

The production of cylindrical tamoxifen citrate-loaded implants via injection molding for the treatment of cancer with poly(sebacic acid-co-ricinoleic-ester anhydride) (70:30 w/w) as matrix carrier was evaluated [108]. This study showed that tamoxifen citrate was in the amorphous state, and that 42 and 62% of the drug was release after 30 days for 10 and 20% w/w drug loaded implants, respectively.

Injection molding in combination with hot-melt extrusion was applied to produce injection implants composed of the biodegradable polymer poly(ε-caprolactone) (PCL) and a broad-sprectrum antiparasitic drug praziquantel (PZQ) [109]. For low drug loadings (6.25 and 12.5% w/w PZQ), the majority of drug crystals were dissolved in the semi-crystalline PCL during processing at 70°C. However, for higher drug loadings (25 and 50% w/w PZQ) drug crystals were detected by SEM and XRD. No significant differences of drug loading on drug release was seen, as similar release profiles were observed for all drug loadings. Furthermore, all implants exhibited a high burst release (± 8% PZQ release after 1 day dissolution) due to immediate dissolution of the drug localized on the surface following immersion in the release medium. Macroscopic evaluation evidenced that two different zones were present in a cross-section of these implants, a peripheral white "drug-free" zone and an inner "pink zone" containing PZQ (figure 9). The boundary (diffusion front) between the two zones moved slowly from the exterior to the interior of the implant during drug dissolution as the drug gradually diffused out of the implant. This was confirmed with mathematical modeling and by quantitative HPLC analysis of both zones. In another study, it was demonstrated that decreasing the diameter (3-8 mm) and increasing the drug loading (25 and 50% w/w) enhanced drug release, whereas end-capping of implants with PCL-sheets in order to avoid drug release from the two ends decreased PZQ release slightly [110].

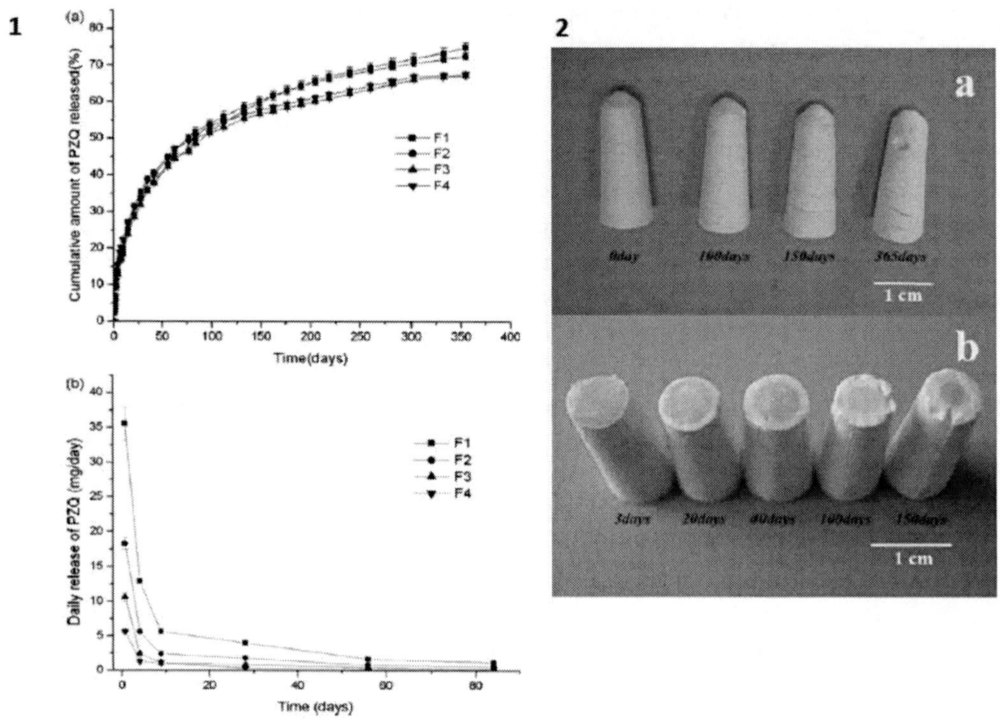

Figure 9. 1: Cumulative praziquantel (PZQ) release from poly(ε-caprolactone) implants containing different amounts of PZQ over a one-year period (a), daily release of PZQ (mg/day) as a function of time (b). 2: Implants after 0, 100, 150 and 365 days dissolution (a), and cross-section of implants after 3,20, 40, 100 and 150 days dissolution (b). Obtained from Ref. 109

3.4. Vaginal Rings/Inserts

Vaginal inserts are primarily used for contraception, the most common active pharmaceutical ingredients include estradiol and norgestrel. These devices aim at a continuous infusion of drugs through the mucosa, thereby preventing first pass metabolisation and hence inefficient therapeutic activity due to surging and ebbing plasma levels.

Sam was the first to review several classes of contraceptive devices, including medicated intra-uterine devices, subdermal implants and intravaginal rings with emphasis on controlled release contraceptive systems such as e.g. Implanon® and Nuvaring® [111]. This study discussed the distinct approaches in designs, comprising the single-rod contraceptive implant, the biodegradable hollow fibre and the two-compartment contraceptive vaginal ring, and manufacturing techniques that are commonly applied (e.g melt extrusion and injection molding).

The effect of drug concentration and solubility with respect to its solid state in extruded ethylenevinylacetate-based coaxial fibers was investigated by van Laarhoven et al. (2002)

[112]. High concentrations (up to 20%) of etonogestrel could be dissolved in the polymeric matrix via the advent of melt extrusion, as the solubility of etonogestrel was considerably higher in EVA at elevated temperatures (figure 10a). Rapid cooling of the extrudate prevented recrystallisation, resulting in the formation of supersaturated solutions. Since drug release is directly proportional to the concentration gradient over the membrane, the amount of drug recrystallizing upon cooling is of crucial importance. These authors found that if the concentration of etonogestrel was below a critical nucleation concentration at room temperature, the dissolved steroid remained in a supersaturated state (±2%). However, if the drug concentration exceeded this critical nucleation concentration, a very slow recrystallization of drug was seen. The amount of dissolved steroid was influenced by the processing conditions: a higher rotation speed of the screw increased the amount of shear while processing and as a consequence more etonogestrel was dissolved (figure 10.b). It was also found that recrystallisation progressed considerably faster when the extrudates were stored at higher temperatures, most probably because of the increased mobility of the polymeric chains and steroid. Therefore, drug release was dependent upon process conditions, drug concentration, and storage temperature.

Another study investigated the in-vitro release properties of etonogestrel and ethinyl estradiol from similar vaginal rings composed of a coaxial fiber made from different EVA grades via co-extrusion. This fiber consisted of a core polymer (EVA28 containing 28% vinyl acetate) that demonstrated a high solubility and permeability for both drugs. This core containing both steroids was enveloped by a thin polymeric membrane (EVA9, 9% vinyl acetate), having a lower drug solubility and permeability [113].

Rathbone et al. (2002) developed an injection molded poly(ε-caprolactone) (PCL) intravaginal insert for the delivery of progesterone as alternative to silicone-based polymeric inserts for estrous control of cattle. Silicon is a thermosetting plastic (i.e. it does not permit reprocessing and is non-biodegradable, therefore require burning or burial for disposal after use. These authors proved that PCL-based insert, engineered into a specific shape for a high retention rate in the vagina of cows and containing 10% w/w progesterone, were bioequivalent to commercially available intravaginal insert [17].

Another study addressed the development of an intravaginal ring for the delivery of oxybutynin in the treatment of urinary incontinence [114]. This drug exhibits a low oral bioavailability and short half-life (<2 h) due to extensive first-pass metabolism resulting in the formation of N-desethyloxybutynin, which is linked to adverse reactions to oral oxybutynin. In this regard, drug delivery of oxybutynin via a non-oral route can result in reduction of both dose and incidence of adverse effects. Implants were manufactured via a reaction injection molding process, allowing the processing of thermosetting polymers and requiring a curing reaction in the mold. Oxybutynin (OXY) (5-60% w/w) was added to a mixture composed of tetrapropoxysilane, silicone elastomer base and stannous octoate catalyst. Blends containing 5 and 40% oxybutyn were reaction injection molded at 50°C to produce cores containing 100 and 800 mg drug, respectively. A mixture containing 60% OXY w/w was too viscous to be reaction molded and was instead extruded at 50°C to produce the active core. These cores were encapsulated via a two-stage reaction injection molding process by a non-medicated silicone sheat. A burst release of oxybutynin (±40 % after 24h) followed by time-dependent drug release was observed for this delivery systems due to the partitioning and diffusion of propan-1-ol, a by-product of the crosslinking of silicone. As propan-1-ol is a good solvent of oxybutynin, the drug dissolved in the core and migrated into the silicone sheat, initiating the burst release. Reducing the length of the core, in combination with a higher drug fraction substantially reduced this burst effect and resulted in slower and time-independent drug release (figure 11).

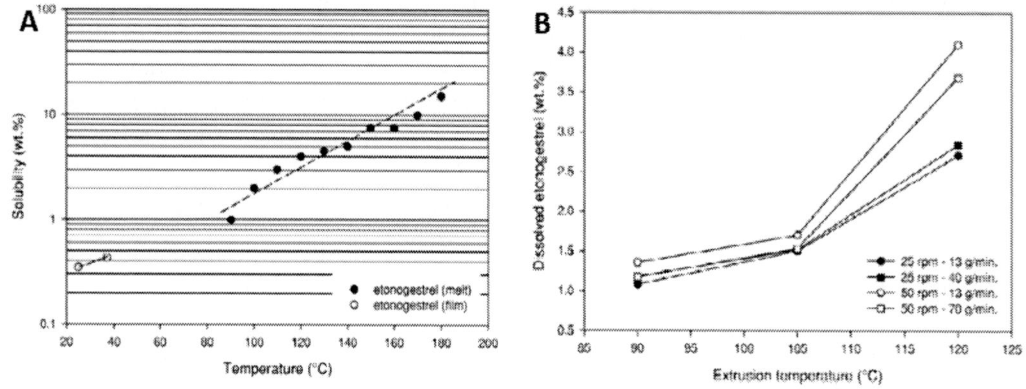

Figure 10. A: Solubility of etonogestrel in ethylenevinylacetate (EVA 28) in function of temperature. B: Influence of the screw speed, throughput rates and temperature on the solubility of etonogestrel in EVA 28 during blend extrusion

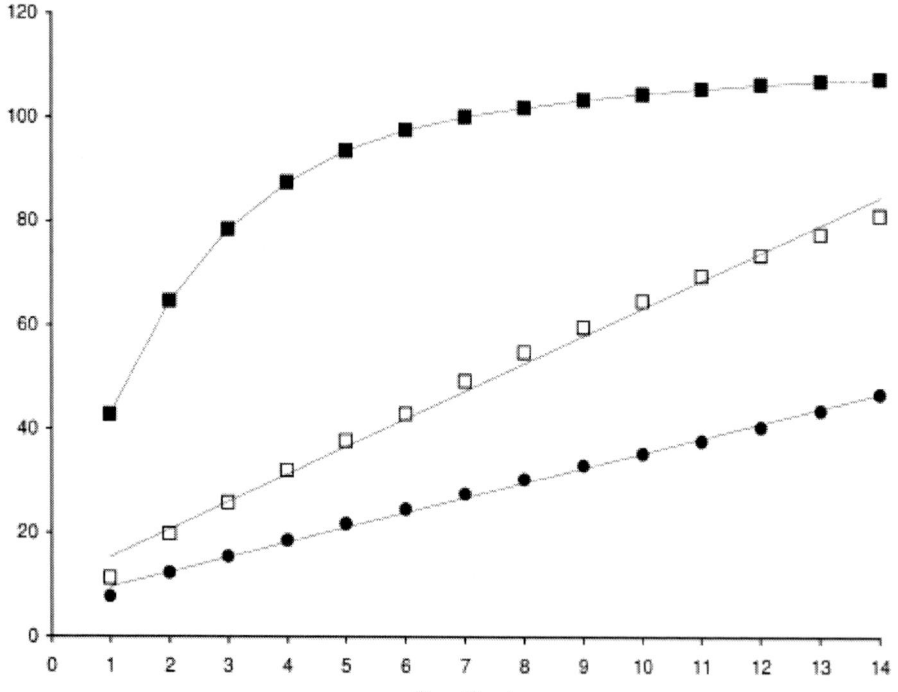

Figure 11. Release of oxybutynin from intravaginal ring devices, formulated as a drug-containing polymer core enveloped by a drug-free silicon layer. The core contained 100 mg drug. (■) Core length: 140.0 mm, drug load: 5%, (□)Core length: 17.5 mm, drug load: 40%, (●) Core length: 11.7 mm, drug load: 60%

3.5. Other (Bio)-Medical Applications

Injection molding is commonly applied to produce parts for medical devices and diagnostic equipment. Examples include catheter tips, luer fittings, syringes, valves, clamps, biodegradable screws in orthopedic surgery [115], microprobes [116], micro-needles [117], bottles, vials and aerosol delivery systems.

Moreover, injection molding can be used in the field of tissue engineering for the manufacturing of scaffolds to restore the structural features and physiological functions of natural tissue in vivo. In most cases, biocompatible, degradable polymers are utilized to induce surrounding tissue ingrowth or to serve as temporary scaffolds for transplanted cells to attach, grow and maintain differentiated functions. Ideally, scaffolds should have highly porous structure (±75%) with an interconnected pore network, for which the optimal pore size for tissue regeneration is dependent on the type of tissue (range 50-750 μm). The material should exhibit good biocompatibility and bio-resorbable with a controllable degradation and resorption rate to match cell/tissue growth. Scaffolds should possess a suitable surface chemistry to allow cell attachment, proliferation and differentiation, and have mechanical properties that matches to those of the tissue at the site of implantation. Convention scaffold manufacturing techniques including woven or non-woven materials from spun fibers, blown films using solvents or propellants, sintered polymer particles, solvent casting and particulate leaching, etc…. . All these techniques have many drawbacks, they make often use of organic solvents, the scaffold lacks structural stability or mechanical strength, or do not address the need for cost-effective manufacturing processes to meet patient needs [39,118]. This favors injection molding as low-cost manufacturing technique, as it capable to mass produce scaffolds with complex geometries due to its repeatability and design flexibility. However, IM can only be successful when a high porosity and interconnectivity can be imparted into the finished product.

Gomes et al. (2001) described the development of scaffolds composed of blends of corn starch with either ethylene vinyl alcohol or cellulose acetate, using a blowing agent to impart porosity in the molded part. In some cases, the scaffolds were reinforced by adding hydroxyapatite. This resulted in the production of scaffolds with a compact skin and a porous core, which had good mechanical and degradation properties. It was possible to tailor the properties and morphology of the end product by simply varying process conditions, material formulations and amount of blowing agent [16].

Another study reported the use of a microcellular injection molding technique which combined supercritical fluid processing and particulate leaching for the production of highly porous and interconnected structures used for tissue engineering. This involved the blending of supercritical carbon dioxide with the polymer melt to create a single-phase polymer-gas solution. The pressure in the barrel is kept around 200 bar in order to maintain this single-phase solution and to ensure homogeneous mixing of gas and polymer prior to injection. Due to the sudden pressure drop during polymer injection the gas emerged from the polymer-gas solution, resulting in the formation of numerous microscale cells. In this way, foamed structures composed of polylactide (PLA) and polyvinyl alcohol (PVA) with sodium chloride as the particulate, were processed without the aid of organic solvents (which can be detrimental to tissue growth). Supercritical carbon dioxide was used as volatile plasticizer and blowing agent, which allowed to process these biodegradable polymers at lower temperatures and to impart porosity into the molded part. As a consequence, microcellular injection molding and subsequent leaching of soluble compounds in water resulted in highly porous PLA foams (>75%) with high interconnectivity [39].

In another study, biodegradable polyurethane scaffolds incorporating tobramycin were prepared by reactive liquid molding to counter infections that compromise the healing of open fractures [119]. Tobramycin release was characterized by a 45-95% burst (tuned by the addition of PEG), followed by up to 2 weeks of sustained release, with a total release 4 to 5-fold greater than equivalent volumes of polymethylmethacyrylate (PMMA) beads. The local antibiotic delivery from PMMA beads is an established clinical treatment of infected fractures, however, surgical removal of the beads is required. This can be avoiding by applying a biodegradable IM scaffold that slowly releases the antibiotic drug.

Another interesting application of injection molding can be found in the domain of bone-analogue composites for temporary or permanent orthopaedic applications. Regarding the bone-analogue concept, the temporary or permanent replacement of hard tissues in load-bearing applications demands mechanically strong biocompatible materials. The attainment of such bone-matching mechanical performance depends on the technical ability of materials to mimic the bone anisotropic character. Biodegradable polymeric composites often reinforced with hydroxyapatite can combine stiffness and strength, resulting in adequate mechanical performance that can be used for implants in high load bearing applications. Sim et al. investigated the properties of high density polyethylene/hydroxyapatite (HDPE/HA) composites via capillary rheometer to evaluate the injection moldability of this composite system. The critical HA concentration was 41%, higher concentrations resulted in voids in the composite [120]. A lubricant was needed to aid processing of blends containing more than 20% HA. Stearic acid reduced the mixing torque allowing a better blending and molding, however it caused a reduction of the elastic modulus. Tetraethoxysilane as coupling agent was ineffective to improve the mechanical properties without microwave heat treatment, however, a slight increase in elastic modulus was observed after heat treatment of composites containing coupling agents.

Sousa et al. (2003) reported the processing of two biodegradable starch-based blends (with ethylenevinyl alcohol copolymer or cellulose acetate) and three high density polyethylene grades via a non-conventional injection molding technique known as shear controlled orientation in injection molding (SCORIM). This was applied to induce a strong anisotropic character to the composites that were reinforced with HA or alternatively in case of HDPE blends with carbon fibres (C fibres) [121]. Mono-sandwich injection molding was applied to develop composites that combine a HDPE/HA outer layer and a HDPE/C fibre reinforced core. These authors concluded that it was possible to produce bone-analogue composites with a mechanical performance comparable to human cortical bone. An excellent review paper that examines the use of different types of polymeric matrix composites (HDPE/HA, biodegradable composites, starch- and ceramic-based polymers, ...) in hard tissue replacement applications is provided by Mano et al. [122].

Conclusion and Outlook

Injection molding represents a relative new, but very promising manufacturing technique for a wide range of drug delivery systems in the pharmaceutical industry. Injection molding has become an attractive alternative to traditional processing technologies since this technique offers many advantages, including a fully automated and continuous production set-up, design flexibility, absence of water or organic solvents, and production of drug delivery systems with a skin/core morphology. It has also been shown that IM permits an accurate control of the design of the drug delivery system, and that the processing conditions can sometimes dictate its morphology and product performance. By carefully choosing the processing parameters, the polymer chain orientation, micro-porous structure and crystallinity of some specific drug delivery system could be altered, which had a profound impact on the drug release behavior. In addition, the homogeneous embedding of drug molecules in carrier matrices via IM is a well-established and successful approach for sustained release applications.

Another main advantage of this technology comprises the fact that IM can offer a solution for drug delivery issues of poorly soluble drugs, since this technique allows the manufacturing of solid solution and dispersions in a controllable and reproducible way. In this regard, it is the belief that IM can be used to "recycle" some promising poorly soluble active pharmaceutical ingredients that have been previously abandoned during drug discovery due to suboptimal formulation/technology approaches, or for new chemical entities formulated as solid dispersion dosage forms, resulting in an acceptable drug bioavailability.

Injection molding for drug delivery purposes, however, was only able to break through after the introduction of a broad selection of (thermoplastic) polymers capable of tailoring drug release. In addition, the availability of different polymeric carriers, plasticizers, processing aids and functional excipients provided new perspectives for formulation research, and resulted in numerous publications and patents issued over the last decade. Although IM is restricted to thermo-insensitive pharmaceutical compounds, some heat- or shear-sensitive drug molecules and peptides have been successfully incorporated into IM or HME dosage forms.

As this review has demonstrated, IM was applied successfully for the production of a wide variety of controlled release drug delivery systems, medical devices, (biodegradable) medical implants and stents, bone-analogue composites, tissue-engineered scaffolds and vaginal rings, clearly illustrating the potential of IM. Moreover, recent advances in the field of micro-molding made mass production of miniaturized parts (often less than 0.1 gram, with a highly accurate dimensional control and precision) feasible, due to extremely short IM cycle times (3-4 seconds) [123]. Numerous applications have emerged from this technology, in particular (bio)-medical devices such as micro-needles, luer-fittings, micro-probes,

There is also a stringent need for in-process control to further fine-tune the properties of injection molded products, thereby improving process understanding and increasing the success rate of the development of a pharmaceutical product that meets all the regulatory requirements and shows excellent product performance. Conventionally product quality in the pharmaceutical industry is assessed via time-consuming and less efficient off-line laboratory testing conducted on randomly collected samples. This provides limited information about the process, making process control more difficult and increasing the risk of batch rejection after manufacturing. The manufacturing processes are often not fully understood and are often inefficient black-boxes. Based on these observations, the Food and Drug Administration launched in 2004 the Process Analytical Technology (PAT) initiative, which forms the basis of the pharmaceutical Good Manufacturing Practice (GMP) rules for the 21st century [124,125]. The ultimate goal of PAT is a better fundamental scientific understanding of manufacturing processes (i.e., the process should not be a black box system) to increase process efficiency. One of the most important statements within the PAT concept is that '*quality should not be tested into products; it should be built in'*. PAT plays a crucial role in design, analysis, and control of manufacturing processes via in-line, on-line and at-line measurements (i.e., during processing) of critical process and material parameters. Via real-time information of all critical process aspects (i.e., continuous process monitoring), it is the objective of PAT to guide the process towards its desired state (i.e., continuous process control), hence ensuring the quality of the end product and possibly allowing real-time release. Continuous gathering of critical process information during production should allow real-time process adjustments to keep product specifications within predefined limits, hence avoiding batch loss. The implementation of PAT in the pharmaceutical industry is strongly encouraged by the pharmaceutical regulatory authorities, but its implementation and the use of in-line monitoring tools in pharmaceutical injection molding processes is still in its infancy. Recently, it has been demonstrated how real-time process and product information can be obtained during hot-melt extrusion. Tumuluri et al. (2008) continuously scanned the surface of a HME film with a Raman probe immediately at the exit of the extruder [126], allowing in-line quantification of the API as well as solid state identification of API and polymer. The implementation of probes allowing continuous monitoring of (spectral) information, without disturbing the process itself, is still a challenge. Recently, Zeaiter et al. (2010) presented a cavity sensor technology enabling in-line monitoring of cavity pressure and temperature during plastic injection molding for real-time evaluation of the weight and dimensions of the final product [127].

A novel research area within the field of HME and IM comprises the reprocessing of materials, which is becoming increasingly important for economical as well as environmental reasons. As the costs of good for active ingredients and polymers are increasing worldwide, especially for petroleum-derived polymers or new drug entities, the reprocessing of rejected batches and IM waste can become interesting. Although thermostable drugs can be processed without problem, there have been no studies published that assessed the effect of multiple HME or IM cycles on drug stability and on the performance of the drug delivery device. However, many concerns will arise how material reprocessing can be successfully implemented in a GMP-regulated production set-up. Especially, as it is known that plastic reprocessing yields products with different properties compared to virgin polymeric material [128]. Javierre et al. (2006) reported that the addition of different percentages of reprocessed material to raw material for injection molding processes changed the rheologic behavior of the material, affecting both mold filling and process conditions [129].

Another possible solution to reduce material loss involves the use of "hot runner systems", in which molten material is transferred from the nozzle of the injection molding machine through heated runners and which is then metered precisely into the cavity of the mold. These systems avoid that material cools down in the runners, which is subsequently discarded with the molded part during the ejection phase as seen with conventional cold runner systems. These systems yield a significant reduction in the amount of material waste during processing. However, hot runners usually make the mold more expensive to manufacture.

The interest in injection molding as pharmaceutical production technology will certainly continue to grow, and this chapter has illustrated that injection molding is an efficient and attractive production technology for a variety of pharmaceutical drug delivery systems.

Retrofitting injection molding machines, increasing numbers of active ingredients, pharmaceutical grade polymers and excipients all provide a promising future for this versatile pharmaceutical production technology.

REFERENCES

[1] Berins, M. L. (1991). *SPI Plastics Engineering Handbook of the Society of the Plastics Industy Inc.* (5th Edition): Injection Molding of Thermoplastics. Springer-Verlag, Berlin, 133-178.

[2] Beck, H. (1952). Injection moulding machine. *German Patent*, *858*, 310.

[3] Ram, A. (1997). Fundamentals of polymer engineering: *Injection molding*. Springer-Verlag, Berlin, 125-133.

[4] Bar-Shalom, D., Slot, L., Lee, W. W. & Wilson, C. G. (2003). Modified-Release Drug Delivery Technology: Development of the Egalet technology. In: M. J., Rathbone, J. Hadgraft, & M. S. Roberts, (Eds.), *Marcel Dekker, Inc*., New York, 263-271.

[5] Verma, R. K., Krishna, D. M. & Garg, S. (2002). Formulation aspects in het development of osmotically controlled oral drug delivery systems. *J. Control. Rel*., *79*, 7-27.

[6] Breitenbach, J. (2002). Melt extrusion: from process to drug delivery technology. *Eur. J. Pharm. Biopharm*., *54*, 107-117.

[7] Crowley, M. M., Zhang, F., Repka, M., Thumma, S., Upadhye, S. B., Kumar, S., McGinity, J. W. & Martin, C. (2007). Pharmaceutical applications of hot-melt extrusion: Part I. *Drug Dev. Ind. Pharm*., *33*, 909-926.

[8] Repka, M., Koleng, J. J., Zhang, F. & Mcginity, J. W. (2002). Encyclopedia of Pharmaceutical Technology: Hot-melt extrusion technology. In: J. Swarbrick, & J. C. Boylan, (Eds.), *Marcel Dekker, Inc*., New York, 1488-1505.

[9] Repka, M., Majumdar, S., Kumar, S., Srirangam, R. & Updadhye, S. B. (2008). Applications of hot-melt extrusion for drug delivery. *Expert Opin. Drug Deliv*., *5*, 1357-1376.

[10] Huang, B., Castro, D., Gale, D. & Hossainy, S. F. A. (2008). *Fabricating polymer stents with injection molding*. US Patent No. 20080001330.

[11] Rothen-Weinhold, A., Besseghir, K., Vuaridel, E., Sublet, E., Oudry, N., Kubel, F. & Gurny, R. (1999). Injection-molding versus extrusion as manufacturing technique for the preparation of biodegradable implants. *Eur. J. Pharm. Biopharm*., *48*,113-121.

[12] McGinity, J. W. & Zhang, F. (2003). Pharmaceutical extrusion technology: Melt Extruded Controlled Release Dosage Forms. In: I., Ghebre-Sellassie, & C. Martin, (Eds.), *Marcel Dekker, Inc*., New York, 183-208.

[13] Langer, R. (1980). Polymeric delivery systems for controlled drug release. *Chem. Eng. Commun*., *6*, 1-48.

[14] Ruprecht, R., Gietzelt, T., Müller, K., Piotter, V. & Haußelt, J. (2002). Injection molding of microstructured components from plastics, metals and ceramics. *Microsyst. Technol*., *8*, 351-358.

[15] Bakar, A., Cheang, P. & Khor, K. A. (2003). Tensile properties and microstructural analysis of spheroidized hydroxyapatite–poly (etheretherketone) biocomposites. *Mater Sci. Eng. A*, *345*, 55-63.

[16] Gomes, M. E., Ribeiro, A. S., Malafaya, P. B., Reis, R. L. & Cunha, A. M. (2001). A new approach based on injection moulding to produce biodegradable starch-based polymeric scaffolds: morphology, mechanical and degradation behavior. *Biomaterials*, *22*, 883-889.

[17] Rathbone, M. J., Bunt, C. R., Ogle, C. R., Burggraaf, S., Macmillan, K. L. & Pickering, K. (2002). Development of an injection molded poly (ε-caprolactone) intravaginal insert for the delivery of progesterone to cattle. *J. Control. Rel.*, *85*, 61-71.

[18] Li, P. & Zhao, L. (2007). Developing early formulations: Practice and perspective. *Int. J. Pharm.*, *341*, 1-19.

[19] Adams, C. P. & Brantner, V. V. (2006). Estimating the cost of new drug development: Is it really $802 million? *Health affairs*, *25*, 420-428.

[20] Pasquali, I. & Bettini, R. (2008). Are pharmaceutics really going critical? *Int. J. Pharm.*, *364*, 176-187.

[21] Langer, R. & Peppas, N. (1983). Chemical and physical structure of polymers as carriers for controlled release of bioactive agents: A review. *J. Macromol. Sci. - Rev. Macromol. Chem. Phys.*, *23*, 61-126.

[22] Ranade, V. V. (1991). Drug delivery systems 5A. Oral drug delivery. *J. Clin. Pharmacol.*, *31*, 2-16.

[23] Kumar, M. N. V. R. & Kumar, N. (2001). Polymeric controlled drug-delivery systems: Perspective Issues and opportunities. *Drug Dev. Ind. Pharm.*, *27*, 1-30.

[24] Nair, L. S. & Laurencin, C. T. (2007). Biodegradable polymers as biomaterials. *Prog. Polym. Sci.*, *32*, 762-798.

[25] Kanjickal, D. G. & Lopina, S. T. (2004). Modeling of drug release from polymeric delivery systems- A review. Crit. Rev.™ in *Therapeutic Drug Carrier Systems*, *21*, 345-386.

[26] Siepmann, J. & Göpferich, A. (2001). Mathematical modeling of bioerodible, polymeric drug delivery systems. Adv. *Drug Deliv. Rev.*, *48*, 229-247.

[27] Crowley, M. M., Zhang, F., Koleng, J. J. & Mcginity, J. W. (2002). Stability of polyethylene oxide in matrix tablets prepared by hot-melt extrusion. *Biomaterials*, *23*, 4241-4248.

[28] Forster, A., Hempenstall, J., Tucker, I. & Rades, T. (2001). The potential of small-scale experiments and the Gordon-Tyler equation to predict the suitability of drug/polymer blends for melt extrusion. *Drug Dev. Ind. Pharm.*, *27*, 549-560.

[29] Forster, A., Hempenstall, J. & Rades, T. (2001). Characterization of glass solutions of poorly water-soluble drugs produced by melt extrusion with hydrophyllic amorphous polymers. *J. Pharm. Pharmacol.*, *53*, 303-315.

[30] Amellal, K., Tzoganakis, C., Penlidis, A. & Rempel, G. L. (1994). Injection molding of medical plastics: a review. *Adv. Polym. Technol.*, *4*, 315-322.

[31] Gogolewski, S. & Mainil-Varlet, P. (1996). The effect of thermal treatment on sterility, molecular and mechanical properties of various polylactides. *Biomaterials*, *17*, 523-528.

[32] Omelzuck, M. O. & Mcginity, J. W. (1992). The influence of polymer glass transition temperature and molecular weight on drug release from tablets containing poly(DL-lactic acid). *Pharm. Res.*, *9*, 26-32.

[33] Quinten, T., De Beer, T., Vervaet, C. & Remon, J. P. (2009). Evaluation of injection moulding as a pharmaceutical technology to produce matrix tablets. *Eur. J. Pharm. Biopharm.*, *71*, 145-154.

[34] Souza Rosa, R. C. R. & Andrade, C. T. (2004). Effect of chitin addition on injection-molded thermoplastic corn starch. *J. Appl. Polym. Sci.*, *92*, 2706-2713.

[35] Mano, J. F., Koniarova, D. & Reis, R. L. (2003). Thermal properties of thermoplastic starch/synthetic polymer blends with potential biomedical applicability. *J. Mater. Sci.-Mater. Med.*, *14*, 127-135.

[36] Tummala, P., Liu, W., Drzal, L. T., Mohanty, A. K. & Misra, M. (2006). Influence of plasticizers on thermal and mechanical properties and morphology of soy-based bioplastics. Ind. Eng. *Chem. Res.*, *45*, 7491-7496.

[37] Tatavarti, A. S., Dollimore, D. & Alexander, K. S. (2002). A Thermogravimetric Analysis of Non-polymeric Pharmaceutical Plasticizers: Kinetic Analysis, Method Validation, and Thermal Stability Evaluation. *AAPS PharmSci.*, *4*, 231-242.

[38] Thumma, S., ElSohly, M. A., Zhang, S., Gul, W. & Repka, M. A. (2008). Influence of plasticizers on the stability and release of a prodrug of Δ^9-tetrahydrocannabinol incorporated in poly(ethylene oxide) matrices. Eur. J. *pharm. Biopharm.*, *70*, 605-614.

[39] Kramschuster, A. & Turng, L. S. (2010). An injection molding process for manufacturing highly porous and interconnected biodegradable polymer matrices for use as tissue engineering scaffolds. *J. Biomed. Mater. Res. Part B*, *92B*, 366-376.

[40] Verreck, G., Decorte, A., Li, H. B., Tomasko, D., Arien, A., Peeters, J., Rombaut, P., Van den Mooter, G. & Brewster, M. E. (2006). The effect of pressurized carbon dioxide as a plasticizer and foaming agent on the hot melt extrusion process and extrudate properties of pharmaceutical polymers. *J. Supercrit. Fluids*, *38*, 383-391.

[41] Lyons, J. G., Hallinan, M. H., Kennedy, J. E., Devine, D. M., Geever, L. M., Blackie, P. & Higginbotham, C. L. (2007). Preparation of monolithic matrices for oral drug delivery using a supercritical fluid assisted hot melt extrusion process. *Int. J. Pharm.*, *329*, 62-71.

[42] Ghebremeskel, A. N., Vemavarapu, C. & Lodoya, M. (2007). Use of surfactants as plasticizer in preparing solid dispersions of poorly soluble API: Selection of polymer-surfactant combinations using solubility parameters and testing the processability. *Int. J. Pharm.*, *328*, 119-129.

[43] De Brabander, C., Van den Mooter, G., Vervaet, C. & Remon, J. P. (2002). Characterisation of Ibuprofen as a nontraditional plasticizer of ethylcellulose. *J. Pharm. Sci.*, *91*, 1678-1685.

[44] Siepmann, F., Le Brun, V. & Siepmann, J. (2006). Drugs acting as plasticizers in polymeric systems: A quantitative treatment. *J. Control. Rel.*, *115*, 298-306.

[45] Repka, M. A., Gerding, T. G., Repka, S. L. & Mcginity, J. W. (1999). Influence of plasticizers and drugs on the physical-mechanical properties of hydroxypropylcellulose films prepared by hot-melt extrusion. Drug Dev. *Ind. Pharm.*, *25*, 625-633.

[46] Tate, S., Chiba, S. & Tani, K. (1996). Melt viscosity reduction of poly(ethylene terephthalate) by solvent impregnation. *Polymers*, *37*, 4421-4424.

[47] Schilling, S. U., Shah, N. H., Malick, A. W., Infeld, M. H. & McGinity, J. W. (2007). Citric acid as a solid-state plasticizer for Eudragit RS PO. *J. Pharm. Pharmacol.*, *59*, 1493-1500.

[48] Zhu, Y., Shah, N. H., Malick, A. W., Infeld , M. H. & McGinity, J. W. (2004). Influence of a lipophilic thermal lubricant on the processing conditions and drug release properties of chlorpheniramine maleate tablets prepared by hot-melt extrusion. *Journal of drug delivery science and technology*, *14*, 313-318.

[49] Chokshi, R. J., Sandhu, H. K., Iyer, R. M., Shah, N. H., Malick, A. W. & Zia, H. (2005). Characterization of physic-mechanical properties of indomethacin and polymers to assess their suitability for hot-melt extrusion process as a means to manufacture solid dispersion/solution. *J. Pharm. Sci.*, *94*, 2463-2474.

[50] Forster, A., Hempenstall, J., Tucker, I. & Rades, T. (2001). Selection of excipients for melt extrusion with two poorly water-soluble drugs by solubility parameter calculation and thermal analysis. *Int. J. Pharm.*, *226*, 147-161.

[51] Hancock, B. C., York, P. & Raymond, C. R. (1997). The use of solubility parameters in pharmaceutical dosage form design. Int. *J. Pharm.*, *148*, 1-21.

[52] Greenhalgh, D. J., Williams, A. C., Timmins, P. & York, P. (1999). Solubility parameters as predictors of miscibility in solid dispersions. *J. Pharm. Sci.*, *88*, 1182-1190.

[53] Maus, M., Wagner, K. G., Kornherr, A. & Zifferer, G. (2008). Molecular dynamics simulations for drug dosage form development: thermal and solubitily characteristics for hot-melt extrusion. *Mol. Simul.*, *34*, 1197-1207.

[54] Giron, D. (1998). Contribution of thermal methods and related techniques to the rational development of pharmaceuticals-part 1., *1*, 191-199.

[55] Munjal, M., Stodghill, S. P., Elsohly, M. A. & Repka, M. A. (2006). Polymeric systems for amorphous Δ^9-tetrahydrocannabinol produced by a hot-melt method. Part I:Chemical and thermal stability during processing. *J. Pharm. Sci.*, *95*, 1841-1853.

[56] Munjal, Elsohly, M. A. & Repka, M. A. (2006). Polymeric systems for amorphous Δ^9-tetrahydrocannabinol produced by a hot-melt method. Part II: Effect of oxidation mechanisms and chemical interactions on stability. *J. Pharm. Sci.*, *95*, 2473-2485.

[57] De Brabander, C., Vervaet, C. & Remon, J. P. (2003). Development and evaluation of sustained release mini-matrices prepared via hot melt extrusion. *J. Control. Rel.*, *89*, 235-247.

[58] Crowley, M. M., Schroeder, B., Fredersdorf, A., Obara, S., Talarico, M., Kucera, S. & McGinity, J. W. (2004). Physicochemical properties and mechanism of drug release from ethyl cellulose matrix tablets prepared by direct compression and hot-melt extrusion. *Int. J. Pharm.*, *269*, 509-522.

[59] Zhu, Y., Shah, N. H., Malick, A. W., Infeld, M. H. & McGinity, J. W. (2002). Influence of thermal processing on the properties of chlorpheniramine maleate tablets containing an acrylic polymer. *Pharm. Dev. Technol.*, *7*, 481-489.

[60] Quinten, T., Gonissen, Y., Cnudde, V., Masschaele, B., Van Hoorebeke, L., Remon, J. P. & Vervaet, C. (2009). Development of injection moulded matrix tablets based on mixtures of ethylcellulose and low-substituted Hydroxypropylcellulose. *Eur. J. Pharm. Sci.*, *37*, 207-216.

[61] Lipinski, C. A., Lombardo, F., Dominyl, B. W. & Feeney, P. J. (1997). Experimental and computational approaches to estimate solubility and permeability in drug discovery and development settings. *Adv. Drug Deliv. Rev.*, *23*, 3-25.

[62] Wadke, D. A., Serajuddin, A. T. M. & Jacobson, H. (1989). Preformulation testing. In *Pharmacuetical dosage forms: Tablets,* Vol. *1*. H. A., Lieberman, L. & Lachman, J. B. Schwartz, (Eds), Marcel Dekker: New York, 1-73.

[63] Fahr, A. & Liu, X. (2007). Drug delivery strategies for poorly water-soluble drugs. Expert Opin. *Drug Deliv.*, *4*, 403-416.

[64] Leuner, C. & Dressman, J. (2000). Improving drug solubility for oral delivery using solid dispersions. Eur. *J. Pharm. Biopharm.*, *50*, 47-60.

[65] Sethia, S. (2003). Solid dispersions: Revival with greater possibilities and applications in oral drug delivery. Crit. Rev.™ in *Therapeutic Drug Carrier Systems*, *20*, 215-247.

[66] Sekiguchi, K. & Obi, N. (1961). Studies on absorption of eutectic mixtures. 1. A comparison of the behaviour of eutectic mixtures of sulfathiazole and that of ordinary sulfathiazole in man. *Chem. Pharm. Bull.*, *9*, 866-872.

[67] Breitenbach, J. & Mägerlein, M. (2003). Pharmaceutical extrusion technology: Melt-extruded molecular dispersions. In: I., Ghebre-Sellassie, & C. Martin, (Eds.), Marcel Dekker, *Inc.*, New York, 245-260.

[68] Verreck, G. & Brewster, M. (2004). Melt extrusion-based dosage forms: excipients and processing conditions for pharmaceutical formulations. *Bulletin Technique Gatefossé*, *97*, 85-94.

[69] Craig, D. Q. M. (2002). The mechanisms of drug release from solid dispersions in water-soluble polymers. *Int. J. Pharm.*, *231*, 131-144.

[70] Serajuddin, A. T. M. (1999). Solid dispersions of poorly water-soluble drugs: Early promises, subsequent problems, and recent breakthrougs. *J. Pharm. Sci.*, *88*, 1058-1066.

[71] Sinha, S., Baboota, S., Ali, M., Kumar, A. & Ali, J. (2009). Solid dispersions: an alternative technique for bioavailability enhancement of poorly soluble drugs. J. Dispersion *Sci.Technol.*, *30*, 1458-1473.

[72] Chokski, R. J., Shah, N. H., Sandhu, H. K., Malick, A. W. & Zia, H. (2007). Stabilisation of low glass transition temperature indomethacin formulations: impact of polymer-type and its concentration. *J. Pharm. Sci.*, *97*, 2286-2298.

[73] Hancock, B. C. & Zografi, G. (1997). Characteristics and significance of the amorphous state in pharmaceutical systems. *J. Pharm. Sci.*, *86*, 1-12.

[74] Hancock, B. C., Shamblin, S. L. & Zografi, G. (1995). Molecular mobility of amorphous pharmaceutical solids below their glass transition temperatures. *Pharm. Res.*, *12*, 835-842.

[75] Wacker, S., Soliva, M. & Speiser, P. (1991). Injection molding as a suitable process for manufacturing solid dispersions or solutions. *Pharm. Ind.*, *53*, 853-856.

[76] Breitenbach, J. & Lewis, J. (2003). Modified-Release Drug Delivery Technology: Two Concepts, One Technology: Controlled-Release and Solid Dispersions with Meltrex. In: M. J., Rathbone, J. Hadgraft, & M. S. Roberts, (Eds.), Marcel Dekker, *Inc.*, New York, 125-134.

[77] Breitenbach, J. (2006). Melt Extrusion Can Bring New Benefits to HIV Therapy: The Example of Kaletra® Tablets. *Am. J. Drug Deliv.*, *4*, 61-64.

[78] Klein, C. E., Chiu, Y. L., Awni, W., Zhu, T., Heuser, R. S., Doan, T., Breitenbach, J., Morris, J. B., Brun, S. C. & Hanna, G. J. (2007). The tablet formulation of Lopinavir/Rotinavir provides similar bioavailability to the sof-gelatin capsule formulation with less pharmacokinetic viriability and diminished food effect. *J. Acquire Immune Defic. Syndr.*, *44*, 401-410.

[79] Chien, Y. W. & Lin, S. (2002). Encyclopedia of Pharmaceutical Technology: Drug delivery-Controlled release. In: J., Swarbrick, & J. C. Boylan, (Eds.), Marcel Dekker, *Inc.*, New York, 811-833.

[80] Colombo, P., Sonvico, F., Colombo, G. & Bettini, R. (2009). Novel platforms for oral drug delivery. *Pharm. Res.*, *26*, 601-611.

[81] El-Agakey, M. A., Soliva, M. & Speiser, P. (1971). Hot extruded dosage forms. Part 1. Technology and dissolution kinetics of polymeric matrices. Pharm. *Acta Helv.*, *46*, 31-52.

[82] Hüttenrauch, R. (1974). Spritzgieβverfahren zur herstellung peroraler retardpräparate. Teil 1: Über das spritzgieβen. *Pharmazie*, *29*, 297-302.

[83] Hüttenrauch, R. & Schmeiss, U. (1976). Spritzgieβverfahren zur herstellung peroraler retardpräparate. Teil 6: Veränderung der äuβeren oberflächengeometrie. *Pharmazie*, *31*, 724-728.

[84] Cuff, G. & Raouf, F. (1996). A preliminary evaluation of injection molding as a technology to produce tablets. Pharm. *Technol.*, *22*, 96-106.

[85] Eith, L., Stepto, R. F. T., Tomka, I. & Wittwer, F. (1986). The injection-moulded capsule. Drug Dev. *Ind. Pharm.*, *12*, 2113-2126.

[86] Stepto, R. F. T. & Tomka, I. (1987). Injection moulding of natural hydrophilic polymers in the presence of water. *Chimia*, *41*, 76-81.

[87] Stepto, R. F. T. (1997). Thermoplastic starch and drug delivery capsules. *Polym. Int.*, *43*, 155-158.

[88] Stepto, R. F. T. (2000). Thermoplastic starch. *Macromol. Symp.*, *152*, 73-82.

[89] Stepto, R. F. T. (2003). The processing of starch as a thermoplastic. *Macromol. Symp.*, *201*, 203-212.

[90] Vilivalam, V. D., Illum, L. & Iqbal, K. (2000). Starch capsules: an alternative system for oral drug delivery. *Pharm. Sci. Technol.* Today, *3*, 64-69.

[91] Watts, P. J. & Illum, L. (1997). Colonic drug delivery. Drug Dev. *Ind. Pharm.*, *23*, 893-913.

[92] Keszie, S., Szabo, A., Marosi, G., Anna, P. & Nagy, S. (2006). Use of thermoplastic starch in continuous pharmaceutical process. *Macromol. Symp.*, *239*, 101-104.

[93] Liu, H., Xie, F., Yu, L., chen, L. & Li, L. (2009). Thermal processing of starch-based polymers. *Prog. Polym. Sci.*, *34*, 1348-1368.

[94] Bar-Shalom, D., Slot, L., Lee, W. W. & Wilson, C. G. (2003). Modified-Release Drug Delivery Technology: Development of the Egalet technology. In: M. J., Rathbone, J., Hadgraft, & M. S. Roberts, (Eds.), Marcel Dekker, *Inc.*, New York, 263-271.

[95] Bar-Shalom, (2004). *Controlled release composition*. US patent 6,787,156.

[96] Bar-shalom, Bukh, N. & Larsen, T. K. (1991). Egalet®, a novel controlled-release system. *Ann. N.Y. Acad. Sci., 618*, 578-580.

[97] Marvola, J., Kanerva, H., Slot, L., Lipponen, M., Kekki, T., Hietanen, H., Mykkänen, S., Ariniemi, K., Lindevall, K. & Marvola, M. (2004). Neutron activation-based gamma

scintigraphy in pharmacoscintigraphic evaluation of an Egalet® constant-release drug delivery system. *Int. J. Pharm.*, *281*, 3-10.

[98] Vaz, C. M., van Doeveren, P. F. N. M., Reis, R. L. & Cunha, A. M. (2003). Development and design of double-layer co-injection moulded soy protein based drug delivery devices. *Polymer*, *44*, 5983-5992.

[99] Vaz, C. M., van Doeveren, P. F. N. M., Reis, R. L. & Cunha, A. M. (2003). Soy matrix drug delivery systems obtained by melt-processing techniques. *Biomacromolecules*, *4*, 1520-1529.

[100] Vaz, C. M., van Doeveren, P. F. N. M., Dias, G. R., Reis, R. L. & Cunha, A. M. (2004). Controlled delivery archieved with bi-layer matrix devices produced by co-injection moulding. *Macromol. Biosci.*, *4*, 795-801.

[101] Deng, J. S., Li, L., Tian, Y., Meisters, M., Chang, H. C., Stephens, D., Chen, S. & Robinson, D. (2001). The relationship between structures and in vitro properties of a polyanhydride implant containing gentamlicin sulfate. *Pharm. Dev. Technol.*, *6*, 541-549.

[102] Deng, J. S., Meisters, M., Li., L., Setesak, J., Claycomb, L., Tian, Y., Stephens, D. & Widman, M. (2002). The development of an injection-molding process for a polyanhydride implant containing gentamicin sulfate. PDA. *J. Pharm. Sci. Technol.*, *56*, 65-77.

[103] Parikh, M., Gross, R. A. & McCarthy, S. P. (1998). The influence of injection molding conditions on biodegradable polymers. *J. Injection Molding Technol.*, *2*, 30-36.

[104] Deng, J. S., Luk, L., Stephens, D., Tian, Y., Harris, F. & Cheng, S. Z. D. (2002). Effect of γ-radiation on a polyanhydride implant containing gentamicin sulfate. *Int. J. Pharm.*, 232, 1-10.

[105] Li, L. C., Deng, J. & Stephens, D. (2002). Polyanhydride implant for antibiotic delivery-from the bench to the clinic. *Adv. Drug Deliv. Rev.*, *54*, 963-986.

[106] Li, J. D., Stephens, D., Tian, Y. & Robinson, D. (2004). Effect of postmolding heat treatment on in vitro properties of a polyanhydride implant containing gentamicin sulfate. Drug Dev. *Ind. Pharm.*, *30*, 341-346.

[107] Soriano, I., Martin, A. Y., Evora, C. & Sanchez, E. (2006). Biodegradable implantable fluconazole delivery rods designed for the treatment of fungal osteomyelitis: influence of gamma sterilization. *J. Biomed. Mater. Res. Part A*, *77A*, 632-638.

[108] Hirematch, J. G., Devi, V. K., Devi, K. & domb, A. J. (2008). Biodegradable poly(sebacic acid-co-ricinoleic-ester anhydride) tamoxifen citrate implants: preparation and in vitro characterization. *J. Appl. Polym. Sci.*, *107*, 2745-2754.

[109] Cheng, L., Guo, S. & Wu, W. (2009). Characterization and in vitro release of praziquantel from poly(ε-caprolactone) implants. *Int. J. Pharm.*, *377*, 112-119.

[110] Li, C., Cheng, L., Zhang, Y., Guo, S. & Wu, W. (2010). Effects of implant diameter, drug loading and end-capping on praziquantel release from PCL implants. *Int. J. Pharm.*, *386*, 23-29.

[111] Sam, A. P. (1992). Controlled release contraceptive devices- A status report. *J. Control. Rel.*, *22*, 35-46.

[112] van Laarhoven, J. A. H., Kruft, M. A. B. & Bromans, H. (2002). Effect of supersaturation and crystallization phenomena on the release properties of a controlled release device based on EVA copolymer. *J. Control. Rel.*, *82*, 309-317.

[113] van Laarhoven, J. A. H., Kruft, M. A. B. & Vromans, H. (2002). In vitro release properties of etonogestrel and ethinyl estradiol from a contraceptive vaginal ring. *Int. J. Pharm.*, *232*, 163-173.

[114] Woolfson, A. D., Malcolm, R. K. & Gallagher, R. J. (2003). Design of a silicone reservoir intravaginal ring for the delivery of oxybutynin. *J. Control. Rel.*, *91*, 465-476.

[115] Burkle, E., Schneebauer, M. & Wurtele, M. (2008). A screw concept for bone substance. *Kunststoffe International*, *98*, 56-59.

[116] Chang, P. C. & Hwang, S. J. (2006). Injection molding of microprobe array parts. *J. Polym. Res.*, *13*, 25-32.

[117] Wang, M. W., Fu, G. L. & Jeng, J. H. (2008). Optimal molding parameter design of PLA micro lancet needles using Taguchi method. Polym.-Plast. *Technol. Eng.*, *48*, 730-735.

[118] Yang, S. F., Leong, K. F., Du, Z. H. & Chua, C. K. (2001). The design of scaffolds in tissue engineering. Part 1. Traditional factors. *Tissue Eng.*, *7*, 679-689.

[119] Hafeman, A. E., Zienkiewicz, K. J., Carney, E., Litzner, B., Stratton, C., Wenke, J. C. & Guelcher, S. A. (2010). Local Delivery of Tobramycin from Injectable Biodegradable Polyurethane Scaffolds. *J. Biomater. Sci.-Polym. Ed.*, *21*, 95-112.

[120] Sim, C. P., Cheang, P., Liang, M. H. & Khor, K. A. (1997). Injection moulding of hydroxyapatite composites. *J. Mater. Process. Technol.*, *69*, 75-78.

[121] Sousa, R. A., Reis, R. L., Cunha, A. M. & Bevis, M. J. (2003). Processing and properties of bone-analogue biodegradable and bioinert polymeric composites. *Compos. Sci. Technol.*, *63*, 389-402.

[122] Mano, J. F., Sousa, R. A., Boesel, L. F., Neves, N. M. & Reis, R. L. (2004). Bioinert, biodegradable and injectable polymeric matrix composited for hard tissue replacement: state of the art and recent developments. Compos. *Sci. Technol.*, *64*, 789-817.

[123] Heckele, M. & Schomburg, W. K. (2004). Review on micro molding of thermoplastic polymers. *J. Micromech. Microeng.*, *14*, R1-R14.

[124] US Food and Drug Administration. *PAT—A Framework for Innovative Pharmaceutical Development, Manufacturing, and Quality Assurance, Guidance for Industry.* Washington, DC: US Food and Drug Administration, Center for Drug Evaluation and Research; 2004.

[125] Balboni, M. L. (2003). Process analytical technology: concepts and principles. *Pharm Technol.*, *27*, 54-66.

[126] Tumuluri, V. S., Kemper, M. S., Lewis, I. R., Prodduturi, S., Majumdar, S., Avery, B. A. & Repka, M. A. (2008). Off-line and on-line measurements of drug-loaded hot-melt extruded films using Raman spectroscopy. *Int. J. Pharm.*, *357*, 77-84.

[127] Zeaiter, M., Knight, W. & Blackwell, B. IFPAC (2010). *twenty-fourth international forum process analytical technology*, January 31 – February 4, 2010, Baltimore, USA.

[128] Henshaw, J. M., Han, W. J. & Owens, A. D. (1996). An overview of recycling issues for composite materials. *J. Thermoplast. Compos. Mater.*, *9*, 4-20.

[129] Javierre, C., Claveria, I., Ponz, L., Aisa, J. & Fernandez, A. (2006). Influence of the recycled material percentage on the rheological behaviour of HDPE for injection moulding process. *Waste Manage.*, *27*, 656-663.

In: Injection Molding: Process, Design, and Applications ISBN: 978-1-61761-307-4
Editor: Phoebe H. Kauffer

Chapter 2

MELT/SOLID WELDLINE IN OVER INJECTION MOLDING: INTERFACIAL CRYSTALLINE STRUCTURES AND ADHESION BETWEEN SEMICRYSTALLINE POLYMER INTERFACES

Hong Wu
Polymer Research Institute of Sichuan University,
The State Key Laboratory of Polymer Materials Engineering, Chengdu, China

INTRODUCTION

Sequential injection molding, also called over-injection molding (OIM), which offers in situ process to join thermopolastics has demonstrated considerable importance in the field of polymer processing due to short manufacturing times and low cost[1-3]. In comparison with the conventional injection molding, sequential overinjection molding includes multi-stage operation: the first part of desired figure is molded by one-stage injection, and then the melt for the second part is subsequently injected over the surface of the first part molded in the previous process. So the bonding characteristics of the interface between the first and second part are very important for the molded products. Some works have been reported on the studies of the relationships between interfacial adhesion and processing parameters used in sequential injection molding[4-8].

Combined with the special thermomechanical environments in the vicinity of the interface formed during injection molding, the complex crystalline structures can be obtained at semicrystalline polymer interfaces due to the fountain flow along injection direction. Therefore, it is important to study the interfacial crystalline structure and its effect on the final adhesion of overinjection molded semicrystalline polymer interfaces.

In this part, the thermomechanical environments, such as melt temperature, mold temperature, injection pressure and injection speed will be changed to investigate the diverse interfacial crystalline structures and their morphology development of PP/PP overinjection molded interface. Based on the experimental results, a formation mechanism of diverse interfacial crystalline structures will be proposed. In addition, the relationship between interfacial crystalline structures and interfacial bonding strength will be discussed.

EXPERIMENTAL

Material

A commercial grade isotatic polypropylene (iPP) F401 with density of 0.9 g/cm^3 and melt flow index of 2.5 g/10 min (230 °C /2.16 kg), was supplied by LanZhou Petrochemical Co., China. The onset melting temperature is 155 °C and the maximum melting temperature is 165 °C by using DSC (Q20, TA Instruments Co., U.S.) at a constant heating rate of 10 °C /min.

Specimens Preparation

The over-injection molded specimens were obtained by a conventional injection molding machine (K-TEC40, Ferromatik Milacron Co., Europe) and a special mold with temperature control system designed in our laboratory. As shown in Figure 1, at the first stage of the injection molding cycle (A), iPP was injected into the empty half-cavity whose molding conditions kept constant, and then the first molded part was prepared. After the first stage, the mold was opened to clear out the runner scraps and took out the first molded part for cooling to room temperature (25 ± 2 °C). At the second stage (B), after rotary joint was rotated 180°, the first molded part was inserted into the holistic-cavity, the second injection was processed and the complete molded specimen was prepared. The dimensions of the specimen were shown in Figure 1(C). The molding conditions for the experimental design and the specimen numbers were summarized in Table 1.

Table 1. Molding conditions for the first and the secondary molding

Specimen No.	Melt temperature, °C	Mold temperature, °C	Injection pressure, MPa	Injection speed, mm/s
A	210	40	60	40
B1/B2	210/250	60	60	40
C1/C2	210/250	40	60	40
D1/D2	210/250	120	60	40
E1/E2	210/250	140	60	40
F1/F2	210	60	40/80	40
G1/G2	210	60	40/80	80

For all specimens: Injection time (s): 4, Holding pressure (MPa): 20, Holding time (s): 10, Cooling time (s): 10.

Note: A is the first molding condition; the numbers from B1 to G2 are the secondary molding conditions.

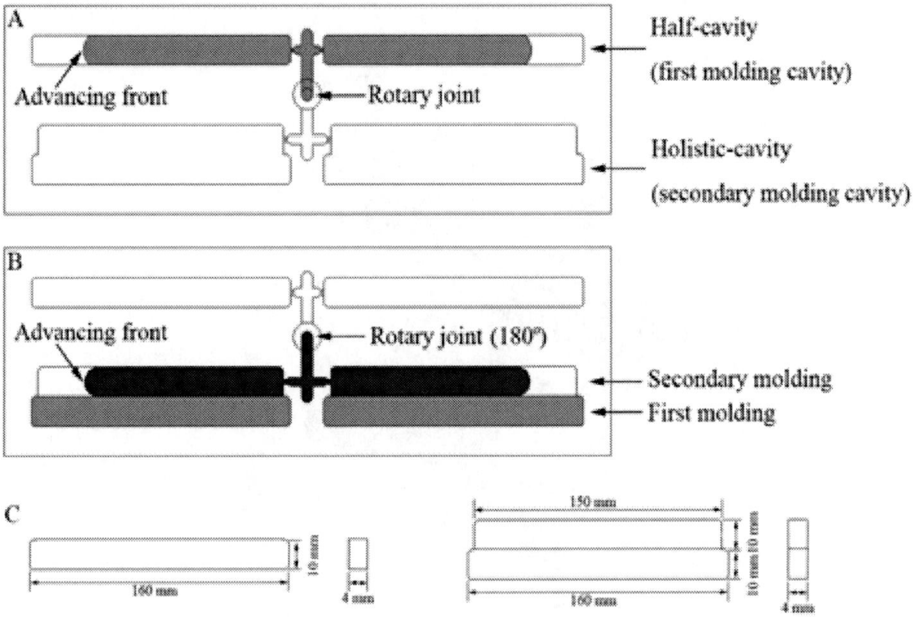

Figure 1. Schematic representation of the OIM (A) the first injection molding (B) the secondary injection molding (C) the dimensions of the first molded part and the complete molded specimen.

CHARACTERIZATION

Polarized Light Microscope (PLM)

Thin slices with 20 μm in thickness were cut by using a microtome (YD-2508, Yidi Medical Appliance Co., China) perpendicular to the flow direction and along the flow direction respectively. Seven cutting procedures were used to slice up the specimens as shown in Figure 2 and the cutting procedure numbers were summarized in Table 2.

Table 2. Specimens cutting procedure numbers

Procedure numbers	Perpendicular to the flow direction (TD×ND)	Along the flow direction (FD×ND)	Distance from the gate (mm)	Distance from the surface (mm)
J	√	-	15	-
Z	√	-	75	-
Y	√	-	135	-
NS	-	√	15	0.3
NC	-	√	15	2
FS	-	√	135	0.3
FC	-	√	135	2

Note: √ is cutting along the direction, - is not the procedure.

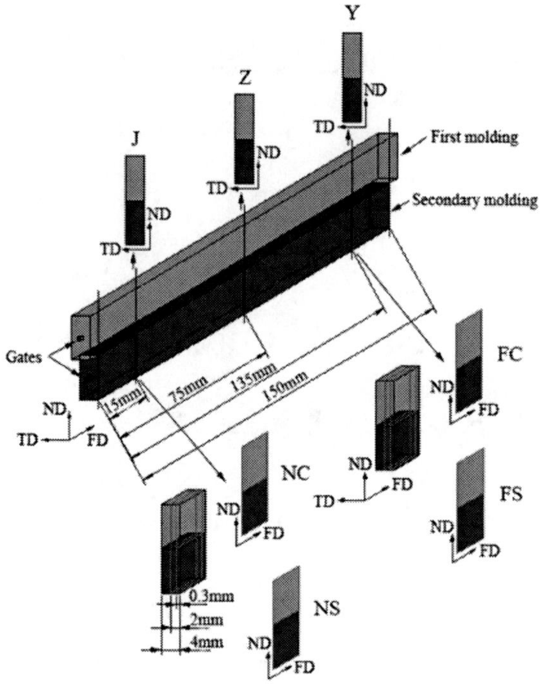

Figure 2. Schematic representations of seven cutting procedures

The thin slices were sandwiched between two microscope cover glasses and placed on the microscope stage. The interfacial crystalline structures during OIM were observed by using a transmission PLM (12POLS, Leica Co., Germany). Main attention has been paid to the crystalline structures of the secondary molded part on the interface in this work.

Meanwhile, the β-iPP formed in the vicinity of the interface was investigated on a PLM (DMIP, Leica Co., Germany) equipped with a hot stage (THMS 600, Linkam Co., U.K.). The slice was heated to 130 °C and held for 2 min to achieve thermal equilibrium. Subsequently, it was heated to 175 °C at a constant rate of 1 °C /min. The proceeding of crystal melting was recorded by taking photomicrographs at appropriate intervals by using a digital camera (POWER SHOT 550, Canon Co., Japan).

Scanning Electron Microscope (SEM)

The specimens, whose surfaces were cut smoothly by a microtome, were etched with a solution [9] prepared by mixing two parts of sulfuric acid (98 %) with one part of phosphoric acid (85%). The etching reagent was then made by dissolving 1 wt% potassium permanganate into the above mixed solvent. After etching, the specimens were washed with distilled water. The surfaces of all the specimens were sputter-coated with a think layer of gold to provide enhanced conductivity. Then the crystalline structures were observed by an SEM (XL30FEG, Philips Co., Netherlands) under an acceleration voltage of 10 kV.

Lap Shear Measurement for Interfacial Bonding Strength

Interfacial bonding strength of the specimen was investigated by using lap shear measurement as shown in Figure 3. Two notches were machined from both surfaces to the interface living a limited bonding interface length of 10 mm ± 0.5 mm, so that the area of the bonding interface under shear condition could be controlled artificially. Bonding strength characterization was measured on a universal materials testing machine (CMT-4104, ShenZhen SANS Testing Machine Co., China) operating in tension mode with a crosshead speed of 10 mm/min at 23 ± 2 °C. The shear strength was calculated by the maximal load dividing the area of the bonding interface (10×4 mm^2) and the given values of strength were calculated as the average over ten specimens.

PART I. INTERFACIAL CRYSTALLINE STRUCTURES AND ADHESION BETWEEN SEMICRYSTALLINE POLYMER INTERFACES

1.1. Morphology Perpendicular to the Flow Direction

Figure 4 shows the optical micrographs in the vicinity of the interface, which were prepared under the secondary molding condition B1 and the cutting procedures J, Z, and Y. It is clearly observed that a bright region and a black region on the interface develop gradually with different distances from the gate. The area of bright region expands and the black region disappears gradually far from the gate, so that there is only bright region on the interface in the position of 135 mm distance from the gate (Y). More evident interfacial crystalline structures can be found in magnified micrographs from two rectangle regions. The interfacial region near the surface of the specimens along TD is defined as region 1 (R1) and the center of the interface is defined as region 2 (R2). Two different morphologies and their development can be shown by comparing R1 with R2. One is the TCL growing perpendicular to the interfacial direction in R1 (J, Z and Y) and R2 (Y). The other is the black region which is hard to find the structure just from the PLM micrographs in R2 (J and Z).

1.2. Morphology along the Flow Direction

The optical micrographs in the vicinity of the interface along the flow direction were obtained under the same molding condition (B1) as shown in Figure 5. The uniform development of the interfacial morphology is manifested with different distances from the gate and the surface. Along the flow direction, there is always TCL growing perpendicular to the interfacial direction near the surface of the specimens, i.e., the direction perpendicular to R1 (from NS to FS). But the morphology develops from the black region to the TCL in the center of the interface, i.e. the direction perpendicular to R2 (from NC to FC). On the other hand, it is clearly seen that the large dimensional shish-kebab structures can be observed in the vicinity of the interface along the flow direction (NS), but this structure can not be found in the other optical micrographs (NC, FS and FC).

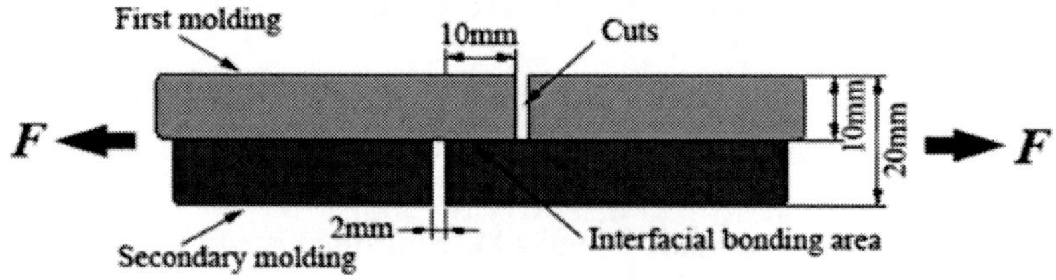

Figure 3. Dimensions of specimen for the lap shear measurement

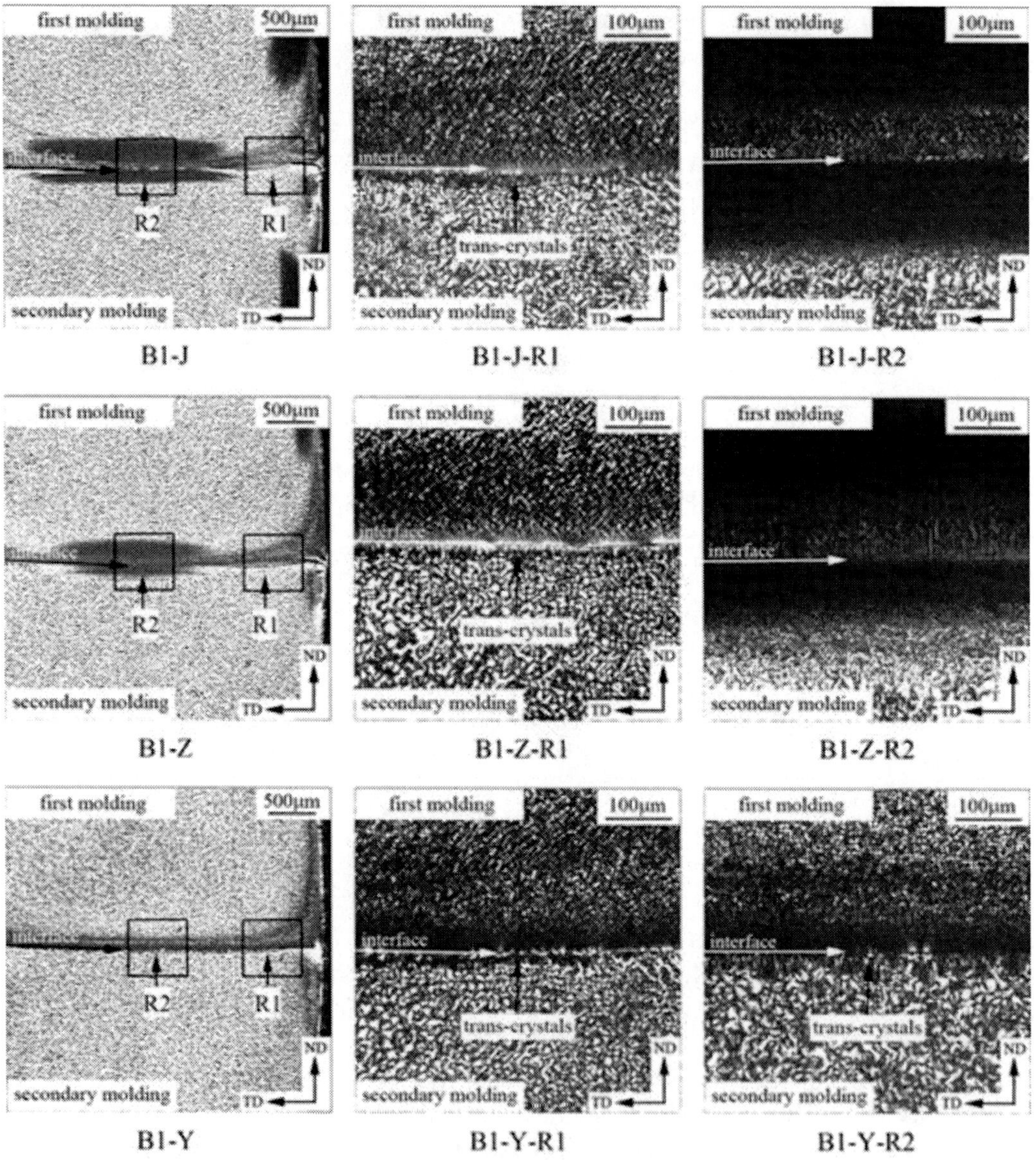

Figure 4. Polarized optical micrographs in the vicinity of the interface at different distances from the gate perpendicular to the flow direction (the secondary molding condition B1 and the cutting procedures J, Z and Y)

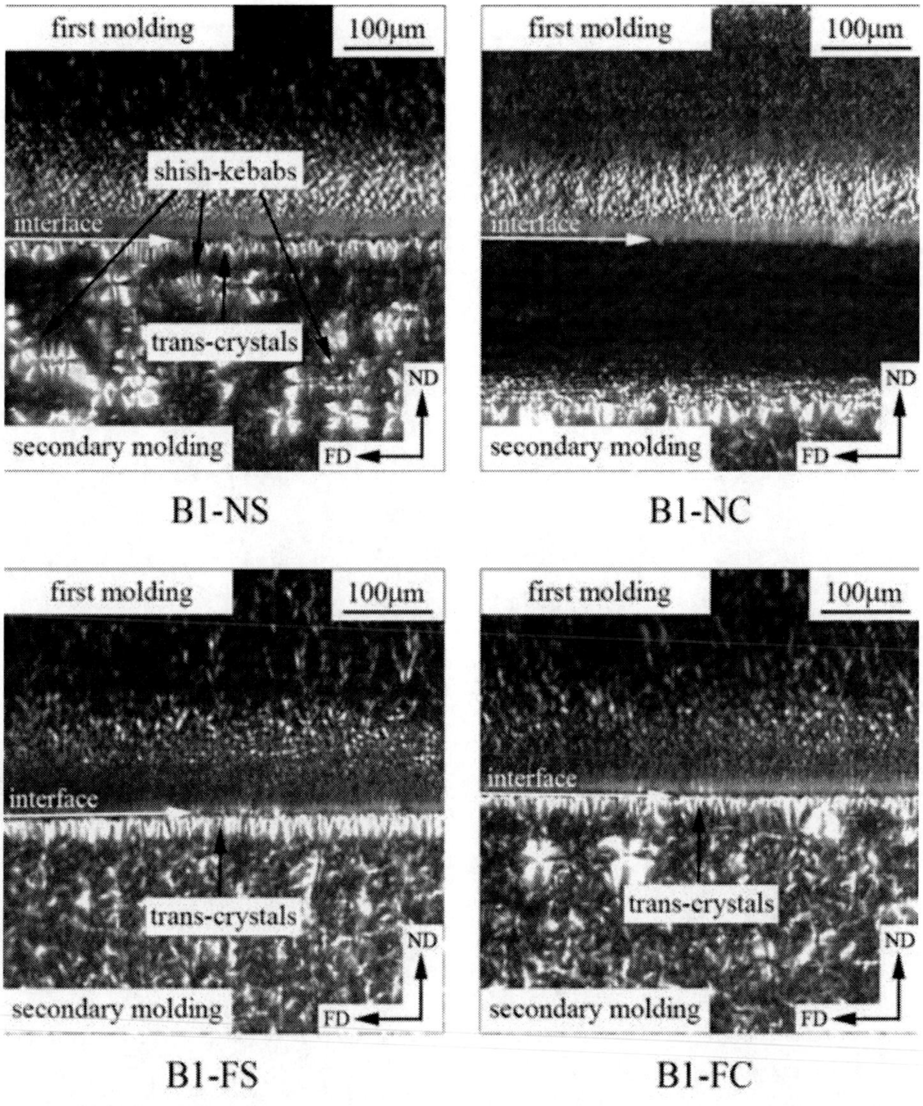

Figure 5. Polarized optical micrographs in the vicinity of the interface at different distances from the gate and the surface along the flow direction (the secondary molding condition B1 and the cutting procedures NS, NC, FS and FC)

Figure 6 shows SEM micrographs in the vicinity of the interface at different distances from the gate and the same distance from the surface (the center of the interface) along the flow direction obtained by etching, i.e. corresponding to Figure 5 (B1-NC and B1-FC). There are lots of tiny dimensional shish-kebab structures growing along the flow direction on the interface in Figure 6 (NC' and NC"), which is different from the black region as shown in Figure 5 (NC). It is difficult to find any structure in the black region because of the low magnification of PLM and the intensive shish-kebab structures. From Figure 6 (FC' and FC"), it is clearly shown that the interfacial crystalline structures are uniform to those presented in Figure 5 (FC). It is further proven that the TCL grows perpendicular to the interfacial direction far from the gate.

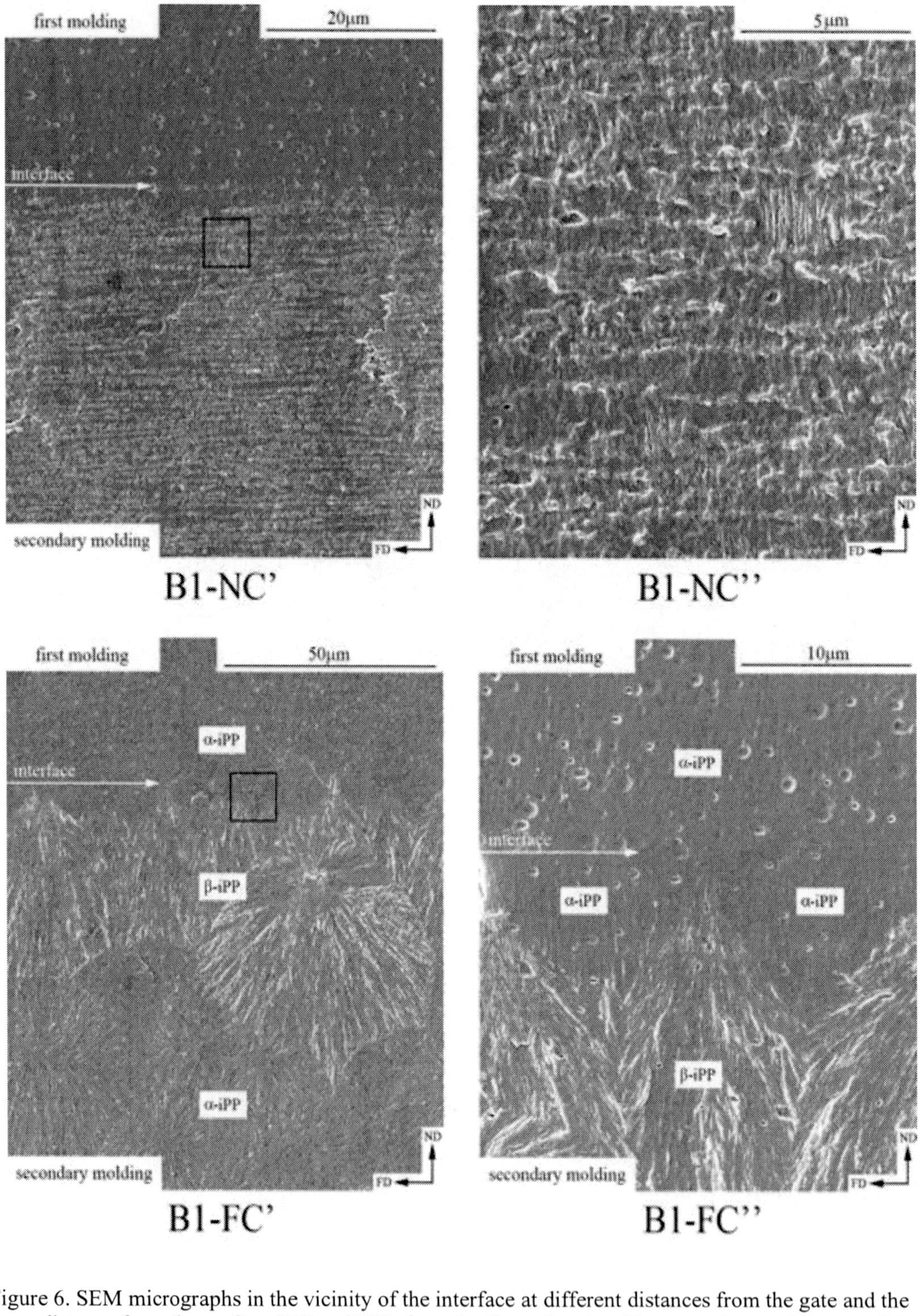

Figure 6. SEM micrographs in the vicinity of the interface at different distances from the gate and the same distance from the surface (NC and FC) along the flow direction (the secondary molding condition B1)

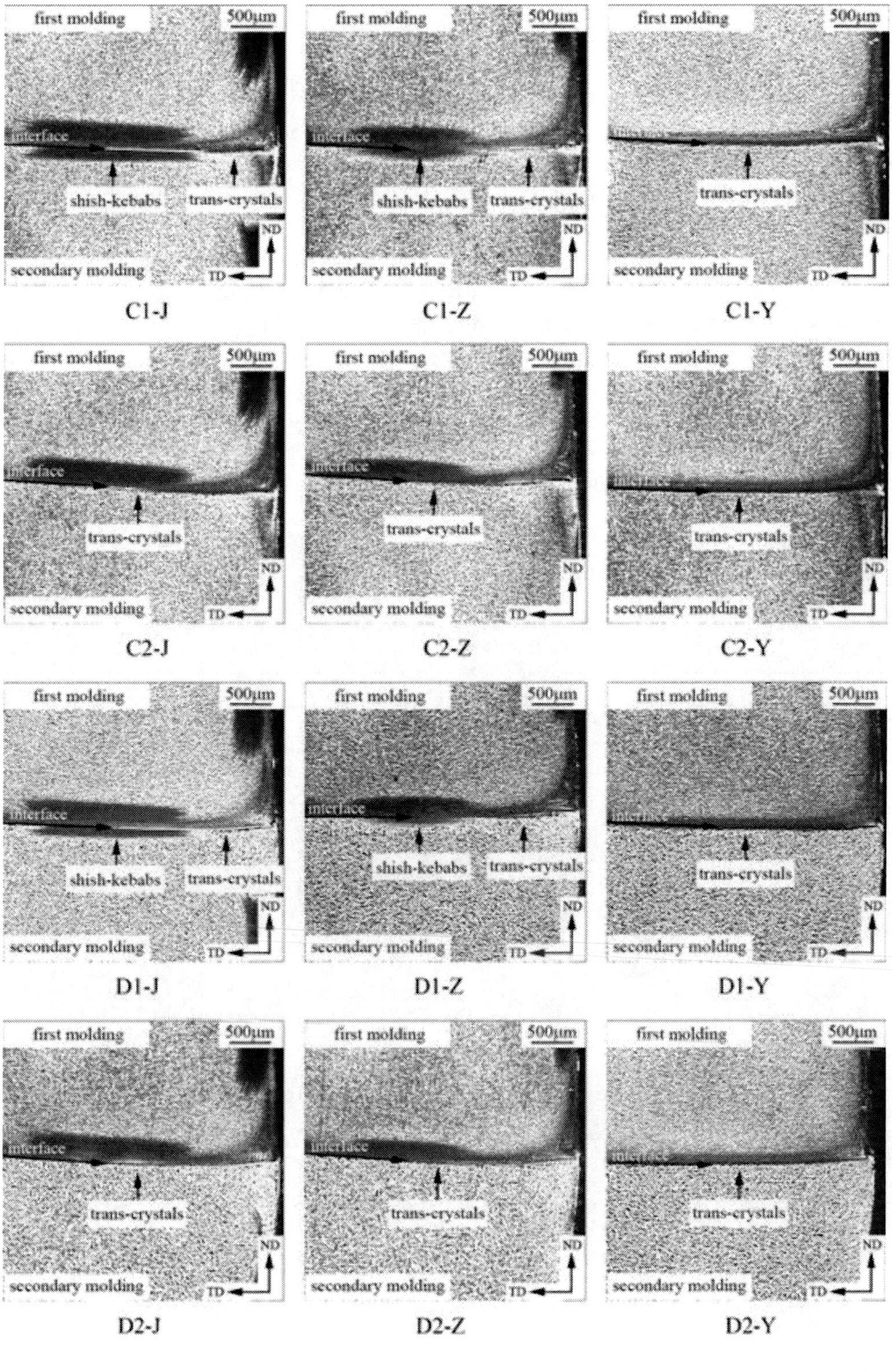

Figure 7. Polarized optical micrographs in the vicinity of the interface prepared by different melt temperature and mold temperature at different distances from the gate perpendicular to the flow direction (the secondary molding conditions C1, C2, D1 and D2 and the cutting procedures J, Z and Y)

2. Effect of Temperature on Interfacial Morphology

Figure 7 shows the effects of melt temperature and mold temperature on the interfacial crystalline structures at different distances from the gate perpendicular to the flow direction (J, Z and Y). The interfacial crystalline structures at low melt temperature (210°C) are apparently different from that at high melt temperature (250°C). The shish-kebab layer (SKL) is more easily developed at low melt temperature and vanishes at high melt temperature comparing C1, D1 with C2, D2. When the high melt temperature is fixed, there is only TCL in the whole range of the interface at different mold temperatures. It can be seen by comparing C1 to D1 that an increase in mold temperature generally results in the shrinkage of the SKL. It is beneficial to expand the TCL and reduce the SKL by increasing melt and mold temperatures. At the same time, the effect of the melt temperature on the interfacial crystalline structures is more obvious than that of the mold temperature. In addition, the interfacial crystalline structures far from the gate (Y) are only composed of the TCL while the SKL can not be found in this region at different melt and mold temperatures.

Figure 8. Magnified optical micrographs in the vicinity of the interface obtained at different melt temperature and mold temperature at the same distance from the gate perpendicular to the flow direction (the secondary molding conditions B1, B2, C1, C2, D1, D2, E1 and E2, the cutting procedure Z)

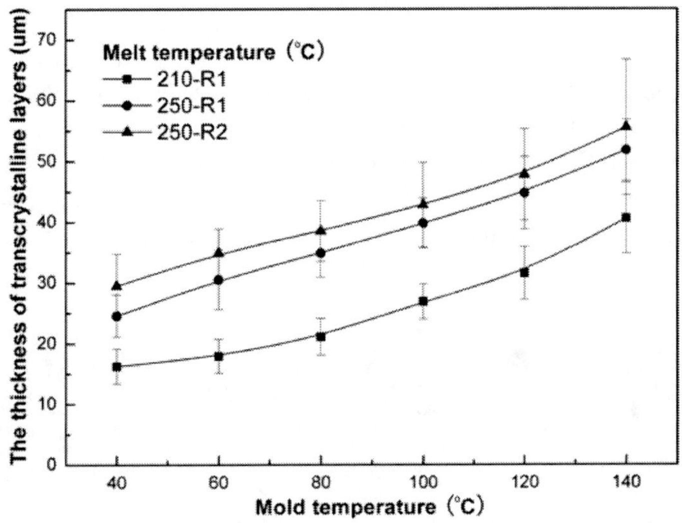

Figure 9. Effects of melt temperature and mold temperature on the thickness of the TCL in different regions (R1 and R2)

To investigate the interfacial crystalline structures in more detail, the magnified optical micrographs in different regions (R1 and R2) are shown in Figure 8, which are obtained by different melt temperature and mold temperature at the same distance from the gate perpendicular to the flow direction. There are two different features of the interfacial crystalline structures from B1 to E2. First of all, the thickness of the TCL gradually increases with raising melt and mold temperatures, which also increases the dimensions of spherulites in the vicinity of the interface of the secondary molded part. Secondly, the different thermal environment composed of melt and mold temperatures lead to the formation of the diverse interfacial crystalline structures, so that the birefringence of the TCL shown in Figure 8 (D2 and E2) differs apparently from the others. According to the previous studies[10-16], the fan-shaped crystalline structures with strong birefringence should be the TCL of β-iPP, as indicated by arrows. Figure 9 depicts the effect of thermal environment on the thickness of the TCL. The thickness gradually increases with the increasing melt and mold temperatures. Furthermore, the thickness in R2 is wider than that in R1 at the same interface perpendicular to the flow direction. It is ascribed to slower cooling rate, i.e. higher crystalline temperature in R2 than that in R1.

3. Effect of Shear Stress on Interfacial Morphology

Figure 10 shows the effects of injection pressure and injection speed on the interfacial crystalline structures at different distances from the gate perpendicular to the flow direction. The wide SKL (black region) can be obtained under high injection pressure (F2 and G2), whereas there is no significant effect on the interfacial crystalline structures by changing the same value of injection speed from 40 to 80 mm/s. From the magnified micrographs, the TCL can be found on the interface under different molding conditions in the same region (R1). At the same time, if the injection pressure keeps high (80 MPa), the SKL gradually expands with increasing injection speed as indicated by arrows in R1 (comparing F2-J-R1 with G2-J-R1). But the SKL can not be found with the change of injection speed in the same region at a fixed low injection pressure (F1-J-R1 and G1-J-R1). On the other hand, the thickness of the TCL has a slight decrease from 18.3 to 16.4 μm with increasing the same injection pressure (increasing 40 MPa) and the same injection speed (increasing 40 mm/s) at the same time. It is concluded that the shear stress controlled by injection pressure and injection speed together has a positive effect on the formation of the SKL and acts a negative effect on the thickness of the TCL.

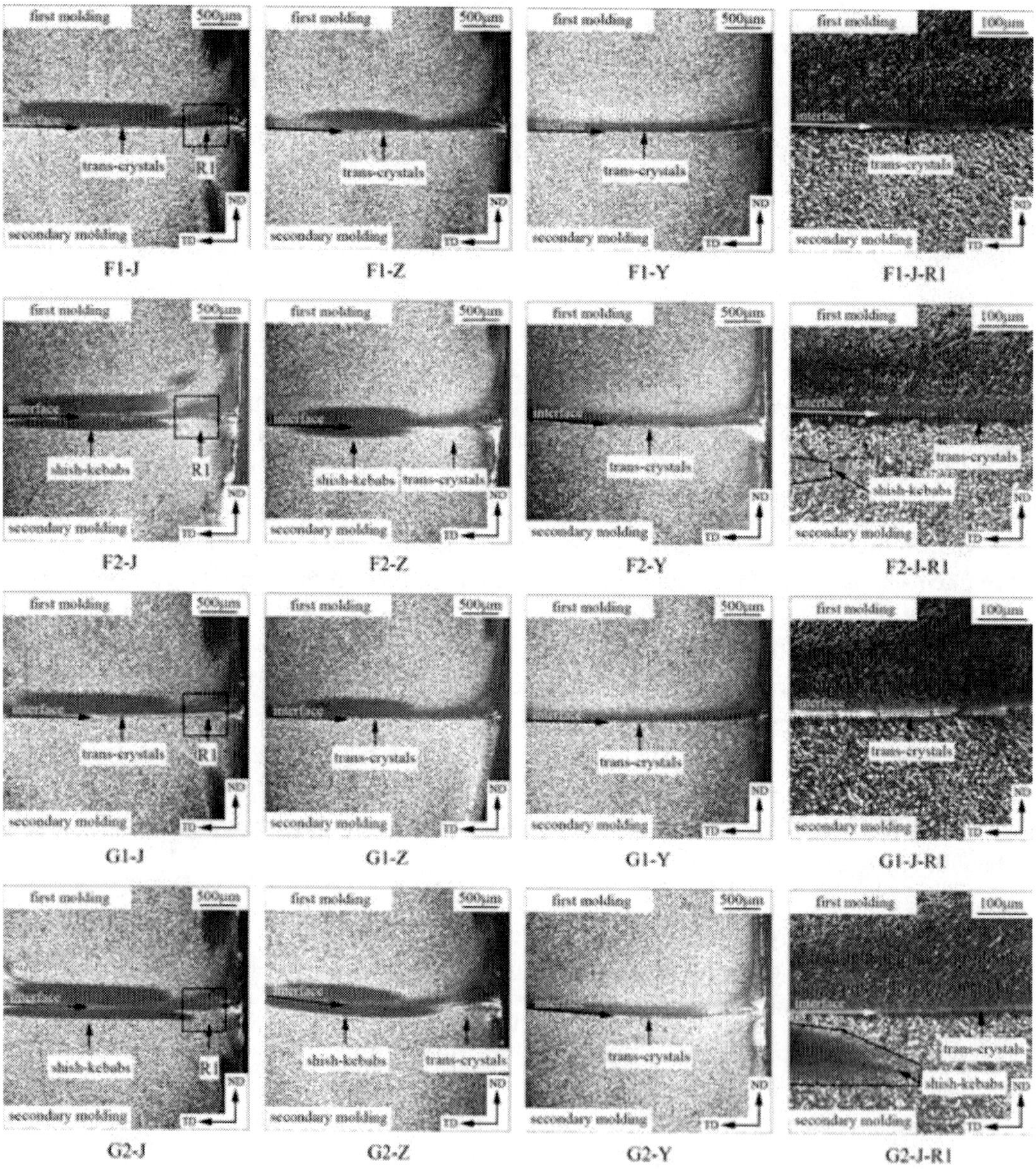

Figure 10. Polarized optical micrographs in the vicinity of the interface prepared by different injection pressure and injection speed at different distances from the gate perpendicular to the flow direction (the secondary molding conditions F1, F2, G1 and G2 and the cutting procedures J, Z and Y)

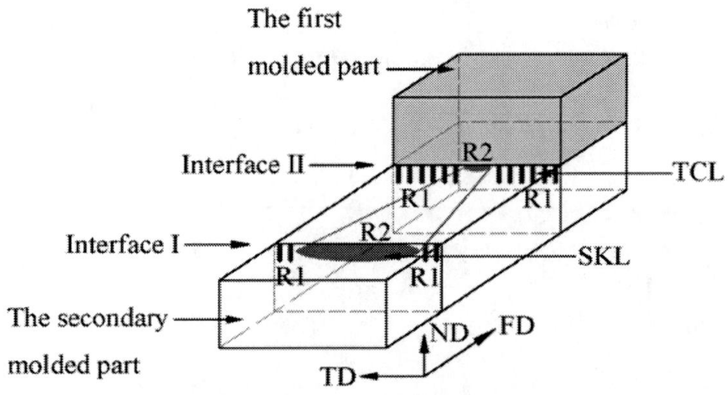

Figure 11. Schematic representations of the morphology development and the diverse interfacial crystalline structures at different regions of different interfaces during OIM

4. The Formation Mechanism of Diverse Interfacial Crystalline Structures

4.1. Competing Interfacial Morphology

As shown in Figures 4, 5 and 6, the SKL and the TCL, competing with each other, constituted the interfacial morphology together during OIM. In order to understand the morphology development and the diverse interfacial crystalline structures more clearly, a three-dimensional model is proposed in Figure 11. It shows the distribution and the evolution rule of the interfacial crystalline structures. During OIM, the complex thermomechanical environments result in different orientation degree of the molecular chains perpendicular to and along the flow direction, thus diverse interfacial crystalline structures are formed at different regions (i.e. R1 and R2 perpendicular to the flow direction) of different interfaces (i.e. I and II along the flow direction). At different regions of every interface, the orientation degree in R2 is higher than that in R1 as a result of "fountain" flow, i.e. the flowing melt is more easily to contact the surface of the first molded part in R2, so that the SKL prefers to form in R2. On the other hand, the shear stress decreases along the flow direction so that the orientation degree reduces gradually. Therefore, the interface I (i.e. near the gate) is dominated by the SKL.

4.2. Interfacial Shish-Kebab Layer (SKL)

There is a significant difference between OIM and conventional injection molding that the shish-kebab structures can be formed on the interface during OIM, as shown in Figure 6 (B1-NC'). To figure out the difference more clearly, an overall view at low magnification is depicted in Figure 12.

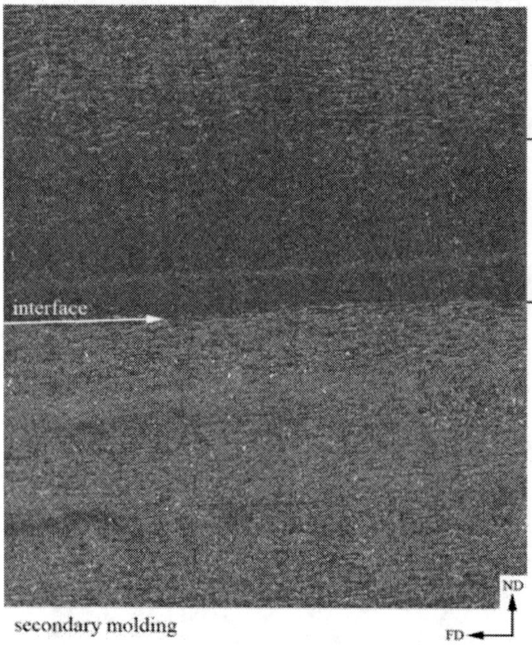

Figure 12. SEM micrographs of the different morphologies in the first molded part and the secondary molded part obtained by conventional injection molding and OIM respectively along the flow direction and near the gate (the secondary molding condition B1 and the cutting procedure NC)

It can be seen that the shish-kebab structures develop in the subskin layer during conventional injection molding, i.e. in the subskin layer of the first molded part for OIM. In addition, the skin layer of the first molded part has not visible structure, which has been reported as a "structureless" skin layer by Kantz et al[17]. However, the shish-kebab structures are obtained on the interface during OIM, i.e. in the skin layer of the secondary molded part. During conventional injection molding, the SKL can be formed due to enhanced orientation degree of molecular chains under high shear between the freezing skin layer and the flowing melt[18-20]. But during OIM, high shear can be induced on the interface between the solid surface of the first molded part and the melt of the secondary molding since the surface of the first molded part is similar to the skin layer in conventional injection molding.

Although the shish-kebab structures are formed in different layers during OIM and conventional injection molding, their formation mechanism is the same. For OIM, the shear stress governed by injection pressure and injection speed, plays an important role in the flow induced orientation of molecular chains and the time for their relaxation before crystallization. High shear stress results in enhanced molecular orientation and short time for the relaxation, thus the thick and wide SKL is formed. On the other hand, thermal environment, which is controlled by melt and mold temperatures, represents an opposite effect on the time for the relaxation. Furthermore, the increased melt temperature, which can decrease the characteristic relaxation time of the molecular chains and prolong the time for the relaxation, is more important than the increased mold temperature, which can only prolong the time for the relaxation.

4.3. Interfacial Transcrystalline Layer (TCL)

The polymer matrix can crystallize on the surface of the substrate as nucleating agent due to sufficiently high density of the nuclei, which will hinder the lateral extension and force growth in one direction, i.e. result in a columnar crystalline layer perpendicular to the interface, known as TCL. So the surface of the substrate, which exhibits strong nucleating ability towards the melt of polymer matrix, is the necessary condition to the formation of the TCL. During OIM, the surface of the first molded part acts as the substrate so that the formation of TCL becomes possible.

The thermomechanical environments during OIM play an important role in determining the nucleation and growth of the TC. The thermal stress, the shear stress and the enhanced mobility of molecular chains in amorphous phase promote the nucleation of the TC on the surface of the first molded part together. Firstly, the thermal stress originates from the shrinkage since the hot melt of the secondary molding contacted the cold solid surface (at room temperature) of the first molded part. Secondly, the shear stress is initiated by the shearing between the flowing melt of the secondary molding and the solid surface of the first molded part. Thirdly, the mobility of molecular chains in amorphous phase at the surface of the first molded part can be enhanced by increasing the interfacial temperature (Ti), which is determined by the initial contact temperature between the solid surface of the first molded part and the melt of the secondary molding. In this work, the surface temperature of the first molded part (T1) keeps invariable at room temperature (25˚C), whereas the melt temperature of the secondary molding (T2) is controlled artificially. So Ti can be estimated from Ti= (T1+T2)/2 and the relative data are listed in Table 3 (T2 is assumed to be constant in the mold during OIM). It is concluded that the crystalline phase at the surface of the first molded part is not melted because the onset melting temperature of the iPP is near 155˚C, which is higher than Ti. Therefore, these three different effects result in a drop of crystallization steric hindrance on the interface so that the high density nucleation can be induced on the interface. On the other hand, the increased crystallization time by improving melt and mold temperatures can promote the growth of the TCL.

It is interesting that the interfacial crystalline structures like the β-iPP were found in some molding conditions during OIM, such as in the position far from the gate as shown in Figure 6 (FC' and FC") and under the conditions of high melt and mold temperatures as shown in Figure 8 (D2 and E2). Figure 13 shows that optical characteristic changes from a strong birefringence to a negative firstly, and then becomes a strong birefringence again, at last the optical characteristic disappears gradually when the temperature is higher than the melting temperature at a constant heating rate of 1˚C /min. The melting test demonstrates that the fan-shaped crystalline structure with strong birefringence is the β-iPP presenting metastability.

Table 3. Temperature parameters of the interface

Melt temperature T2 (°C)	Interfacial temperature Ti (°C)	Thermal gradient △T=T2-Ti (°C)
210	117.5	92.5
250	137.5	112.5

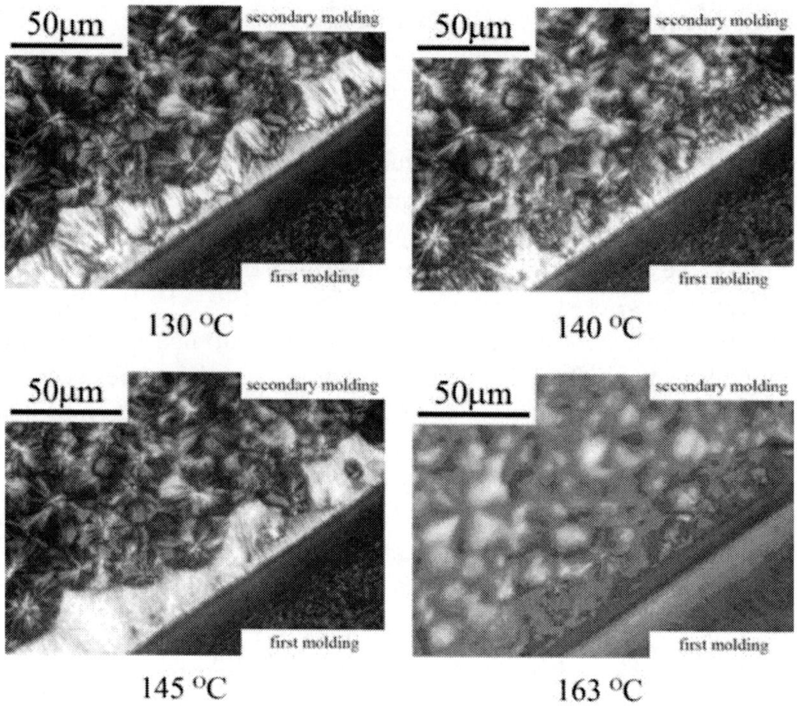

Figure 13. Melting test by using PLM equipped with a hot stage at a constant heating rate of 1˚C /min

Many investigations have reported that β-iPP was metastable and could be formed under special conditions, such as high thermal gradient, shearing or elongation of the melt and the addition of β-nucleating agents[9-14, 21, 22]. During OIM, the complex nonisothermal gradient, which is dependent on melt and mold temperatures, is the most important factor for the formation of the β-iPP. Varga et al[9, 23] have reported that the linear growth rate of the β-iPP from the melt was higher than that of the α-iPP within the temperature range of $T\alpha\beta$ (α to β transition) ≈ 100 ˚C and $T\beta\alpha$ (β to α transition) ≈ 140 ˚C, but it was lower than that of the α-iPP outside this temperature range. Furthermore, β-iPP was easily formed under high thermal gradient. As listed in Table 3, the thermal gradient near the interface (△T) is related to Ti and T2. Ti was limited in the range from 100 ˚C to 140 ˚C, in which the growth rate of β-iPP was faster than that of the α-iPP, so that β-iPP easily began to grow along the oriented α-iPP front as shown in Figure 6 and Figure 8. When T1 is fixed, the increased mold temperature can only extend the time for crystallization but can not affect Ti. However, the increased melt temperature not only has the same effect of extending the time for crystallization but also can increase Ti and △T as shown in Table 3, which form a suitable temperature range for the growth of the β-iPP. On the other hand, the effect of shearing and elongation of the melt on the formation of the β-iPP during OIM has not been discussed in this work. The relative study and the orientation of the molecular chains of the diverse interfacial crystalline structures during OIM will be investigated by using 2D WAXS in the future work.

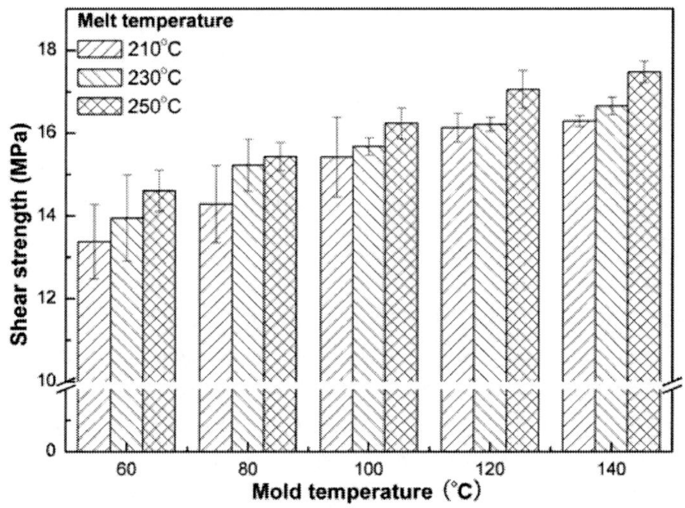

Figure 14. Effects of melt temperature and mold temperature on interfacial bonding strength by using lap shear measurement.

5. The Interfacial Adhesion of the Interface

The formation of the interfacial crystalline structures determines the interfacial mechanical property during OIM, i.e. the interfacial bonding strength. Figure 14 shows the effects of melt and mold temperatures on the interfacial bonding strength by using lap shear measurement. The interfacial bonding strength increases gradually with increasing melt and mold temperatures. And the maximal bonding strength, the value is 17.5 MPa, can be obtained by highest melt and mold temperatures. Actually, the fracture position has not been along the interface when the melt temperature was 250˚C and the mold temperature was higher than 120˚C. As shown in Figure 7 and Figure 8, the diverse crystalline structures are the decisive factor for the change of the interfacial bonding strength. With the increased thickness of the TCL, the interfacial bonding strength will gradually be augmented. Oppositely, the bonding strength will become worse with the enlargement of the SKL.

The relationship between the competing interfacial morphology and the interfacial bonding strength should be understood, so that the bonding strength can be predicted by the development of the interfacial morphology and can be controlled by the thermomechanical environments artificially. Firstly, increased temperature is favorable to obtain better interfacial bonding strength which has been proved in this work. Secondly, the interfacial bonding strength would be decreased under high injection pressure and injection speed. Similar conclusions have been obtained from the report that the PET films acted as the first molded part and another PET was used as the secondary molding during film insert molding[24]. Thirdly, the interfacial bonding strength far from the gate is improved more obviously than that near the gate. Actually, the specimens near the gate will fracture along the interfacial direction due to the impact effect from the mechanical vibration during cutting the specimens, but this phenomenon will not take place on the specimens far from the gate. According to the prediction of the interfacial bonding strength of the interface, optimized interfacial bonding strength could be obtained by using simple measures, which is not necessary to reheat the surface of the first molded part or to increase the interfacial temperature above its melting temperature.

CONCLUSIONS

For OIM, the diverse interfacial crystalline structures, which consisted of the SKL and the TCL competing with each other, formed at different regions (perpendicular to the flow direction) of different interfaces (along the flow direction). The interfacial morphology developed from the SKL to the TCL along the flow direction. At the same time, the SKL preferred to form in region 2 (R2) and the TCL was more easily obtained in region 1 (R1) perpendicular to the flow direction. The thermomechanical environments controlled by molding conditions artificially played a significant role in the development of the interfacial morphology and the formation of the diverse interfacial crystalline structures. The thermal environment consisted of melt temperature and mold temperature had a positive effect on the formation of the TCL. However, the SKL could be obtained under high shear stress controlled by injection pressure and injection speed. In addition, the SKL formed in the position near the gate generally, whereas the TCL generated easily in the position far from the gate.

It was significantly different that the shish-kebab structures could be formed on the interface during OIM but in the subskin layer during conventional injection molding. The formation of the SKL was ascribed to high shear stress resulted from enhanced orientation of molecular chains and short time for their relaxation before crystallization. On the other hand, the nucleation of the TC was influenced by the thermal stress, the shear stress and the enhanced mobility of molecular chains in amorphous phase. For the TCL, the increased interfacial temperature and the increased thermal gradient near the interface could promote β-iPP growth at a suitable temperature range.

The formation of the TCL could improve the interfacial bonding strength obviously, but the existence of SKL played a negative effect on it, so that thermal environment enhanced the interfacial bonding strength, whereas high stress had a negative effect on it.

PART II. ENHANCEMENT OF INTERFACIAL ADHESION AT ASYMMETRIC POLYMER INTERFACE VIA IN-SITU REACTIVE COMPATIBILIZATION

Introduction

In part II, the effect of processing parameters on interfacial crystalline structures and adhesion between symmetric semicrystalline polymer interfaces formed in a sequential injection molding was investigated based on overinjection molded PP/PP interface. However, most of the polymer pairs are thermodynamically immiscible, the interfacial adhesion is very poor due to the lack of interactions at the interfaces. Therefore, how to enhance the interfacial adhesion between different components is a key issue to extent the applications of this technology. Commonly, the addition of premade copolymers[25-32] or in situ formation of those[33-42] are employed to improve the compatibility at polymer-polymer interface. Both routes to reinforce the interfacial properties and control the morphologies of polymer blends have been studied extensively. In particular, reactive compatibilization which forms the copolymers at the interface by the reaction of functionalized polymers was believed to be more efficient than the method of adding premade copolymers in compatibilization. It has attracted more and more interests from both academic and industrial viewpoints.

In this part, the interfacial adhesion between polyethylene (PE) and polyamide 6 (PA6) reinforced by in situ reactive compatibilization in a sequential two-staged injection molding process has been studied. Two routes were designed to achieve in situ reactive compatibilization between PE and PA6 interface. One is the addition of maleic anhydride grafted PE (PE-MAH) into PE interface directly, the other is the introduction of a tie layer comprising PE/PE-MAH blend at PE/PA6 interface. The effects of PE-MAH content and processing parameters, such as injection pressure, injection speed, melt temperature, and mold temperature, on the interfacial adhesion of PE/PA6 were investigated experimentally. The interfacial morphology and fracture mechanism were also studied to correlate with the enhanced interfacial adhesion. In addition, the enhanced interfacial adhesion and the amount of copolymers formed in situ was compared in two different PE/PA6 interfaces overinjection-molded through two different routes.

Experimental

The materials and its' properties used in this study were summarized in Table 1.

Preparation of Specimens

In order to study the effect of reactive compatibilization on the interfacial adhesion of PE/PA6 interface, the mixtures of dried HDPE and different weight percent of HDPE-*g*-MAH were put into a co-rotating twin-screw extruder (diameter of screw = 20, length/diameter ratio = 40/1) for melt blending with a screw speed of 100 rpm and a barrel temperature profile of 160/200/200/190°C from hopper to die. The resulted pellets were injected into the first mold cavity (150 mm × 10 mm × 4 mm), shown in Figure 1(a). During the second injection molding, the first injection molded parts was transferred into the second cavity of the mold quickly. Equilibrating the temperature at least 5 min, and then the melt of PA6 were injected over it.

To compare simply, the injection molding conditions for the first part were same and listed in Table 2. The injection molding conditions for the second part were varied to study the effect of processing parameters on the enhanced interfacial adhesion of PE/PA6.

Table 1. The materials and its' properties used in the experiment

Materials	Code	Properties
HDPE	6055AG	$M_n = 6.5 \times 10^4$, PDI = 5.5
HDPE-MAH	CMG9804	$M_n = 5.1 \times 10^4$, maleic anhydride grafted percent 0.5 wt.-%
PA6	M3400	RI = 6 g/10 min (230°C/2.16 Kg)

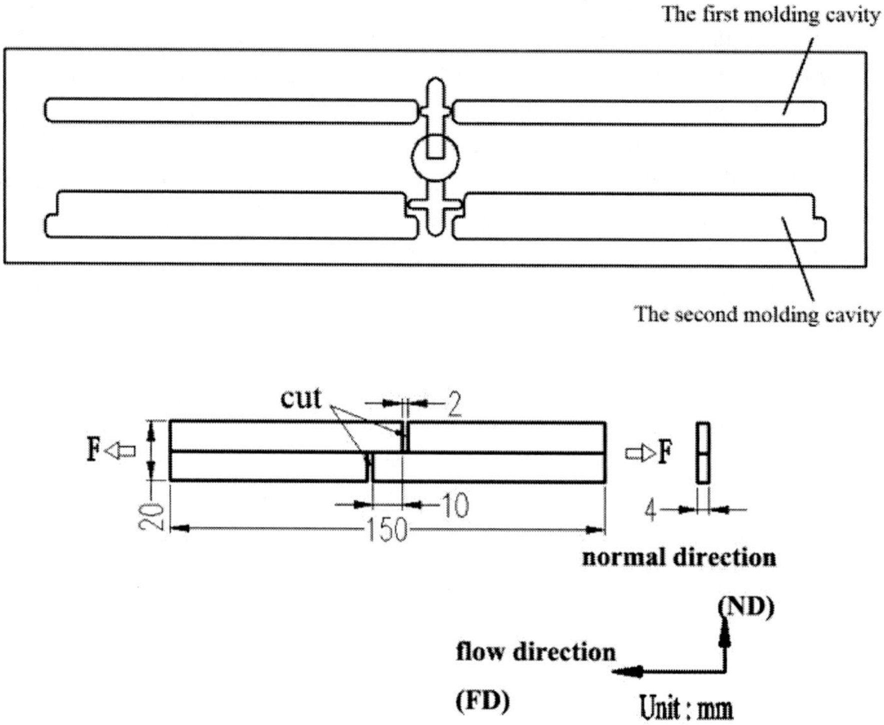

Figure 1. Schematic of mold used for OIM (a) and samples for lap shear measurement (b)

Table 2. Injection molding conditions for the first part (P1)

	PE/PE-MAH blend as P1	PA6 as P1
Injection melt temperature (°C)	200	240
Mold temperature (°C)	40	40
Injection speed (mm/s)	40	40
Injection pressure (MPa)	60	60
Cooling time (s)	10	10

Characterization

Interfacial Adhesion

Interfacial adhesion strength of the specimens was characterized by lap shear test. All the specimens were cut in a lap shear joint geometry (depicted in Figure 1(b)). Stress-strain behaviors were measured with an Instron 5567 machine (Canton, MA, USA) operating in tension mode at 23°C. A crosshead speed of 10 mm/min was used in the test. The shear strength was calculated as the maximal load divided by lap area of shear plane. To minimize error, ten replicates were tested for each set of processing conditions.

Interfacial Morphology

Interfacial morphology of specimens prepared at different conditions was observer by polarized optical micrography in a polarized transmission mode. For this purpose, a trans-section with a thickness of 25 μm was produced using an ultra-thin microtome.

Determination of the Density of Copolymers at the Interface

To determine the areal density of copolymers at the interface, the adhesive interface was treated by the method established by Boucher's. Then chemical analysis on the fracture surface was conducted on the X-ray photoelectron. The spectra near 285 and 400 eV were collected for the 1s peaks of carbon (C 1s) and nitrogen (N 1s) on a KRATOS XSAM800 spectrometer (Kratos Corp., U.K) using MgKa source (hv = 1253.6 ev). A pressure of 5×10^{-7} Pa was maintained during the analysis.

Interfacial Fracture Mechanism

To study the failure mode and locus of failure, the fracture surfaces of the samples after lap shear measurement were coated with gold and then observed using a Hitachi S3400 + EDY scanning electron microscope (SEM, Japan) at an accelerating voltage of 20 kV.

In order to understand more precisely the fracture mechanism for the specimens overinjection molded at different mold, The chemical composition of the surfaces of the fracture interface was analyzed by ATR-IR qualitatively and by DSC quantificationally, respectively.

The infrared spectra from 400 to 4000 cm^{-1} were obtained on a Nicolet 170X spectrometer (ThermoNicolet Corporation, USA) at a resolution of 2 cm^{-1} with 32 scans.

Melt and crystalline behaviors of the interface of the overinjection molded specimens were measured by a TA-Q20 (USA) thermal system purged with nitrogen. The sample for DSC analysis, which has a thickness of 25 μm, was cut from the facture interface by an ultra-thin microtome carefully. The heating and cooling rate is 10°C/min.

SECTION I. THE INTERFACIAL ADHESION AND FRACTURE MECHANISM OF PE/PA6 IN SITU REACTIVE COMPATIBILIZED BY THE ADDITION OF PE-MAH INTO PE INTERFACE

1. Effect of PE-MAH Content and Processing Parameters on the Interfacial Adhesion

Figure 2 shows the effect of the PE-MAH content on the interfacial shear strength between PE and PA6 at three different second injection melt temperature ($T_{melt\ 2}$). It can be seen that the interfacial shear strength increases with the content of PE-MAH at the same $T_{melt\ 2}$. For each PE-MAH concentration series, higher $T_{melt\ 2}$ resulted in higher interfacial shear strength.

The anhydride group of PE-MAH can react with the amine end-group of PA6 and form in-situ grafted copolymer containing imide linkage at PE/PA6 interface. Therefore, the enhancement of interfacial adhesion between mixed PE/PE-MAH blend and PA6 is expected to be due to the in-situ formation of copolymers. Some experiment[44-46] indicated that the chemical reactions occurring at the interface are considered to be diffusion controlled. Small differences in melt temperature will greatly affect the diffusion constant, and the total interpenetration depth distance at HDPE/PA6 interface can be increased at high $T_{melt\ 2}$. This means more functionalized chains can react in the interfacial region, which is beneficial to the improving of interfacial shear strength.

Figure 3 shows the interfacial shear strength as a function of injection speed of the second melt (PA6). The interfacial strength firstly increases with injection speed and shows a maximum at around 60 mm/s, then decreases at high injection speed.

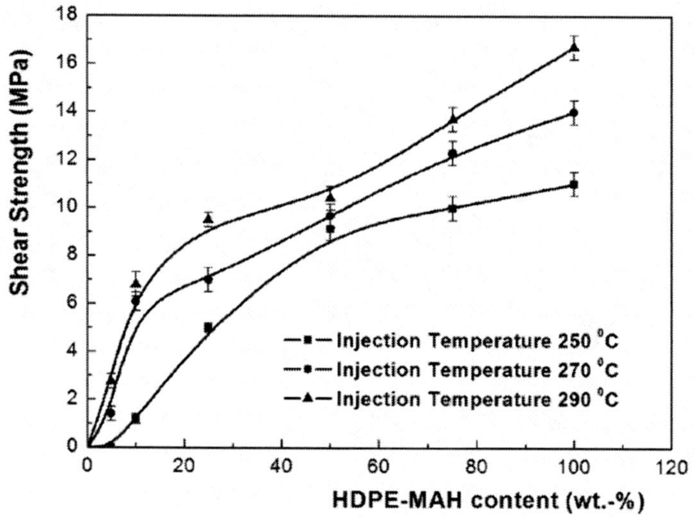

Figure 2. Effect of the PE-MAH content and the second injection melt temperature on interfacial shear strength of overinjection molding PE/PA6 interface

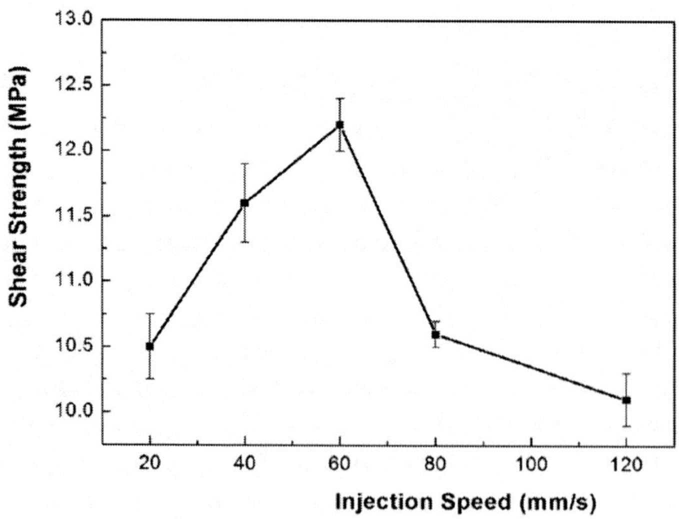

Figure 3. Effect of the second injection speed on the interfacial shear strength of overinjection molding PE/PA6 interface

Figure 4 and Figure 5 are interfacial shear strength as a function of injection pressure (p) and cooling time respectively.

Interfacial shear strength shows a weakly dependence on p (Figure 4), which is attributed to intimate contact and wetting between PA6 melt and PE surface during injection molding. With regard to cooling time, the interfacial strength is hardly influenced by that. Due to cooling during injection molding, the interfacial reaction will be suspended when the interfacial temperature is lower than the crystallization temperature of either interface, The result, shown in Figure 5, indicate that 15 s would be adequate for interfacial reaction in our cases. After consideration, 75 MPa and 60 s were chose as our processing conditions for later experiment.

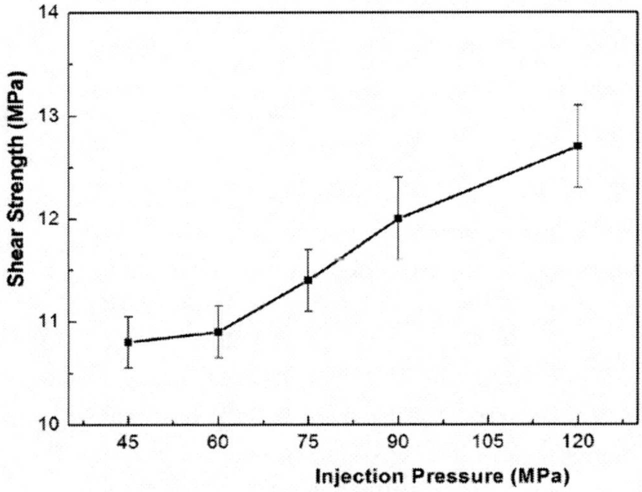

Figure 4. Effect of injection pressure on the interfacial shear strength of overinjection molding PE/PA6 interface

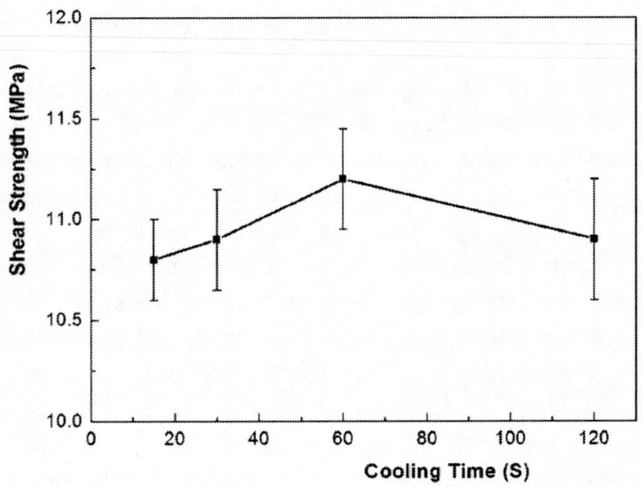

Figure 5. Effect of cooling time on the interfacial shear strength of overinjection molding PE/PA6 interface

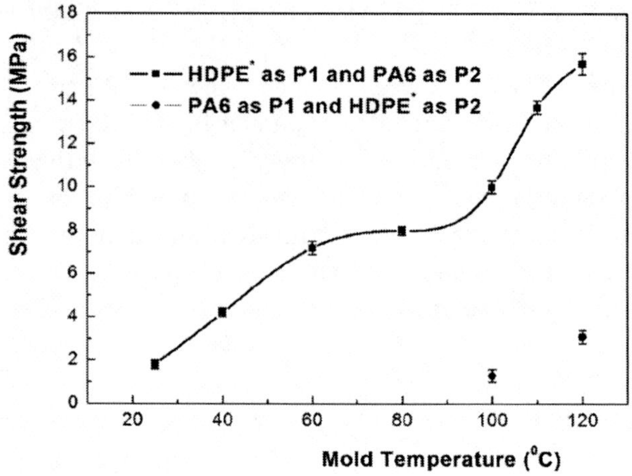

Figure 6. Effect of mold temperature on the interfacial shear strength of overinjection molding PE/PA6 interface

Figure 6 shows measured interfacial shear strength for specimens injected at constant melt temperature (270°C) as a function of mold temperature (T_{mold}). From the results, it was found that the interfacial strength is obviously improved with increasing T_{mold}. For the reaction which goes on between the groups of semicrystalline polymer at solid-melt interface, the second injected melt should provide enough enthalpy to melt a small thickness of solid layer (the first part) while remaining in the melt state, therefore, the extent of reaction significantly depends on the interface temperature and the heat transfer across the interface.

We estimated the interface temperature (T_i) by standard heat transfer equations[47-48]for asymmetric polymers put in contact at different starting temperature.

$$T_i = (T_1 \cdot b_1 + T_2 \cdot b_2) / (b_1 + b_2)$$

Where, T_1 is the temperature of P1, which is equal to T_{mold}, and T_2 is the second melting injection temperature. b_1 and b_2 stand for the thermal effusivities of both polymers. $b_i = (\rho k C_p)^{1/2}$, ρ, k, C_p correspond to density, thermal conductivity, heat capacity of the polymer. The calculated results were listed in Table 3.

Obviously, T_i increases with T_{mold}. The kinetics of coupling reaction held at the immiscible polymer interface was depended on the reactivity of the functional groups and the diffusion of the reactive chains. The increasing interfacial temperature was beneficial to the diffusion of reactive chains. It means that the reaction kinetics of the interfacial reaction was accelerated. In order to validate the role of heat transfer across the interface in adhesion process, we exchanged the injection sequence, i.e. PE/PE-MAH blend melt (as P2) was injected over the solid PA6 (as P1) at the same melt and mold temperature listed in Table 3. The calculated interface temperature was also listed in Table 3. It can be found that the calculated interface temperature of this situation is far lower than that of former at same injection molding parameters. Accordingly, the interface was debonded at low tensile strength, as shown in Figure 6. No effective bonding was formed until T_{mold} was 100°C, but interestingly, the interface could be reinforced when T_i fell below the melt temperature of solid PA6 interface.

Table 3. Interfacial strength and calculated interface temperature for samples prepared at different mold temperature and injection sequence

P1- PE/PE-MAH blend, P2-PA6				P1- PA6, P2- PE/PE-MAH blend			
T_{mold} (°C)	T_{melt} (°C)	T_i (°C)	Interfacial Strength (MPa)	T_{mold} (°C)	T_{melt} (°C)	T_i (°C)	Interfacial Strength (MPa)
270	25	174	1.8	270	25	118	debond
270	40	180	4.2	270	40	130	debond
270	60	187	7.2	270	60	142	debond
270	80	195	8	270	80	154	debond
270	100	203	10	270	100	166	1.3
270	120	211	15.7	270	120	178	3.1

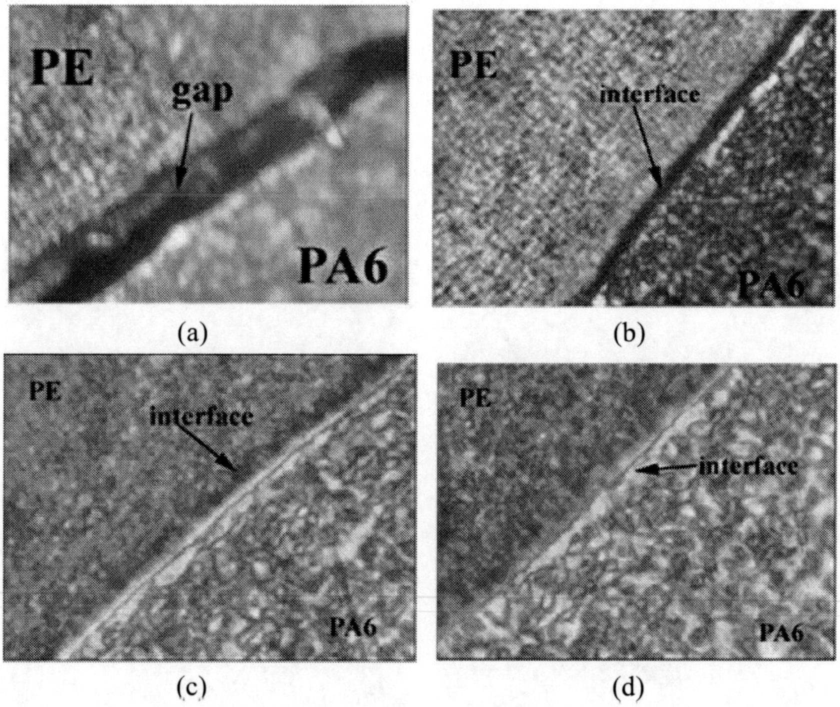

Figure 7. Trans-sectioned polarized optical micrograph of the interface region for the PE+PE-MAH/PA6 adhesive joints prepared at different mold temperature (a) 40°C, (b) 60°C, (c) 80°C, (d) 120°C

2. Interfacial Morphology

Interfacial morphology of specimens prepared at different mold temperature is shown in Figure 7. Crystalline layers are visible on both the PE/PE-MAH and PA6 sides of interface. As shown in Figure 7a, it could be found that the adhesive interface overinjected at mold temperature of 40°C was broken during microtome. There existed the visible gap between PE/PE-MAH and PA6 sides, indicating the weak interfacial interaction. As the mold temperature increased to 40°C, even obvious interface between two sides can be observed, the adhesive interface was intact, which means the interfacial adhesion was reinforced. The interfacial region become very bright and blurry when the mold temperature up to 80°C, some crystalline structure similar to transcrystalline are visible, especially for the sample overinjection molded at mold temperature of 120°C. In part II, the transcrystalline structure was observed in the overinjection molded PP/PP and PA6/PA6 interfaces[4], which is found to be helpful to enhance the interfacial adhesion. From the trans-sectioned polarized optical micrograph of the interface region shown in Figure 7, it is not enough to prove the formation of the transcrystalline structure at overinjection molded PE+PE-MAH/PA6 interface. however, it can be obtained that the adhesive joint between PE/PE-MAH and PA6 side was improved with increasing mold temperature.

3. Interfacial Fracture Mechanism

To investigate the effects of mold temperature on the interfacial fracture mechanism, The PE and PA6 surfaces of a fracture PE+PE-MAH/PA6 adhesive interface prepared at different T_{mold} were observed with SEM as shown in Figure 8 and Figure 9.

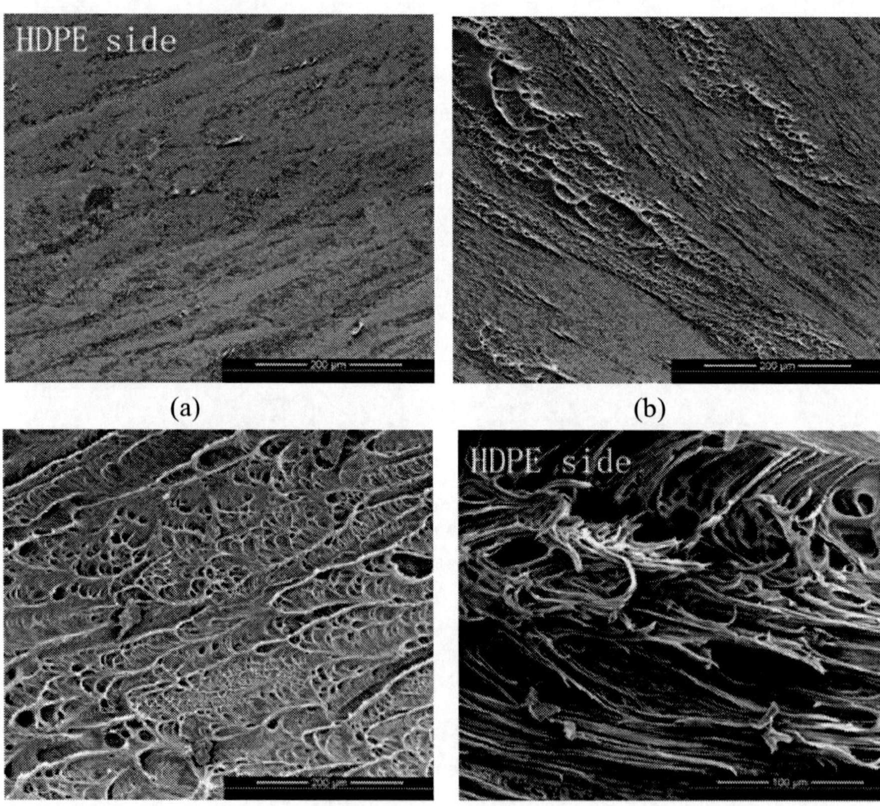

(a) (b) (c) (d)

Figure 8. Scanning electron micrographs of fracture PE surfaces for PE+PE-MAH/PA6 adhesive interface prepared at different mold temperature (a) 40°C, (b) 60°C, (c) 80°C, (d) 120°C

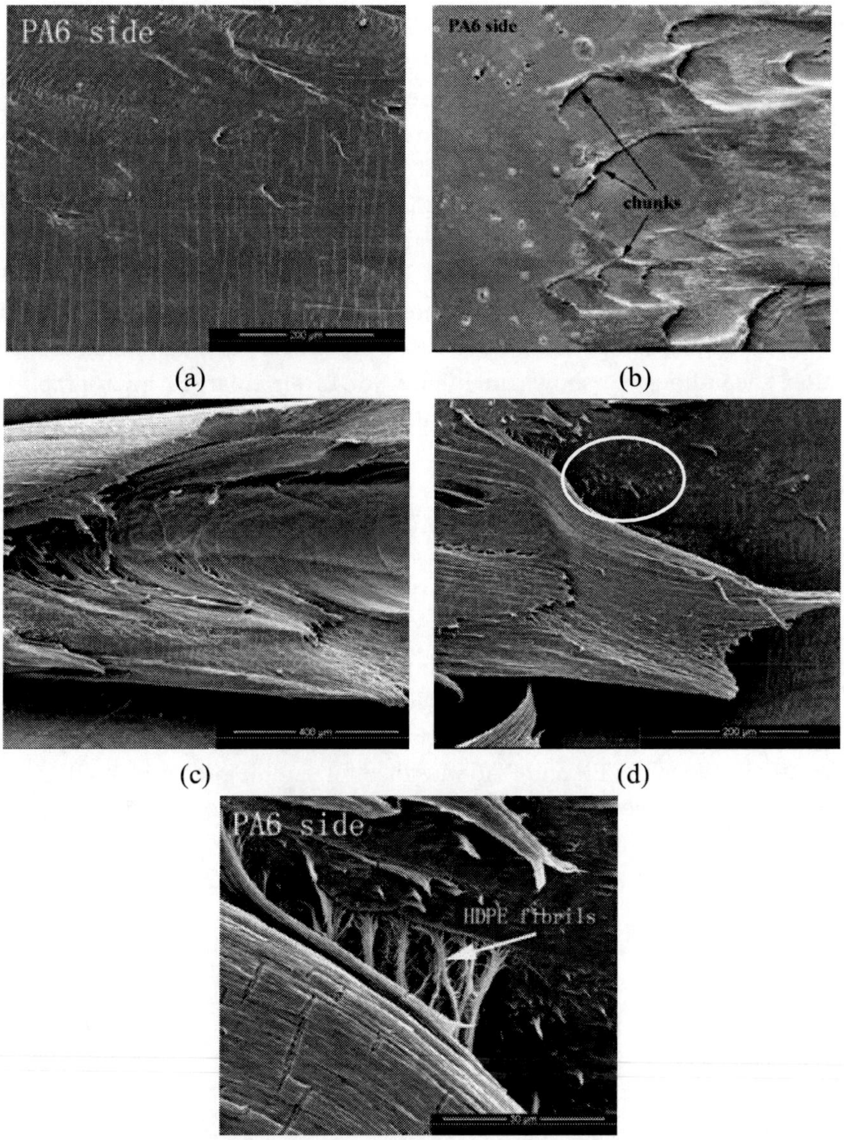

(a) (b) (c) (d) (e)

Figure 9. Scanning electron micrographs of the fracture surfaces PA6 for PE+PE-MAH/PA6 adhesive interface prepared at different mold temperature (a) 40˚C; (b) 60˚C; (c) 80˚C; (d) 120˚C; (e) 120˚C (× 4000)

The in-situ formed copolymers at the interface will build up some stress during lap shear measurement. However, the built-up stress of the specimen prepared at low T_{mold} is not enough to induce the plastic deformation of PE (Figure 8a), resulting in interface adhesive failure at low strength. As T_{mold} increases, the stress developed during shear fracture is able to initiate obvious plastic deformation of HDPE. Figure 8 b and c show the SEM micrographs of the fractured PE surface for specimens formed at a mold temperature of 80˚C and 100˚C respectively. Higher T_{mold}, more deformed PE was observed, which means that more copolymers were formed at the interface with increasing T_{mold}. Therefore, higher interfacial strength can be obtained. At a much higher T_{mold}, e.g. 120˚C, large fibrils of HDPE are formed along the direction of shear force, shown in Figure 8d, and fibril scissions along the shear direction will then follow. Of course, the appearance of a large amount of deformed HDPE fibrils is caused by an increase of the areal density of copolymers interlinking the two sides across the interface. A sharp increase of interfacial strength for the specimens prepared at this mold temperature is, therefore, expected.

For the matching PA6 surfaces of a fracture PE+PE-MAH/PA6 interfaces prepared at different T_{mold}, just shown in Figure 9a, the fracture surface is very smooth at low T_{mold}, indicated low interfacial adhesion. When the T_{mold} is 60°C, the surface becomes coarse and some chunks are found on the side. This phenomena becomes more obvious with increasing T_{mold}, especially at a mold temperature of 120°C, a piece of layer is remained at the surface of PA6 side after shear fracture, shown in Figure 9d. In an enlarged micrograph of circular region at PA6 side, which was shown in Figure 9e, the interface between PA6 and PE are found to be linked by some unbroken fibrils.

The same fracture PE and PA6 surfaces were investigated by FTIR and DSC in order to understand more precisely the fracture mechanism for the specimens overinjection molded at different T_{mold}.

Figure 10 shows the ATR-IR spectra of the fracture surfaces for the PE/PA6 adhesive interface prepared at a mold temperature of 40°C.

The infrared analysis revealed only polyamide characteristic peaks (υ_{-NH}, 3300 and $\upsilon_{C=O}$, 1640 cm^{-1}) on PA6 surface and only polyethylene characteristic peaks ($\upsilon_{as,\ C-H}$, 2920, $\upsilon_{s,\ C-H}$ 2850, δ_{C-H}, 1462 cm^{-1}) on the PE surface. This indicates that the interfacial crack tended to propagate strictly between the PE and PA6 interface for sample prepared at 40°C.

However, distinct polyethylene peaks ($\upsilon_{as,\ C-H}$, 2920, $\upsilon_{s,\ C-H}$ 2850, δ_{C-H}, 1462 cm^{-1}) appeared on the PA6 side, still no polyamide characteristic peaks appeared on the PE side for the specimens overinjected molding at mold temperature of 100°C, just shown in Figure 11.

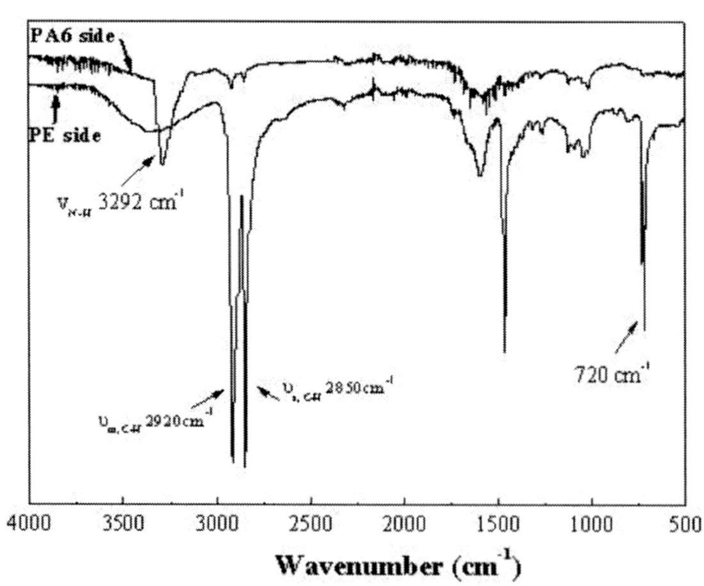

Figure 10. ATR-IR spectra of the fracture surfaces for PE/PA6 adhesive interface prepared at a mold temperature of 40°C

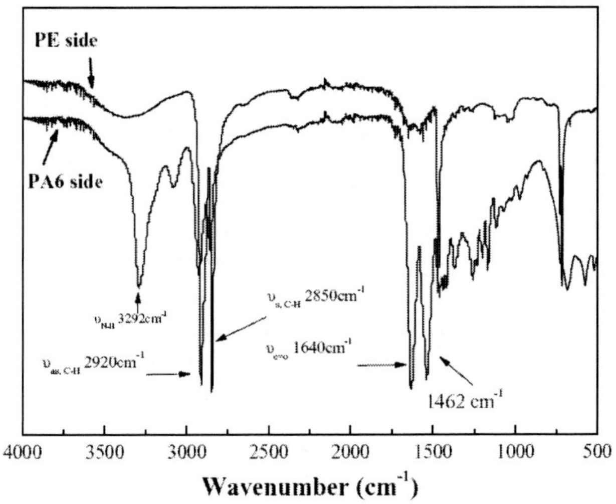

Figure 11. ATR-IR spectra of the fracture surfaces for PE/PA6 adhesive interface prepared at a mold temperature of 100°C

Combining SEM observation and infrared analysis on the fracture surfaces, it can be concluded that the remaining fragment on PA6 side is PE chunk. Since the yield stress of HDPE is much lower than that of PA6, the fracture tends to happen in the bulk of the material with low yield stress, i.e. PE, when the strong interface is fractured. The interface failure will occurs through cohesive failure from PE bulk. It can be concluded that, at the low T_{mold}, litter copolymer are formed at the interface, the interfacial interaction can not induce the plastic deformation . The interfacial crack tends to propagate between the PE and PA6 interface, resulting in adhesive failure. With the increasing T_{mold}, more PE chains were anchored on the surface of PA6 due to the formation of more copolymers at interfacial region. When the interfacial adhesion is high enough to initiate the plastic deformation of PE, the crack will prefer to propagate in the bulk of PE, and cohesive failure is observed instead of adhesive failure, consequently, the interfacial strength was improved.

In order to analyze the interfacial fracture mechanism quantificationally, the melting and crystallization behaviour of the fracture PE and PA6 side were measured using DSC. The heating thermograms of PE side were given in Figure 12.

For all specimens tested, it was found that a melting peak around at 131°C, which belongs to PE crystalline, appears in the melting thermograms, while, no PA6 melting peak appears.

The heating and cooling thermograms of PA6 side were given in Figure 13 and Figure 14, respectively.

It can be found that, when the T_{mold} is higher than 40°C, a melting peak (T_m) around at 131°C appears in the melting thermograms of PA6 surface. It means that no PE molecular chains are pulled out when the mold temperature is 40°C, the interfacial fracture mechanism should be chain scission of copolymers at interface. When mold temperature higher than 40°C, there are some PE molecular chains are pulled out from PE interface due to interfacial adhesion, and resided on PA6 side. The interfacial fracture mechanism should be chain pull or crack propagation. The result of DSC analysis supports the above interfacial fracture mechanism for specimens obtained by SEM and FTIR.

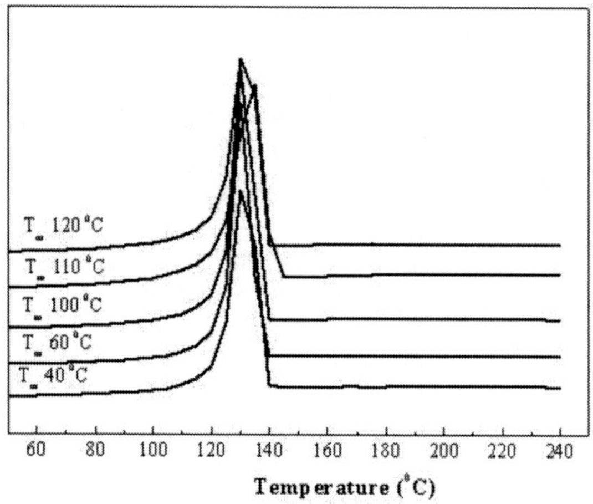

Figure 12. DSC thermograms of the PE surface after shear fracture

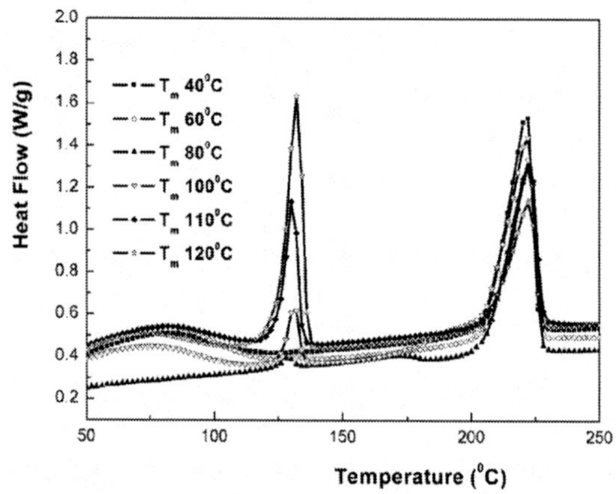

Figure 13. DSC first heating thermograms of the fracture PA6 surfaces for PE/PA6 adhesive interface prepared at different mold temperature

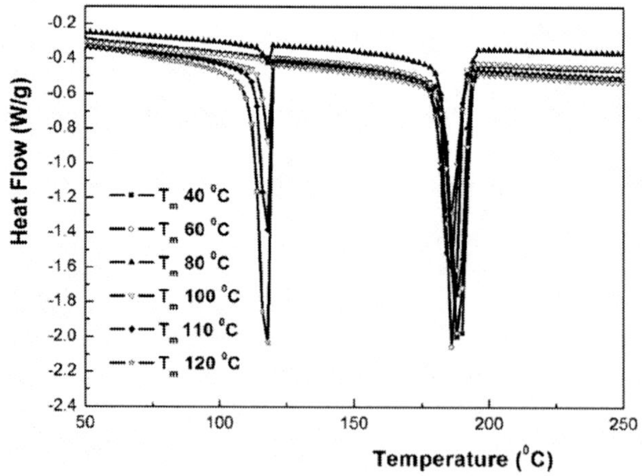

Figure 14. DSC cooling thermograms of the fracture PA6 surfaces for PE/PA6 adhesive interface prepared at different mold temperature

From Figure 13 and 14, it also can been seen that the melting enthalpy (ΔH_c) of PE component, which resided on the PA6 surfaces after fracture, increases with T_{mold}, however, the crystallization temperature (T_c) of that hardly changes. It means the change of the specific melting enthalpy of PE is attributed to the increasing amount of PE in samples which analyzed by DSC, not the degree of crystallinity of PE component. In other word, the amount of PE component resided on PA6 side after fracture increases with the increasing T_{mold}. Whether ΔH_c of the remaining HDPE component on PA6 surface can be correlated with the interfacial shear strength of specimens. To answer this question, the data of ΔH_c and interfacial shear strength were plotted on a log-log scale as shown in Figure 15, fitting the entire set of data yields an exponent of 1.76.

Boucher et al. [50, 51] found that the fracture energy of adhesive interface were varied as Σ^2 for reactive reinforced PP/PA6 interface (Σ is the areal density of copolymers at the interface). We considered that whether the correlation of interfacial strength to melting enthalpy of HDPE can be used to determine the amount of copolymer chains in the interfacial region. When the interface is fractured through cohesive failure, the HDPE chains are pulled out from the HDPE bulk and resided on the PA6 surface. More amount of copolymer chains are formed in the interfacial region, more HDPE chains will be pulled out from the HDPE bulk. So the change of ΔH_c of HDPE is relative to the amount of copolymer chains in the interfacial region. However, the exponent in our experiment is lower than 2.

In order to explain that, a simple model was presented to illustrate reactive compatibilization process at a solid-melt interface, just shown in Figure 16. When being contacted with the second injected melt, (PA6 in the experiment), the solid skin layer of the first injected PE will become molten within small thickness due to the heat transfer from PA6 melt. Therefore, the PE-MAH molecules in the molten skin layer can react with the amine groups of PA6 molecules at the interface randomly to form PA6-HDPE copolymers (Figure 16b). As the interface temperature slowly decreases during the cooling, the partial molecules in PA6 interface will tend to crystallize separately and form individual lamellas. When the interface temperature decreases further to the crystalline temperature of PE, lamellas of PE will tend to be formed. Partial PA6 and PE blocks in the copolymer molecular chains may crystallize separately to form the corresponding PA6 and PE lamellas. Consequently, the PA6 and PE lamellas are connected by the covalent bond of the copolymer in the interfacial region, just shown in Figure 16 c. The interfacial adhesion is enhanced. PA6 and PE are semicrystalline polymers, the amorphous parts of the formed copolymer also contribute to the enhancement in the interfacial adhesion through molecular entanglements in the interfacial region. So the contribution to the enhanced interface adhesion due to in-situ formed copolymer is composed of the amorphous and crystalline parts. The fact that the data from DSC measurement can only represent the contribution of crystalline parts of the formed copolymers is the reason why the exponent of our experiment is only 1.76, though more experimental data should be obtained to confirm our view.

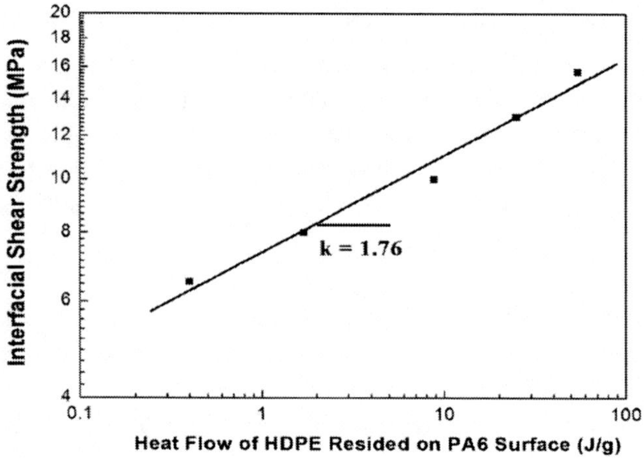

Figure 15. Relationship between the interfacial shear strength of PE/PA6 adhesive interface and ΔH_c of the remaining HDPE component on fracture PA6 surface plotted on a log-log scale

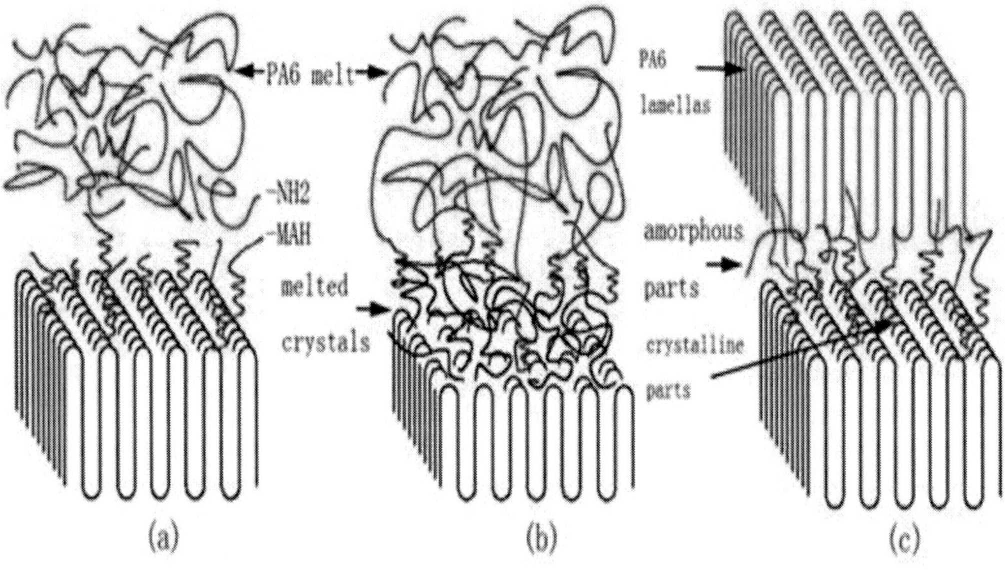

Figure 16. Schematic representations for the formation of copolymers at a reactive solid-melt interface

In this section, the adhesion at solid-melt PE/PA6 interface is enhanced via in-situ reactive compatibilization during a sequential two-staged injection molding process. The interfacial strength is governed by several parameters such as the content of functional polymers, processing conditions and injection sequence. Interfacial temperature and heat transfer across PE/PA6 interface after the second melt injected were found to play a significant role in the interfacial reinforcement. With increasing mold temperature, the interface failure occurs firstly through adhesive failure, chain scission is main fracture mechanism. When the interfacial interaction can initiate the plastic deformation of PE, the crack propagates in the bulk of PE but very near the interface, and cohesive failure dominates the fracture mechanism.

SECTION II. EFFECT OF GRADIENT COOLING ON THE REACTIVE REINFORCEMENT IN A SEQUENTIAL INJECTION MOLDING

It is well known that the chemical reactions between small molecules are usually under diffusion control due to the reduced mobility of reactive sites. When both the reactive agent and the reaction media are highly entangled polymeric chains, the reaction kinetics and the mechanism become even more complicated.

In the past several years, there are two main theories which were used to depict the reaction kinetics and the mechanism of the coupling reaction at polymer-polymer interface: one is diffusion-controlled (DC) models suggested by Fredrickson[52-54] and the other is reaction-controlled (RC) models established by O'Shaughnessy and Sawhney[55-59].

For DC reaction, the formation of the copolymers at the interface is controlled by the mass transfer of the functionalized polymer chains. In the early stages of the reaction, when the density of reactive chains in the interfacial region can be considered to be the same as the value in the bulk, coupling reaction is controlled by the quasilocal reaction rate coefficient.

$$\rho_i(t) \approx \rho_0$$

$$\sigma(t) \approx K_0 \rho_0^2 t$$

In the intermediate regime, a depletion hole of reactants in the interfacial region is formed with the progression of reaction,

$$\rho_i(t) \approx \rho_0 \left(\pi t / \tau_\rho\right)^{-1/4}$$

$$\sigma(t) \approx \left(2/\pi^{-1/2}\right)\sigma^* \left(t/\tau_\sigma\right)^{1/2} = \left(2/\pi^{-1/2}\right)\rho_0 \left(D_0 t\right)^{1/2}$$

Therefore, the reaction is controlled by center-of-mass diffusion of the reactive chains to fill the depletion hole.

Finally, the potential barrier arising from previously formed copolymers at the interface suppressed the reaction in the last regime, therefore, the diffusion of the functionalize chains across the barrier plays a very important role in the further coupling reaction at the interface. The copolymer grows slowly as

$$\sigma(t) \sim \sigma^* \ln^{1/2}\left[N^{1/2} t / \left(\tau_\sigma \ln N\right)\right]$$

$\sigma(t)$ is the number of copolymer chains per area of interface that have been formed in the time interval t after initiating the reaction. ρ_0 is the concentration of the reactive chains with function groups A and B. D_0, and N are the diffusion coefficient and degree of polymerization of a reactive chain, respectively.

O'Shaughnessy and Sawhney considered that the reaction depends on the reactivity of the functional groups and the stage of the reaction. They predicted that the reaction kinetics obeys mean field theory in the reaction between functional groups with weak reactivity ($Q \ll Q^*$):

$$k(t) = Qa^3h \tag{9}$$

where Q, a, and h were the local functional group reactivity, the group size, and the interfacial thickness, respectively.

When the functional groups are sufficiently reactive ($Q \gg Q^*$), $k(t)$ would be controlled by the diffusion of the reactive chains.

In a sequential injection molding, a steep temperature gradient is built up across the interface, which would result in a different diffusion rate and time of reactive chains, as well as the reactivity between functionalized groups, within the locations of steep temperature gradient. Hence, a question is begged. How does the interfacial reaction depend on the gradient cooling in a sequential injection molding.

In this section, in order to investigate the difference of the reactive compatibilization between overinjection molded and isothermal annealing interface, the relationship between the enhanced interfacial adhesion (*IS*) and the initial concentration of reactive chains (ρ) was compared experimentally in two systems.

1. The Relationship between IS and ρ in a Sequential Injection Molding Process

Different weight percent of PE/PE-MAH blend was injected as the first part, and then PA6 melt was overinjected to for the adhesive interface. Figure 1 shows the effect of the concentration of reactive PE-MAH on IS of adhesive interface. Obviously, the interfacial adhesion was enhanced due to the in situ reactive compatibilization. The degree of enhancement increases with increasing concentration of reactive PE-MAH, with 100 wt% PE-MAH sample having IS value up to 25.7 MPa, which is approximately equal to the yield strength of bulk PE.

In order to obtain the relationship between IS and ρ in this interface, a log-log plot of date from Figure 1 is shown in Figure 2. Fitting the entire set of date yields an exponent of 1.07. It means the enhanced interfacial adhesion through in-situ reactive compatibilization approximately increases linearly with the initial concentration of reactive chains in a sequential injection molding.

2. The Relationship between IS and ρ in Isothermal Annealing Process

The same interface was annealed at 215°C for 300 min. Figure 3 shows the interfacial adhesion of PE/PA6 with different PE-MAH concentration.

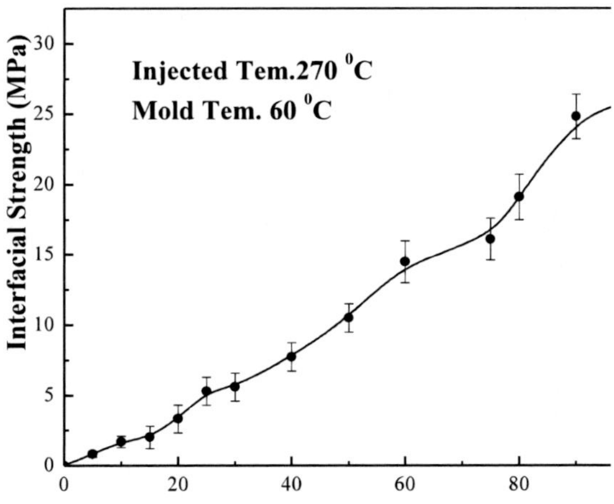

Figure 1. Effect of PE-MAH content on the interfacial adhesion of PE+PE-MAH/PA6 formed in a sequential injection molding

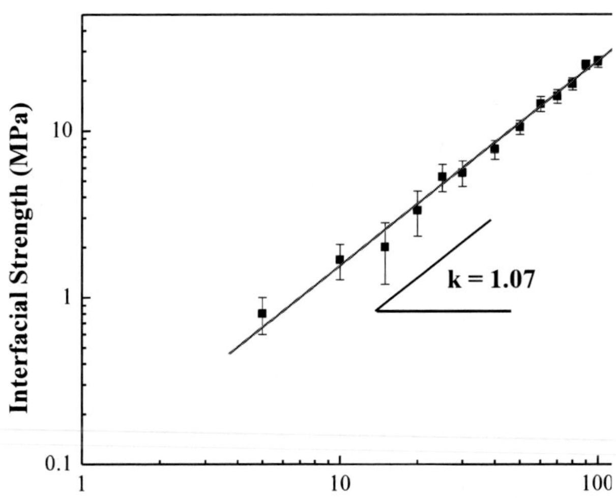

Figure 2. The date shown in Figure 1 plotted on a log-log scale to show a linear behavior

Similar to the interface formed in sequential injection molding processing, the peel strength of PE+PE-MAH/PA6 samples rise rapidly with increasing PE-MAH content. The date was also plotted on a log-log scale, just shown in Figure 4, The result demonstrates that a line of slope two does fit the data, Cole et. al[36] also obtained the interfacial fracture energy is square-law depended on the initial concentration of PP-MAH and annealing time when they study the in-situ reactive compatibilization on the enhanced interfacial adhesion of PP/PA6. Based on Brown model[61], the interfacial fracture energy was found to vary as the square of the areal density of copolymer formed at interface. Therefore, they considered that the reaction controlled kinetics can be used to describe the interfacial reaction during static annealing.

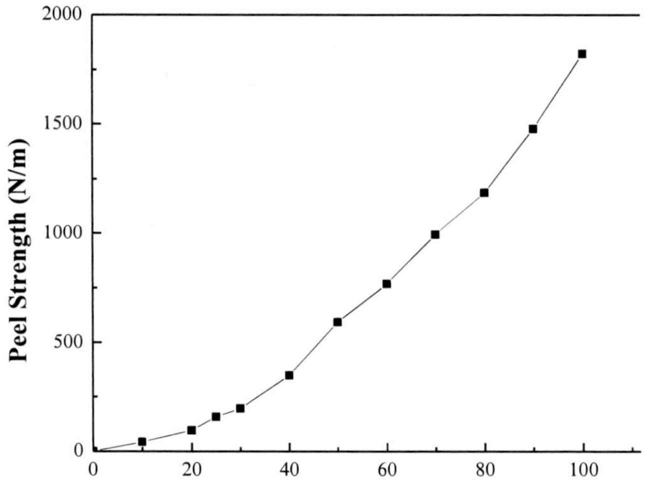

o

Figure 3. The peel strength of PE+PE-MAH/PA6 bilayer samples raise rapidly with increasing PE-MAH content

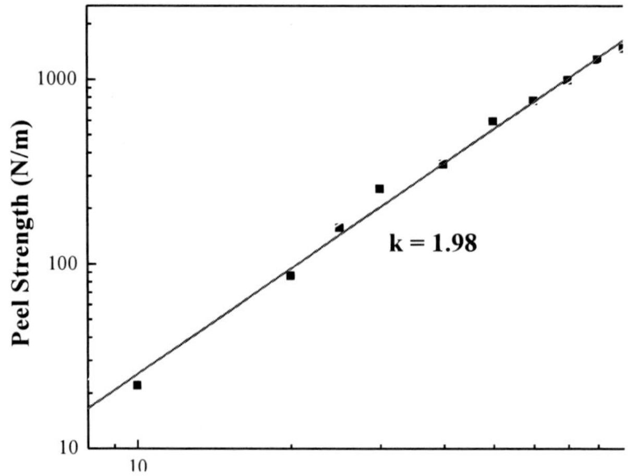

Figure 4. Relationship between interfacial adhesion and the addition of PE-MAH shown in Figure 3 on a log-log scale to show the square-law behavior

From the above result, we can obtain that the gradient cooling in sequential injection molding has alter the enhanced interfacial adhesion compared with isothermal annealing process. The relationship between interfacial adhesion and the initial reactive chains changed from square law to linear due to the gradient cooling in sequential injection molding. Small differences in melt temperature will greatly affect the diffusion constant of reactive chains. Therefore, the gradient cooling causes the coupling reaction at overinjection molding interface is mainly controlled by the diffusion of reaction chains. In order to prove this conclusion, the adhesive interfaces formed in overinjection molding were annealed at some conditions to eliminate the inadequate diffusion of reactive chain at interface.

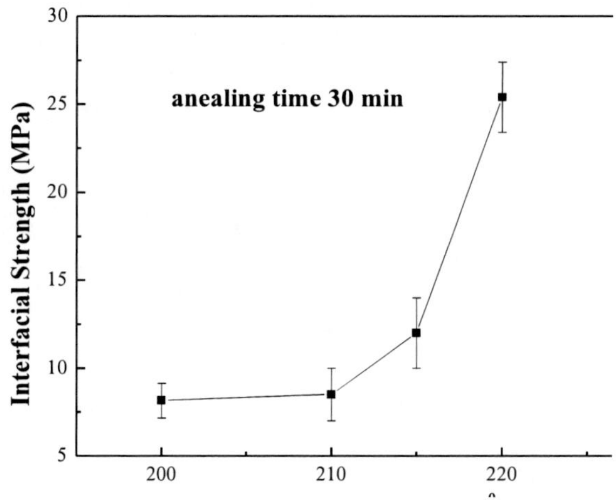

Figure 5. Interfacial strength of overinjection molded PE+PE-MAH/PA6 interfaces as a function of annealing temperature

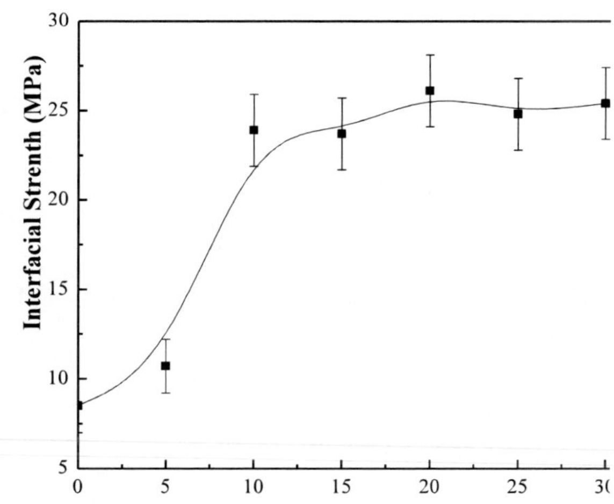

Figure 6. Interfacial strength of overinjection molded PE+PE-MAH/PA6 as a function of annealing time

Firstly, we investigated the effect of annealing conditions on the interfacial adhesion. Figure 5 and 6 present the interfacial adhesion of overinjected PE+PE-MAH/PA6 adhesive joints as a function of the annealing temperature and time, respectively.

It can be seen that the interfacial adhesion has not obvious increase after 30 min annealing when the annealing temperature was lower than the melting temperature of PA6. The striking increase in interfacial adhesion is observed at annealing temperature of 220°C. As the increasing of annealing time, the interfacial adhesion was rapidly improved until it reaches a plateau at approximate 10 min.

As the annealing temperature is well above the melting temperature of PA6, the crystals of PA6 melt and the mobility of polymer chains will become high, the coupling reaction between functionalized groups proceeds. As the annealing time prolongs, more copolymer can be formed at the interface, hence, interfacial adhesion is enhanced. The interfacial adhesion shows a saturation at annealing time of 10 min, mainly is attributed to the interfacial fracture mechanism. When the interfacial adhesion strength is higher than the bulk of PE, the interfacial will crack in the bulk of PE, which has been discussed in the former section in detail.

According to the results shown in Figure 5 and Figure 6. The overinjection molded PE+PE-MAH/PA6 adhesive joints with different PE-MAH content was annealed at 220°C for 10 min. The comparison of interfacial strength before and after annealing is presented in Figure 7.

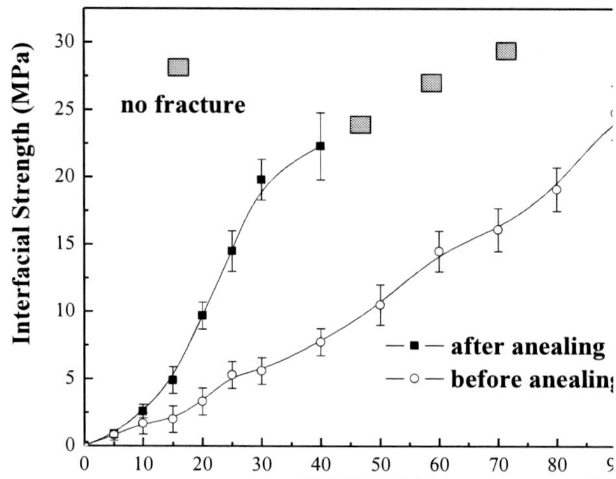

Figure 7. Comparison of the interfacial strength of overinjection molded PE+PE-MAH/PA6 adhesive joints before and after annealing

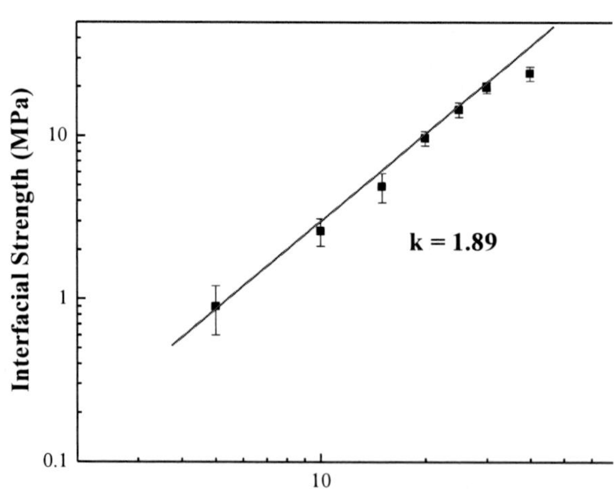

Figure 8. Relationship between interfacial strength of annealing PE+PE-MAH/PA6 adhesive joints and the addition of PE-MAH content plotted on a log-log scale

As expected, the interfacial adhesion dramatically increases after annealing due to the progress of coupling reaction during annealing. It was also experimentally found that the interface will crack in the PE bulk when the PE-MAH content exceed to 50 wt% and all the interfacial strength does not change with the increasing PE-MAH content, indicating the adhesive strength of PE+PE-MAH/PA6 is higher than the bulk strength of PE. The date of interfacial adhesion after annealing was also plotted on a log-log scale, shown in Figure 8.

It appears that the annealing interfaces approximately have an interfacial strength proportional to the square of the initial reactive chains, i.e., PE-MAH content. The relationship between interfacial adhesion of overinjection molding PE+PE-MAH/PA6 and the initial PE-MAH content was again restored upon annealing.

Based on the present experimental result, it is not adequate to get a conclusion that the reaction kinetics of in-situ reactive compatibilization at overinjection molded interface is controlled by the diffusion of the reactive chains. However, due to the gradient cooling, the diffusion of the reactive chains across the interface is altered, resultantly, the interfacial adhesion strength is weaker than the same interface which is annealed in isothermal annealing process. When the effect of diffusion of reaction chains is eliminated by annealing the interfaces at high temperature, the interfacial adhesion of overinjection molded samples is restored to that of annealing process. In addition, the interfacial adhesion was shown to scale linearly with the initial reactive chains, providing the possibility to predicate the interfacial adhesion formed in sequential injection molding.

SECTION III. EFFECT OF A TIE LAYER ON THE ENHANCED INTERFACIAL ADHESION BETWEEN PE AND PA6 IN A SEQUENTIAL INJECTION MOLDING

From the former two sections, it was obtained that the interfacial adhesion between PE and PA6 can be enhanced through in-situ reactive compatibilization in a sequential injection molding when a certain amount of the PE-MAH is incorporated into PE component. However, in order to obtain effective welding strength, a mass of PE-MAH should be incorporated since the available reaction time and interfacial area are limited in process. A majority of compatibilizer in PE side can not diffuse to the interface where the reaction between PE-MAH and PA6 is conducted due to gradient cooling in overinjection molding. In addition, the mechanical and rheological properties of PE component will be influenced with the addition of functionalized PE.

In this section, a tie layer composed of PE and PE-MAH was introduced between PE and PA6 interface in sequential injection molding. The effects of the composition and thickness of the tie layer and processing temperature on the interfacial adhesion were investigated, The relationship between the interfacial adhesion and interfacial fracture mechanisms was also discussed.

The tie layer was prepared in two steps. The first step involved the melt mixing of HDPE-g-MAH into the HDPE using a corotating twin-screw extruder (diameter of screw=20 mm, length/diameter ratio=40/1) with a screw speed of 100 rpm and a barrel temperature profile of 160/200/200/190°C from hopper to die. The resulted pellets were dried at 80°C for

24 h. Then completely dried pellets were blown into film on a HAAKE PolyLab OS System (Thermo Fisher Scientific, Germany), connected to a film blowing machine at 210°C. The extrusion speed was fixed at 40 rpm, and the films with a thickness of 5-80 μm were obtained by varying blow-up ratio and frost-line height.

The preparation of the measured specimens also includes two steps: Firstly, HDPE substrates with dimensions of 150 mm × 10 mm × 4 mm were injection-molded. Then HDPE substrates were wrapped by a layer of the tie layer. The parts composed of HDPE substrate and tie film were denoted as P1. In order to avoid the slide of the tie film during the second injection molding process, both ends of the tie film were fixed using adhesive tape. During the second injection molding process, PA6 was over-injected on P1. So the samples were made in the "sandwich" geometry, shown in Figure 1.

1. Effect of Processing Parameters on Interfacial Adhesion of Overinjection Molded PE/Tie Layer/PA6 Interface

Figure 2 presents IS of PE/tie layer/PA6 adhesive interface as a function of the PE-MAH content in the tie layer.

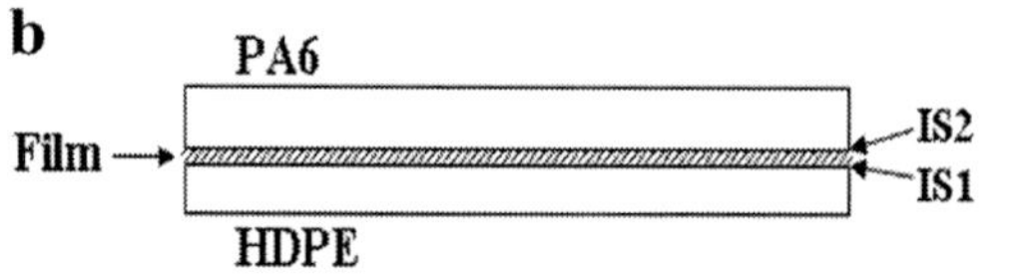

Figure 1. Schematic of sandwich PE/tie layer/PA6 overinjection molded samples

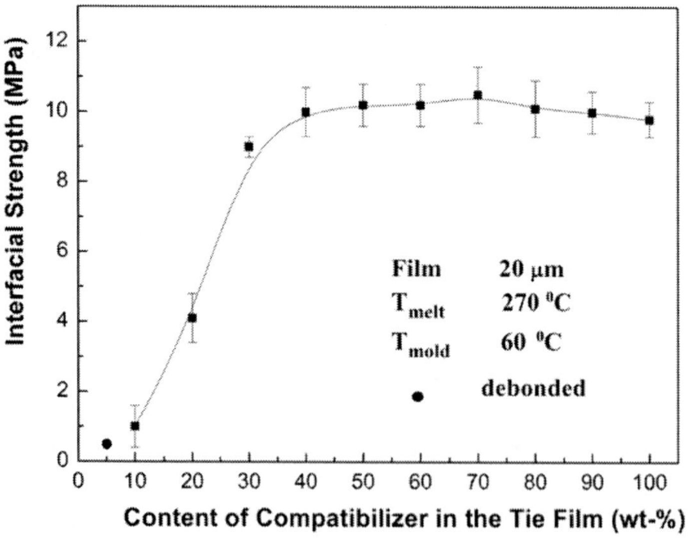

Figure 2. Interfacial strength of PE/tie layer/PA6 adhesive interface as a function of PE-MAH content in the tie layer

It was shown that a small amount of PE-MAH, 5 wt%, does not result in good adhesive strength and the interface was debonded when the samples were cut for lap shear measurement. As the PE-MAH concentration increases, IS increases dramatically and then levels off. From the result obtained in section II, $IS \square \rho$, higher PE-MAH concentration in the tie layer, more covalent bonds were able to be formed in the interfacial region, consequently, the interfacial adhesion was enhanced. The appearance of saturated IS values at about 40 wt% PE-MAH in the tie layer will be discussed in detail later.

Figure 3 shows IS of PE/tie layer/PA6 adhesive interface as a function of the thickness of the tie layer. it can be seen that the interfacial adhesion between PA6 and PE can be improved even when 5 μm-thick film was used as a tie layer. When the thickness of the tie film increases to 10 μm, IS increases about twice, and becomes slightly level off until the thickness of the tie film increases to 30 μm. Then the interfacial adhesion between PA6 and HDPE decreases obviously with increasing thickness of the tie layer. When the thickness of the tie film exceeds 60 μm, the interface between PA6 and PE can not be jointed effectively.

Obviously, the amount of functionalized PE in the interfacial region increases with increasing thickness of the tie layer, the early increase in the interfacial adhesion with the thickness of the tie layer is mainly ascribed to an increase in the number of intermolecular reactions at the interface. However, IS of PA6/tie layer/PE adhesive joint depends not only on the amount of copolymer at interface but also on the interfacial fracture mechanism. The latter decrease of IS with the thickness of the tie layer should be ascribed to the change of interfacial fracture mechanism, which also be discussed later.

Figure 4 and 5 show IS of PE/tie layer/PA6 adhesive joints as a function of the second melt injection and mold temperature, respectively.

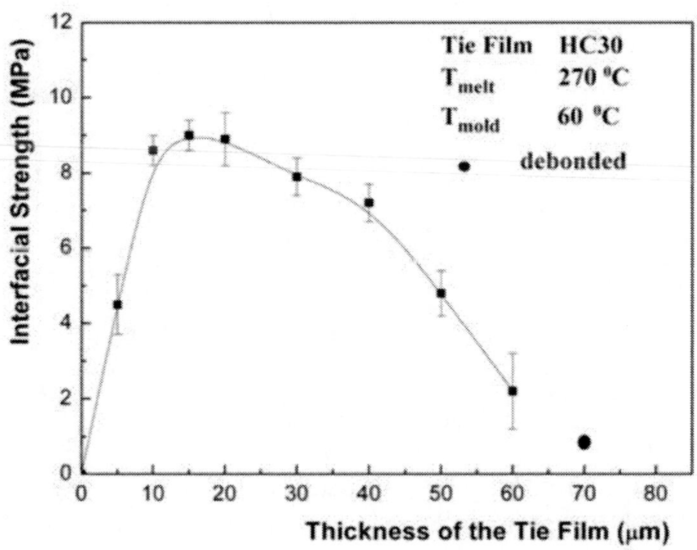

Figure 3. Interfacial strength o PE/tie layer/PA6 adhesive interfaces as a function of the thickness of the tie layer

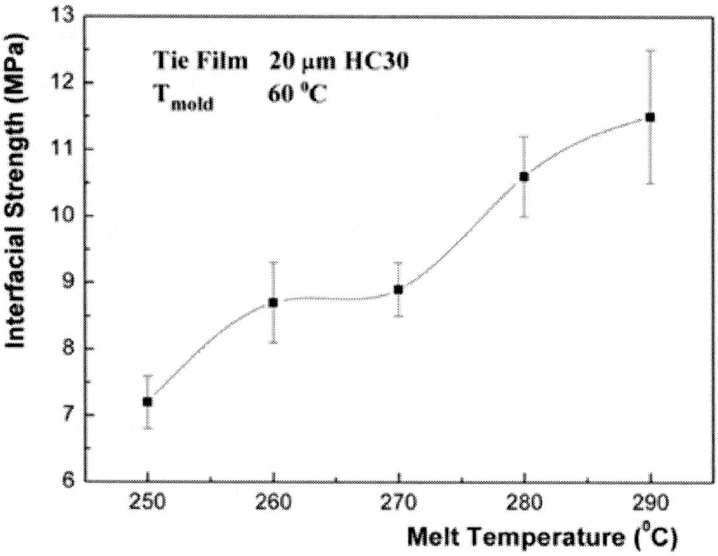

Figure 4. Interfacial strength of PE/tie layer/PA6 adhesive interface as a function of the second melt injection temperature.

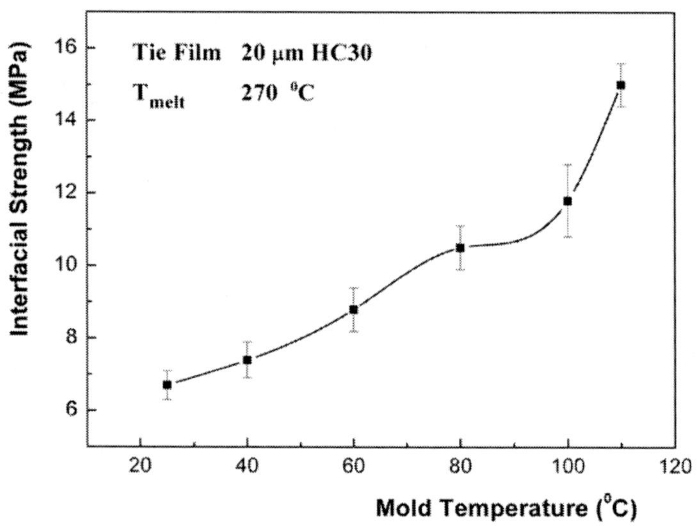

Figure 5. Interfacial strength of PE/tie layer/PA6 adhesive interface as a function of the mold temperature.

It can be seen that IS of PE/tie layer/PA6 adhesive joints both increase with the second melt injection and mold temperature. From the result obtained in section I, the enhanced extent of the interfacial adhesion during overinjection molding process significantly depends on the interface temperature and the heat transfer across the interface. Obviously, the interfacial temperature will be both increased as the increasing second melt injection and mold temperature, resultantly, the interfacial adhesion is also improved.

2. Interfacial Morphology of PE/Tie Layer/PA6 Adhesive Interface

Interfacial morphology of PE/tie layer/PA6 adhesive interface overinjection molded at different mold temperature was observed using polarized optical microscope. The micrograph of the interface region is shown in Figure 6.

Two interfaces are visible in PE/tie layer/PA6 adhesive joints overinjection molded at a mold temperature of 40°C, i.e., the interface between PE and tie layer designated as Interface 1, the interface between tie layer and PA6 designated as Interface 2, shown in Figure 6a. When the mold temperature increases to 60°C, still two interfaces can be observed, but the thickness of the tie layer becomes thin. As the mold temperature increasing, Interface 1 almost disappears, only a very short segment can be observed at interface overinjected at 80°C. Interface 1 fully disappears, only one interface exists in sample overinjected at 120°C, as shown in Figure 6b - d. No special crystalline structure is observer at the interfacial region of samples overinjection molded at whatever the mold temperature. The interfacial morphology of samples overinjection molded at different mold temperature indicates that the tie layer can be molten with increasing mold temperature, and the interface between PE and tie layer is healing gradually until PE substrate and tie layer becomes a whole.

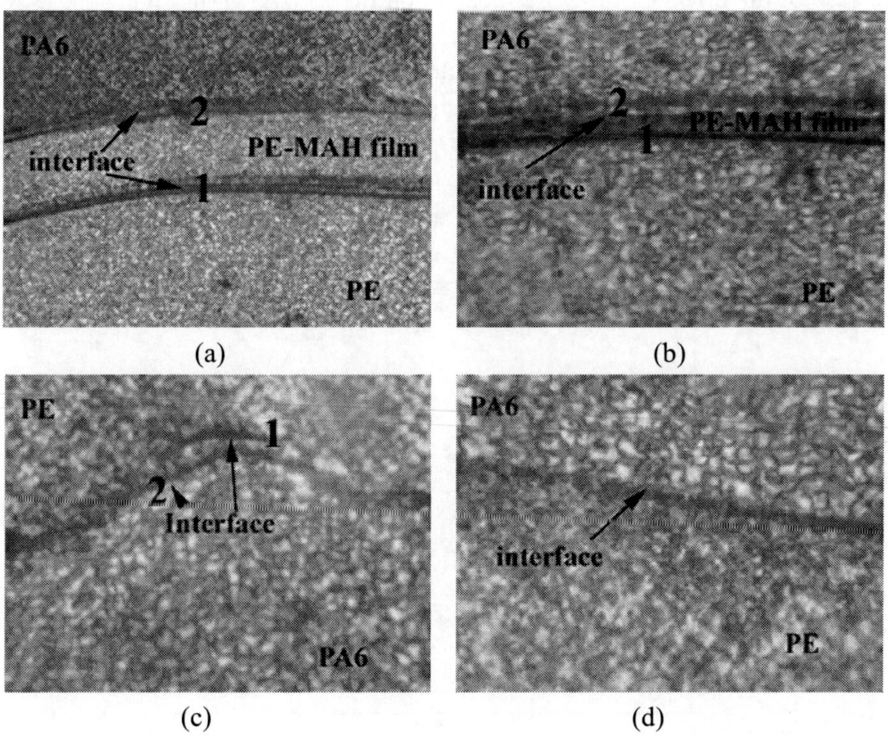

Figure 6. Trans-sectioned polarized optical micrograph of the interface region for the PE/tie film/PA6 adhesive joints prepared at different mold temperature. (a) 40°C, (b) 60°C, (c) 80°C, (d) 100°C

3. The Relationship between the Interfacial Adhesion and Fracture Mechanism

In order to correlate the interfacial adhesion of PE/tie layer/PA6 overinjection molded at different processing parameters with the interfacial fracture mechanism, the fracture surfaces on each side of PE/tie layer/PA6 adhesive interfaces were microscopically examined by SEM and DSC analysis.

The interfacial adhesion is determined by the adhesive strength at the interface and by the cohesive strength of both polymers. The tensile strength of the PE, tie layer and PA6 was 25, 23, and 65 MPa, respectively. Depending on the variation of the adhesive and cohesive strength at the interface, the interfacial fracture prefers the weaker interface. Here, we designated the interfacial strength between the tie film and PE substrate as IS1, and that between the tie film and PA6 as IS2, just shown in Figure 1. The value of IS1 depends on the heat and heat transfer rate across the tie layer/PE interface. For IS2, it is definitely related to the amount of the copolymer formed in-situ at the tie layer/PA6 interface.

PE-MAH content in tie layer

In the case of the effect of PE-MAH content on the IS (shown in Figure 2), IS increases with PE-MAH content in tie layer and reaches a saturated value. This can be explained as follows. As shown in Figure 7, the second melt injection temperature and mold temperature were constant, which means that the heat in the second injection process was same for all samples. In addition, the cooling condition and the thickness of the tie layer was same, which indicates that the heat transfer rate was also same, therefore, we believed that IS1 was a constant value for all samples.

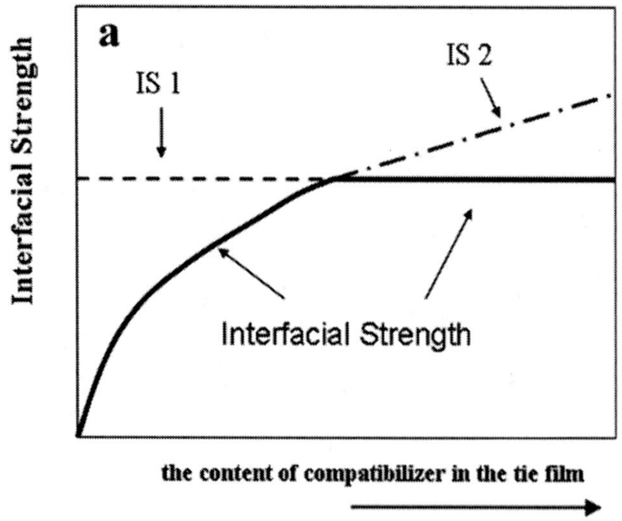

Figure 7. Schematic representation of the locus of failure of PE/tie layer/PA6 interface as a function of different PE-MAH content in the tie layer. Dash lines (IS1) denote the interfacial strength between HDPE/tie films. Dash dot lines (IS2) denote the interfacial strength between tie film/PA6. Black bold lines implicate the overall path of the shear fracture

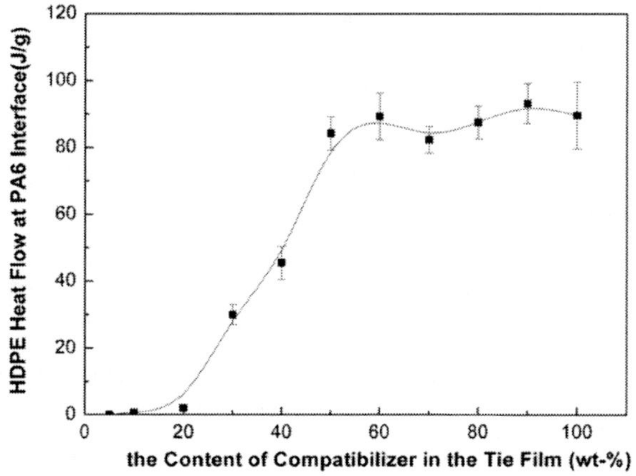

Figure 8. ΔH_c of the remaining HDPE component on fracture PA6 surfaces as a function of the content of PE-MAH in the tie layer

In section I, we obtained that IS2 increased with more PE-MAH added due to the formation of more copolymers in the interfacial region during overinjection molding. When a small amount of PE-MAH was incorporated into the tie layer, the concentration of copolymers formed at the interface was not sufficient to produce a strong adhesion between PA6 and tie layer. So IS2 was lower than IS1, the interfacial fracture occurs firstly through the tie layer/PA6 interface. With more PE-MAH was added, IS2 became higher than IS1 due to the formation of a large number of copolymers at the interface, the interfacial fracture will occur through PE/tie layer interface, and the interfacial strength would not change any more with the content of PE-MAH added further into the tie film.

According to DSC analysis conducted on fracture PE and PA6 sides, a melting peak around at 131°C, which belongs to PE crystalline, appears in the melting thermograms of PA6 side, while no PA6 melting peak is found in melting thermograms of PE side, indicating the fracture occurs very near the PE interface or in the bulk of PE, and the fracture remnant resided on the PA6 surfaces is PE component. Figure 8 shows ΔH_c of the remaining PE component on fracture PA6 surface as a function of the content of PE-MAH in the tie layer.

It was also found that ΔH_c of the remaining PE increases with the PE-MAH concentration and then reaches a plateau, indicating more PE component is pulled from PE side firstly and then does not change much with increasing PE-MAH concentration. The result is consistent with the above-mentioned interfacial fracture mechanism, and the dependence of PE-MAH content on interfacial adhesion can be explained adequately.

The thickness of tie layer

If the thickness of the tie layer is variable, the amount of PE-MAH was also increased in the interfacial region with increasing thickness of the tie layer, which resulted in an increased number of reaction at the tie layer/PA6 interface, therefore, IS2 still increased with the thickness of the tie film at the same injection melt and mold temperature in overinjection molding process, just shown in Figure 9. But the quantity of heat which transferred across the tie layer to the PE/tie layer interface would decrease due to the barrier effect of the thicker layer. Poor heat transfer across the interface caused the lack of entanglements between the tie layer and PE substrate, resultantly, IS1 should decrease with the thickness of the tie layer.

So in the case of the film thickness dependence, IS firstly increases with the thickness of the tie layer and shows a maximum around 15μm, then decreases at thicker tie layer. When the tie layer was too thick to transfer heat across the interface, no effective entanglements at PE/tie layer interface could be formed, such as 70μm in our experiment, The interfaces prefer to propagate toward the weaker interface, and the overall interface could not be reinforced in OIM process. Hence, there is a maximum in the variation of the overall interfacial strength (IS_{max}) with the thickness of the tie layer, as shown in Figure 3.

The SEM micrographs of the fractured PE sides of a series of typical specimens with different thickness of the tie layer are shown in Figure 10.

It can be seen that the introduction of 5μm tie film is able to induce the formation of some plastics deformation on PE side, which indicates that the interfacial adhesion between PE and PA6 can be improved effectively through our processing technique. As the thickness of the tie film increases (Figure 10b - c), the fracture PE surfaces all exhibit significant plastics deformation, and the hills and valleys of the deformed region become evident, indicating the interfacial adhesion increases. However, when the thickness of the tie film exceeds to 20μm, the fracture PE surfaces are seemingly to become less deformed, and the deformed zone on PE side decreases with the thickness of the tie film and eventually very litter deformed zone is observed for the sample with 60μm tie film (Figure 10d - f). Without doubt, increasing the thickness of the tie film increases the number of reaction at the interface hence the interfacial adhesion, which is correctly reflected by the fracture surface of the samples in Figure 10a - c. When the tie film is thicker than 20μm, the interfacial strength between PA6 and the tie film may still increases with the thickness of the tie film, but the interfacial strength between the tie film and PE becomes poor due to the heat transfer across the interface. So the fracture tends to propagate toward the weaker PE/tie film interface, and the fracture PE surfaces exhibit less deformed with the thickness of the tie film. The change of fracture morphology of PE sides with the thickness of tie layer is exactly indication of the interfacial fracture mechanism shown in Figure 9.

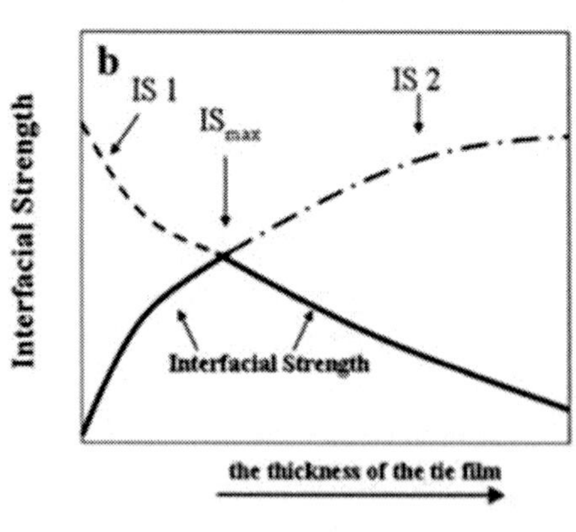

Figure 9. Schematic representation of the locus of failure of PE/tie layer/PA6 interface as function of different thickness of the tie layer. Dash lines (IS1) denote the interfacial strength between HDPE/tie films. Dash dot lines (IS2) denote the interfacial strength between tie film/PA6. Black bold lines implicate the overall path of the shear fracture

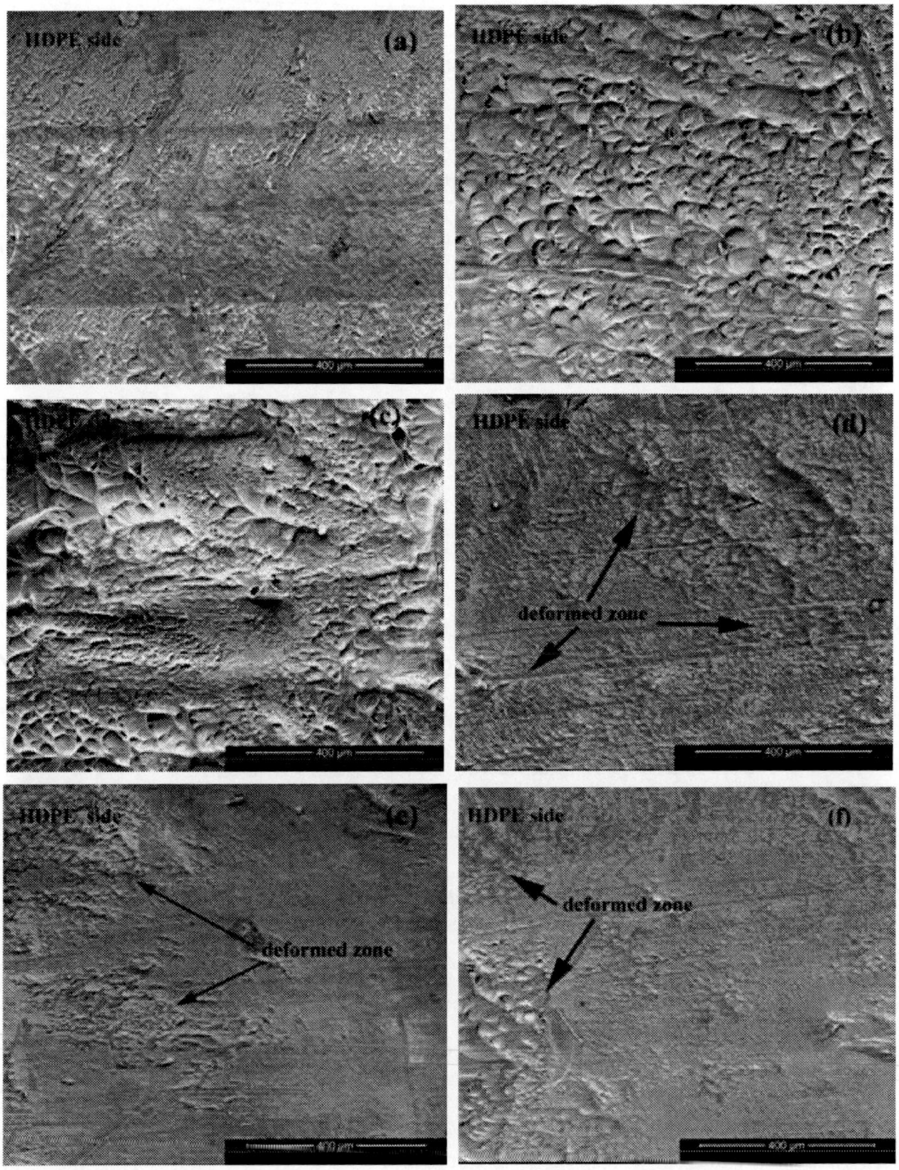

Figure 10. Scanning electron micrographs of the fracture surfaces of PE for the HDPE/tie layer/PA6 adhesive joints prepared with different thickness of the tie film. (a) 5μm, (b) 10μm, (c) 15μm, (d) 30μm, (e) 50μm, (f) 60μm

To prove the dependence of the thickness of the tie layer on the interfacial fracture mechanism, the fracture PA6 surface were analyzed using FTIR, The ATR-IR spectra of the fracture PA6 surfaces for the PE/tie layer/PA6 adhesive interfaces prepared with three different thickness of tie layer is given in Figure 11. The spectras of samples with 20 and 50μm tie layer are very similar. polyamide characteristic peaks (υ_{-NH}, 3300 and $\upsilon_{C=O}$, 1640 cm^{-1}) and PE characteristic peaks ($\upsilon_{as, C-H}$, 2920 cm^{-1}, $\upsilon_{s, C-H}$ 2850 cm^{-1}) were found on PA6 side, indicating the interfacial fracture occurs between tie layer and PA6 interface, PE component was pulled out from PE interface and resided on the PA6 side. However, the

ATR-IR spectra of PA6 surface for the sample with 70μm tie layer is remarkably different. Surprisingly, polyamide characteristic peaks disappears on PA6 side. PE characteristic peaks ($\upsilon_{as, C\text{-}H}$, 2920 cm^{-1}, $\upsilon_{s, C\text{-}H}$ 2850 cm^{-1}, 720 cm^{-1}) becomes very strong, and a distinct peak at 1790 cm^{-1}, which belongs to maleic anhydride group appears. It means the PA6 side is covered by a layer composed of PE/PE-MAH, i.e., the tie layer in the experiment. Therefore, the fracture crack should be propagated toward the interface between tie layer and PE substrate in the case of sample with 70μm tie layer, the interfacial fracture mechanism presented in Figure 9 was further proved experimentally.

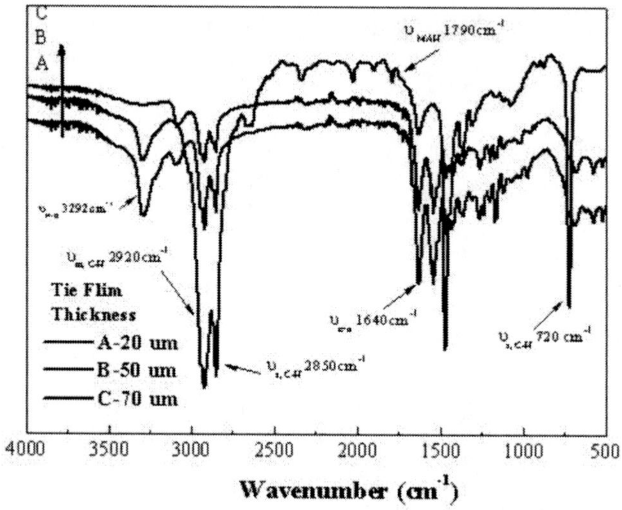

Figure 11. ATR-IR spectra of the fracture PA6 surfaces for the PE/tie layer/PA6 adhesive interfaces prepared with different thickness of tie film

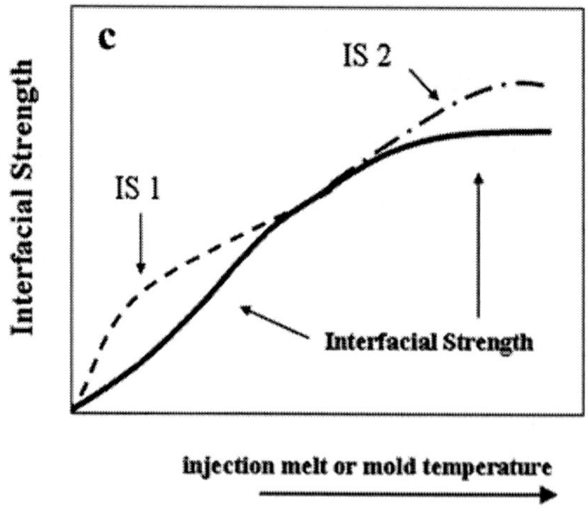

Figure 12. Schematic representation of the locus of failure of PE/tie layer/PA6 interface as a function of different injection melt or mold temperature in OIM process. Dash lines (IS1) denote the interfacial strength between HDPE/tie films. Dash dot lines (IS2) denote the interfacial strength between tie film/PA6. Black bold lines implicate the overall path of the shear fracture

Melt injection temperature and mold temperature

In section I, the effect of interfacial temperature on interfacial adhesion and fracture mechanism was investigated. Obviously, the interfacial temperature will increase with increasing second melt injection temperature or mold temperature. IS1 and IS2 both increased due to the interfacial reaction and entanglements or co-crystallization in the interfacial region. Though it is not easy to decide which strength is high, we can conclude that, at least, the overall interfacial adhesion was improved with increasing second injection melt temperature as well as mold temperature, just presented by the black bold line in Figure 12.

Figure 13 and Figure 14 give the ΔH_c results for samples overinjection molded at different melt injection and mold temperature, respectively.

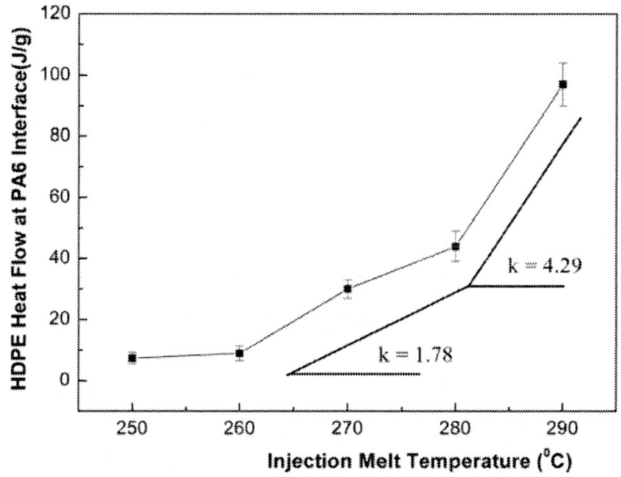

Figure 13. ΔH_c of the remaining PE component on fracture PA6 surfaces as a function of the second injection melt temperature

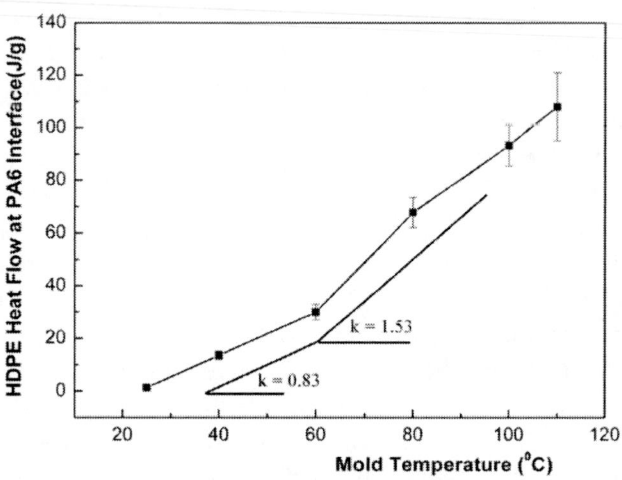

Figure 14. ΔH_c of the remaining HDPE component on fracture PA6 surfaces as a function of the mold temperature

It shows that as the second melt injection temperature and mold temperature increase, ΔH_c of the remaining HDPE component both increases, indicating more PE molecules were pulled out from PE side during shear fracture. With increasing interfacial temperature, the interface between PE substrate and tie layer will disappeared gradually, therefore, the propagation of crack should change from the interface between tie layer and PA6 to the bulk of PE due to strong interfacial adhesion, which can induce the plastic deformation of PE bulk.

Conclusion, The experimental results in this section showed that the interfacial adhesion could be significant enhanced by the introduction of the tie layer, and the interfacial strength depends on not only the interfacial adhesion of PA6/tie layer, but also that of tie layer/PE substrate. It was found that interfacial strength increases with the weight fraction of PE-MAH in the tie layer and becomes slightly leveled off. The interfacial strength shows a maximum with a tie film at the thickness of about 15μm, and then decreases with thicker tie film. Increasing the second injection melt and mold temperature is beneficial to improve the interfacial adhesion. The different dependence of the interfacial strength on the experimental parameters is ascribable to the change of the interfacial fracture mechanism.

SECTION IV. A COMPARISON OF TWO METHODS FOR REACTIVE REINFORCEMENT OF PE/PA6 INTERFACE IN A SEQUENTIAL INJECTION MOLDING PROCESS

In the former sections, two different methods were used to enhance the interfacial adhesion of PE/PA6 via reactive compatibilization in a sequential injection molding process, the addition of PE-MAH into PE component (PE+PE-MAH/PA6 bilayer system) versus the insert of a tie layer comprising of PE-MAH (PE/tie layer/PA6 sandwich system). Here, the efficiency of enhanced interfacial adhesion through two methods was compared by evaluating the interfacial strength, the amount of copolymers formed at the interface. The interfacial adhesion of reinforced interfaces through two methods was also correlated with the interfacial fracture mechanism. In PE+PE-MAH/PA6 bilayer system, 50 wt% PE-MAH was incorporated into PE interface to introduce reactive compatibilization. Equally, the tie layer in PE/tie layer/PA6 sandwich system was also comprised of 50 wt% PE-MAH and 50 wt% PE. The injection molding parameters were the same for all samples.

Figure 1 presents the interfacial strength of the bilayer and sandwich specimens prepared at different mold temperature.

It can be seen that interfacial strength is obviously improved with increasing mold temperature for bilayer and sandwich samples due to increasing interfacial temperature. It also can be found that the interfacial strength of sandwich sample is better than that of bilayer sample at the same mold temperature, especially at low temperature region.

It is known that the interfacial adhesion via reactive reinforcement is usually influenced by the amount of copolymer in situ formed at the interface. In order to study the dependence of the interfacial adhesion on the copolymers formed at the interface, we measured the copolymer concentration of the samples by XPS, which has been proved to be a reliable technique for determining the areal density of copolymer chains (Σ) at the reactive interface[26].

To determine Σ at reactive interface, the PA6 bulk was completely removed using the method established by Boucher's. The N 1s signal came exclusively from PA6 chains grafted onto the PE surface, and we normalized the N 1s singles by C 1s signals, to filter out fluctuations in the intensity of the X-ray source. Σ can be evaluated directly from the measured N 1s and C 1s signals using eq.

$$\Sigma = -(N_a \rho / M_n)\Lambda In(1 - \frac{(N/C)}{(N/C)_\infty})$$

where $(N/C)\infty$ is the normalized N1s signal for pure PA6; Λ accounts for the escape length of the electrons through the layer, here we used the result of Boucher[26] (Λ = 19 Å), N_a is Avagadro's number, ρ is the density of PA6 and M_n is the number-average molecular weigh of PA6.

Figure 2 presents XPS spectra in the N 1s region for bilayer and sandwich samples prepared at mold temperature of 40 and 100°C. The peak located at 399.8 ev belonged to N 1s of amides. The percent of the area of N 1s peak was listed in Table. Based on eq. , the calculated results were also listed in Table 1.

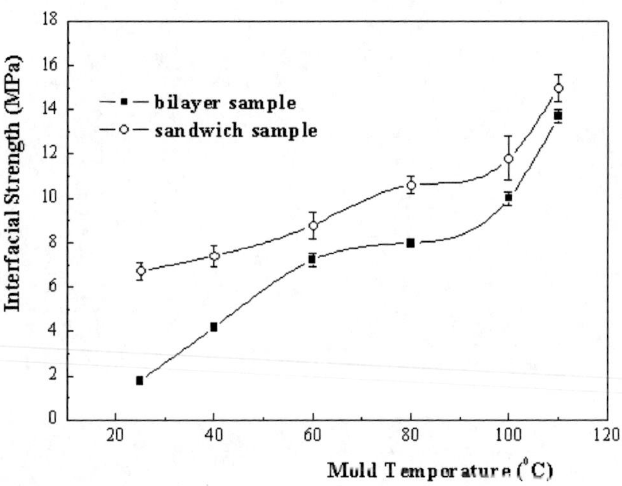

Figure 1. Comparison of the interfacial strength of bilayer versus sandwich samples prepared at different mold temperature

Table 1. The results of N1 concentration and the areal density of copolymer at different overinjection molded interfaces

Mold temperature (°C)	N (%)		Σ ($nm^{-2} \times 10^{-3}$)	
	bilayer	sandwich	bilayer	sandwich
40	0.42	1.08	1.5	3.4
60	0.73	1.29	3.2	5.0
80	0.86	1.44	3.9	7.1
100	1.41	1.56	6.5	7.8

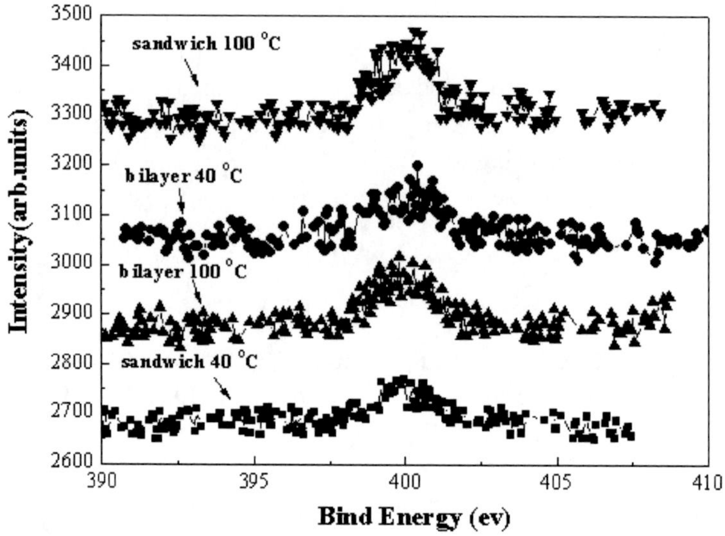

Figure 2. XPS in the N 1S region for bilayer and sandwich PE/PA6 adhesive joints overinjection molded at mold temperature of 40 and 100°C

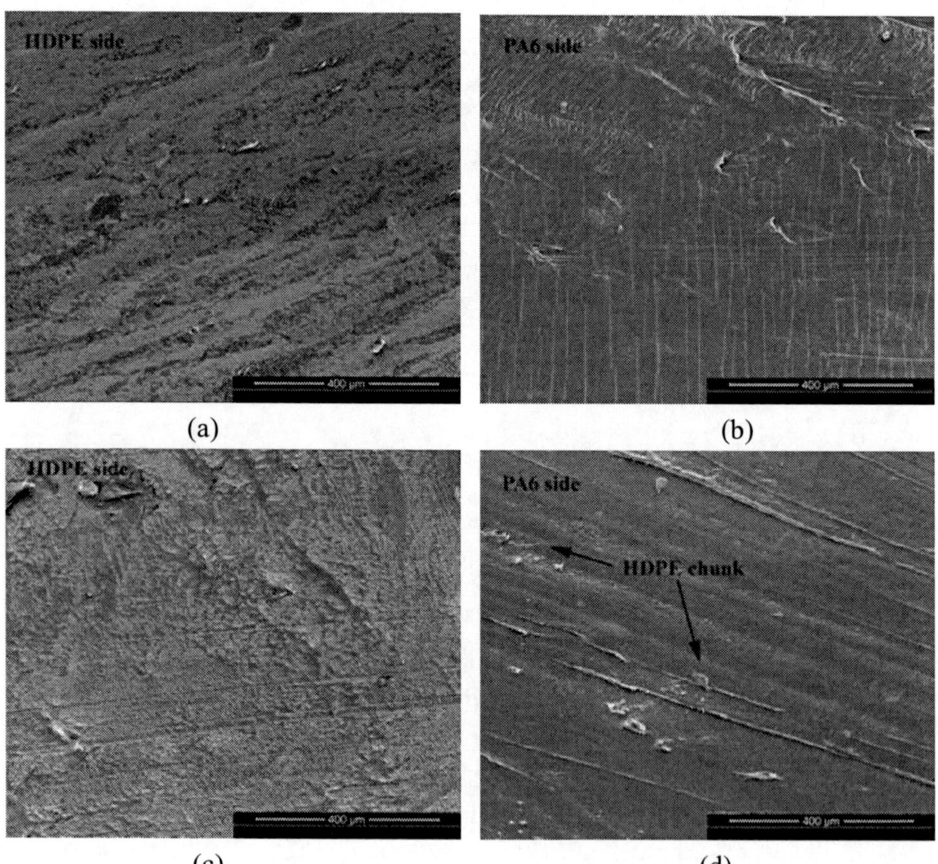

Figure 3. Scanning electron micrographs of the fracture surfaces of PE and PA6 for bilayer (a, b) and sandwich (c, d) adhesive joints prepared at mold temperature of 40°C

Table 2. ΔH_c of the remaining PE component on fracture PA6 surfaces for bilayer and sandwich sample overinjection molded at a mold temperature of 40 and 100°C.

Mold temperature (°C)	ΔH_c (J/g)	
	bilayer	sandwich
40	0	13.6
100	8.8	93.3

As expected, it is clearly shown that the amount of nitrogen on PE side increases with increasing molding for a series of bilayer and sandwich specimens. For the specimens overinjection molded at the same mold temperature, the composition percent of nitrogen at sandwich interface is all higher than that at bilayer interface, implying that more copolymers were formed at reactive interface of sandwich samples compared with bilayer interface. The different amount of copolymers in situ formed at interface should be the reason that caused the different degree of enhanced interfacial adhesion of bilayer and sandwich interfaces overinjection molded at the same conditions.

Figure 3 shows SEM photographs of the fractured surface on the PE and the matching PA6 side for bilayer and sandwich specimens overinjection molded at 40°C.

For the bilayer samples, a rugged fracture surface is observed on PE side. The matching PA6 fracture surface, shown in Figure, exhibits relatively smooth feature, indicating poor interfacial adhesion. In the case of PE/tie layer/PA6 sandwich system, the PE fracture surface becomes rougher and rugged than bilayer system, and the zone of obvious plastic deformation is observed. Some chunks are remained on the matching PA6 side after shear fracture, just shown in Figure 3d. The same PA6 fracture surfaces were analyzed by DSC, the special melting enthalpy (ΔH_c) of PE component, which resided on the PA6 surfaces after fracture was measured, and the experimental results were listed in Table 2.

From the DSC result, when the mold temperature of overinjection molding was 40°C, ΔH_c of the remaining PE component on fracture PA6 surfaces is zero for bilayer interface, while it changes to 13.6 for sandwich interface. This means the interfacial fracture mechanism is chain scission of copolymers for bilayer interface, but it changes to chain pull or crack propagation for sandwich interface. The strong interfacial adhesion at sandwich interface can induce plastic deformation of PE side, and resultantly, some PE fragments are pulled out and resided on PA6 side after fracture.

When the specimens were overinjection molded at mold temperature of 100°C, the morphology of the fractured surface for PE and PA6 side is remarkably different, as shown in Figure 4. Highly deformed PE fibrils are formed along the direction of shear force on HDPE side for bilayer sample. The broken fibrils also can be observed on the matching PA6 side. Obviously, the appearance of large amount of deformed fibrils is ascribed to an increase of the copolymers formed in-situ at the interfaces. The stress developed at the interface is able to initiate the significant plastics deformation of PE, indicating that the interface is enhanced evidently. The fractured surfaces for sandwich sample demonstrate different morphological structure. As shown in Fig.4b, a thin layer of deformed HDPE is drawn along the shear direction besides some striations or fibrils on PE side. The scissions of the thin layer is also observed on the matching PA6 side, shown in Fig.4d. The characteristic of these fractured surfaces indicates the interfacial adhesion of the sandwich sample has increase to such a degree that the interfacial strength is much greater than the cohesive strength of PE, and the cracks tend to run through a thin layer in PE domain rather than along the interface. In comparison of bilayer sample, total cohesive failure is observed in the case of sandwich sample instead of adhesive failure. The result of ΔH_c of the remaining PE component on fracture PA6 surfaces also indicated that more plastics deformation was taken place at sandwich interface, as presented in Table.

Based on the experimental result obtained in this section, two different routes were investigated to enhance the interfacial adhesion of PE/PA6 in a sequential molding process, the addition of PE-MAH into PE interface (PE+PE-MAH/PA6 bilayer system) versus the insert of a tie layer comprising of PE-MAH (PE/tie layer/PA6 sandwich system). When the processing conditions and the PE-MAH percent were the same, It was found that more copolymers can be formed at sandwich PE/tie layer/PA6 interface compared with bilayer PE+PE-MAH/PA6 interface, the different amount of copolymers in situ formed at bilayer and sandwich interface induce different interfacial fracture mechanism of reinforced interfaces, As a result, the interfacial adhesion of sandwich interface is stronger than that of bilayer interface.

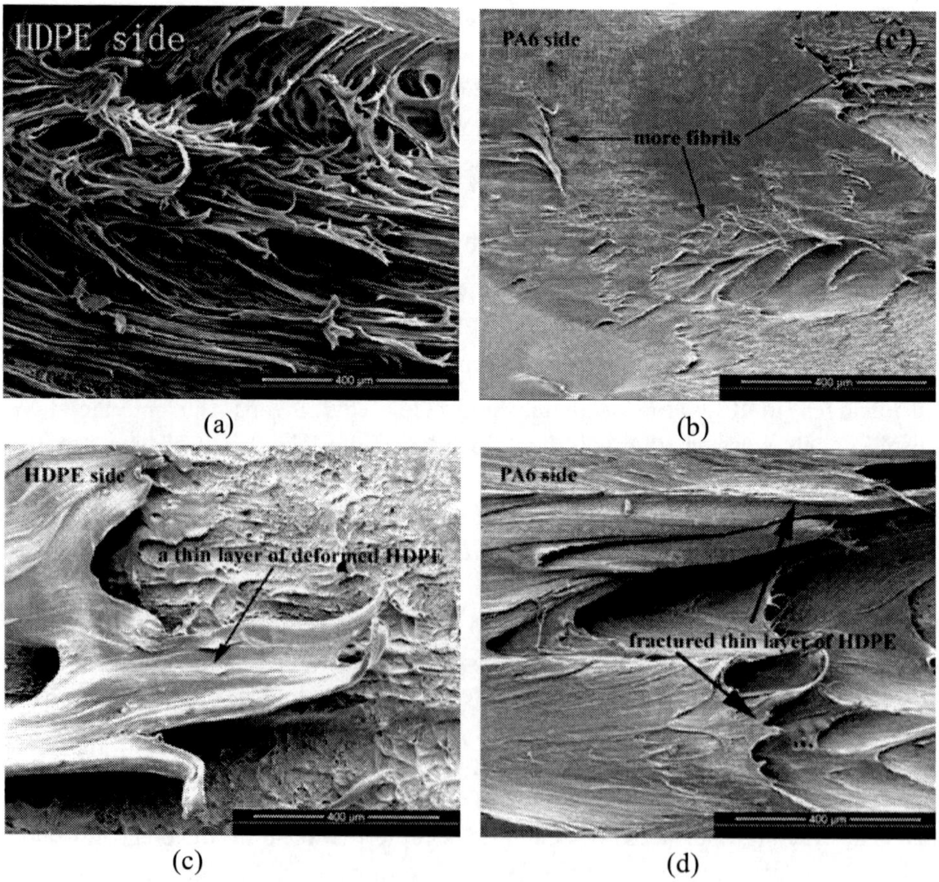

Figure 4. Scanning electron micrographs of the fracture surfaces of PE and PA6 for bilayer (a, b) and sandwich (c, d) adhesive joints prepared with at mold temperature of 100°C

SUMMARY

Technical feasibility of enhancement of PE/PA6 interface via in situ reactive compatibilization in a sequential injection molding process was investigated. Some conclusions were obtained:

(1) The interfacial adhesion of PE/PA6 was improved by the addition of PE-MAH into PE interface. Interfacial temperature and heat transfer across PE/PA6 interface after the second melt injected were found to play a significant role in the interfacial reinforcement. At the same melt injection temperature, more copolymers can be formed in situ at PE/PA6 interface with increasing mold temperature, the interfacial fracture mechanism changes from chain scission between PE and PA6 interface to chain pull and plastic deformation on PE side. Consequently, strong interfacial adhesion is obtained.

(2) Due to gradient cooling in sequential injection molding, the diffusion of the reactive chains across the interface is altered, resultantly, the interfacial reaction kinetics is mainly influenced by the diffusion of reactive chains, the interfacial adhesion strength is weaker than the same interface which is annealed in isothermal annealing process. The relationship between interfacial strength (IS) and the initial content of reactive chains (ρ) changes from $IS \sim \rho^2$ for isothermal annealing interface to $IS \sim \rho$ for overinjection molding interface.

(3) The interfacial adhesion of PE/PA6 also can be improved by insert of a tie layer composed of PE and PE-*g*-MAH in a sequential injection molding process. The interfacial adhesion of PE/tie layer/PA6 sandwich interface is determined by the interfacial strength between PE and tie layer and the interfacial strength between tie layer and PA6, depending on the variation of the composition and thickness of the tie layer, the second injection melt and mold temperature.

(4) At the same processing parameters and PE-MAH content, more copolymers can be formed in situ at sandwich PE/tie layer/PA6 interface compared with bilayer PE+PE-MAH/PA6 interface. Macroscopical interfacial adhesion of sandwich is higher than that of bilayer, which is the indication of microcosmic interfacial interaction and interfacial fracture mechanism.

REFERENCES

[1] Harte, A. M. & Mc Namare, J. F. (2007). *J. Mater. Proc. Technol., 182 12.*

[2] Harte, A. M. & Mc Namare, J. F. (2006). *J. Mater. Proc. Technol., 173 376.*

[3] Weng, D., Andries, J., Morin, P., Saunders, K. & Politis, J. (2000). *Injection Molding Technol., 4 22.*

[4] Huang, D. Y. & Chen, R. S. (1999). *Polym. Eng. Sci., 39 2159.*

[5] Carella, A. R., Alonsol, C., Merino, J. C. & Pastor, J. M. (2002). *Polym. Eng. Sci., 42 2172.*

[6] Aurrekoetxea, J., Castillo, G., Cortes, F., Sarrionandia, M. A. & Urrutibeascoa, I. (2006). *J. Appl. Polym. Sci., 102 261.*
[7] Arzondo, L. M., Pino, N., Carella, J. M., Pastor, J. M., Merino, J. C., Póveda, J. & Alonso, C. (2007). Polym. Eng. Sci., *44 2110.*
[8] Dondero, M., Pastor, J. M., Carella, J. M. & Perez, C. J. (2009). *Polym. Eng. Sci., 49 1886.*
[9] Olley, R. H. & Bassett, D. C. (1982). *Polymer, 23 1707.*
[10] Varga, J. & Karger-Kocsis, J. (1995). *Polymer, 36 4877.*
[11] Wu, C. M. & Chen, M. (1999). J. Karger-Kocsis, *Polymer, 40 4195.*
[12] Li, H., Jiang, S., Wang, J., Wang, D. & Yan, S. (2003). *Macromolecules, 36 2802.*
[13] Li, H., Zhang, X., Duan, Y., Wang, D., Li, L. & Yan, S. (2004). *Polymer, 45 8059.*
[14] Li, H., Zhang, X., Kuang, X., Wang, J., Wang, D., Li, L. & Yan, S. (2004). *Macromolecules, 37 2847.*
[15] Sun, Y., Li, H., Huang, Y., Chen, E., Zhao, L., Gan, Z. & Yan, S. (2005). *Macromolecules, 38 2739.*
[16] Liu, J., Wang, J., Li, H., Shen, D., Zhang, J., Ozaki, Y. & Yan, S. (2006). *J Phys Chem B, 110 738.*
[17] Kantz, M. R., Newman, J. R. & Stigale, F. H. (1972). *J Appl Polym Sci, 161249.*
[18] Tadmor, Z. (1974). *J Appl Polym Sci., 18 1753.*
[19] Fleischmann, E., Zipper, P., Janosi, A., Geymayer, W., Koppelmann, J. & Schurz, J. (1989). *Polym Eng Sci., 29 835.*
[20] Zipper, P., Janosi, A., Geymayer, W., Ingolic, E. & Fleischmann, E. (1996). *Polym Eng Sci., 36 467.*
[21] Boucher, E., Folkers, J. P., Creton, C., Hervet, H. & Leger, L. (1997). *Macromolecules, 30 2102.*
[22] Lovinger, A. J., Chua, J. O. & Gryte, C. C. (1977). *J Polym Sci Polym Phys Ed, 15 641.*
[23] Varga, J., Karger-Kocsis, J. & Polym. J. (1996). Sci. *Part B Polym. Phys., 34 657.*
[24] Leong, Y. W. & Hamada, H. (2008). *Int. Polym. Proc., 2 183.*
[25] Utracki, L. A. (1982). *Polym. Eng. Sci., 22*, 1166.
[26] Chun, S. B. & Han, C. D. (2000). *Macromolecules, 33*, 3409.
[27] Fayt, R. & Jérôme, R. (1989). Ph. Teyssié, *J. Polym. Sci., Polym. Phys. Ed., 7*, 775.
[28] Auschra, C., Stadler, R. & Voigt-Martin, I. G. (1993). *Polymer, 34*, 2081.
[29] Harrats, C., Fayt, R. & Jérôme, R. (2002). *Polymer, 43*, 5347.
[30] Fleischer, C. A., Marales, A. R. & Koberstein, J. T. (1994). *Macromolecules, 27*, 379.
[31] Brown, H. R., Char, K., Deline, V. R. & Green, P. F. (1993). *Macromolecules, 26*, 4155.
[32] Scott, C. & Macosko, C. W. (1994). *J. Polym. Sci., Part B: Polym. Phys., 32*, 205.
[33] Anastasiadis, S. H., Gancarz, I. & Koberstein, J. T. (1989). *Macromolecules, 22*, 1449.
[34] Harrats, C., Dedecker, K., Groeninckx, G. & Jérôme, R. (2003). *Macromol. Symp., 198*, 183.
[35] Liu, N. C. & Baker, W. E. (1992). *Adv. Polym. Technol., 11*, 249.
[36] Cartier, H. & Hu, G. H. (2001). *Polymer, 42*, 8807.
[37] Son, Y. & Weiss, R. A. (2002). *Polym. Eng. Sci., 42*, 1322.
[38] Jiang, C., Filippi, S. & Magagnini, P. (2003). *Polymer, 44*, 2411.
[39] Hong, S. M., Hwang, S. S., Seo, Y., Chung, I. J. & Kim, K. U. (1997). *Polym. Eng. Sci., 37*, 646.

[40] Kim, H. J., Lee, K J., Seo, Y., Kwak, S. & Koh, S. K. (2001). *Macromolecules*, *34*, 2546.

[41] Jeon, H. K. & Kim, J. K. (1998). *Macromolecules*, *31*, 9273.

[42] Isayev, A. I. & Hong, C. K. (2003). *Polym. Eng. Sci.*, *43*, 91.

[43] Li, J., Guo, S. Y., Slezák, R. & Hausnerová, B. (2005). *Macromol. Chem. Phys.*, *206*, 2429.

[44] Chen, G. S., Guo, S. Y. & Li, H. L. (2002). *J. Appl. Polym. Sci.*, *86*, 23.

[45] Harton, S. E., Stevie, F. A., Spontak, R. J., Koga, T., Rafailovich, M. H. & Sokolov, J. C. (2005). Ade, H. *Polymer*, *46*, 10173.

[46] Harton, S. E., Stevie, F. A. & Ade, H. (2005). *Macromolecules*, *38*, 543.

[47] Arzondo, L. M., Pino, N., Carella, J. M., Pastor, J. M., Merino, J. C., Póveda, J. & Alonso, C. (2007). Polym. *Eng. Sci.*, *44*, 2110.

[48] Agassant, Avenas, Segent, Carreau, Polymer Processing, *Principles and Modelling*, Hanser Publishers, (1991)

[49] Qiu, J. H., Tsuboi, A., Izumi, K., Wu, H., Guo, S. Y. & Huang, Y. D. (2007). Polym. *Eng. Sci.*, *47*, 2164.

[50] Boucher, E., Folkers, J. P., Hervet, H., Léger, L. & Creton, C. (1996). *Macromolecules*, *29*, 774.

[51] Boucher, E., Folkers, J. P., Creton, C., Hervet, H. & Léger, L. (1997). *Macromolecules*, *30*, 2102.

[52] Fredrickson, G. H. (1996). Phys. Rev. *Lett.*, *76*, 3440.

[53] Fredrickson, G. H. & Milner, S. T. (1996). *Macromolecules*, *29*, 7386.

[54] Fredrickson, G. H. & Leibler, L. (1996). *Macromolecules*, *29*, 2674.

[55] O'Shaughnessy, B. & Sawhney, U. (1996). Phys. Rev. *Lett.*, *76*, 3444.

[56] O'Shaughnessy, B. & Vavylonis, D. (1999). Europhys. *Lett.*, *45*, 638.

[57] O'Shaughnessy, B. & Sawhney, U. (1996). *Macromolecules*, *29*, 7230.

[58] O'Shaughnessy, B. & Vavylonis, D. (1999). *Macromolecules*, *32*, 1785.

[59] O'Shaughnessy, B. & Vavylonis, D. (2000). *Phys. Rev. Lett.*, *84*, 3193.

[60] Cole, P. J. (2002). *Polymer-polymer adhesion in melt-processed multilayered structures PhD Thesis*, University of Minnesota.

[61] Brown, H. R. (1991). *Macromolecules*, *24*, 2752.

In: Injection Molding: Process, Design, and Applications
Editor: Phoebe H. Kauffer

ISBN: 978-1-61761-307-4

Chapter 3

SMOOTHED PARTICLE HYDRODYNAMICS AND ITS APPLICATION TO NON-NEWTONIAN MOULDING FLOW

X. J. Fan**, *R. I. Tanner and R. Zheng
School of Aerospace Mechanical & Mechtronic Engineering,
University of Sydney, NSW, Australia

ABSTRACT

Smoothed Particle Hydrodynamics (SPH) is a meshless and fully Lagrangian method. It has been widely applied in simulating fluid flows with complex moving boundaries, such as free surfaces, and dynamic response of materials involving large deformation, fracture and fragmentation. In the simulation of flow problems this method usually solves the conservation equations explicitly and the stable results can be obtained only for momentum dominated flows with low viscosity and pressure. However, in polymer processing, the fluid is non-Newtonian with the viscosity being as high as $O(10^3)$ to $O(10^4)$ *Pa-S*, and under pressure which can be as high as $O(10^6)$ to $O(10^{10})$*Pa*. The standard SPH algorithm is infeasible in this case.

In this chapter, we give a concise review of SPH with emphasis on the SPH formulation, stability problems and enforcing boundary conditions; and describe how to apply SPH to the simulation of injection molding flows. The viscosity of the fluid is modeled by a power-law function. The fluid is considered compressible under high pressure. The pressure of fluid is calculated in terms of realistic state equations for polymer melts, such as the Tait equation, Sun-Song-Yang's equation or a weakly compressible equation. To relieve the stability constraints on the time step, we propose a semi implicit SPH formulation for high viscous non-Newtonian flow, so that a reasonable time step can be adopted in simulations. An iterative algorithm is used to solve the resulting system equations of the SPH formulation. This method is completely matrix-free and robust.

It has been known that inconsistency and tensile instability are the intrinsic weakness of standard SPH. For a highly viscous flow under high pressure, the SPH system becomes highly unstable. To make the system more stable, we use an artificial pressure to stabilize the SPH system thus avoiding the instability.

The present method is firstly tested in simulating simple flows: the transient slit flow and simple shear flow. The simulated flow profiles are in good agreement with the analytical solutions. Secondarily, the fountain flow is simulated. The simulated free surfaces for the Newtonian and shear-thinning fluids are in agreement with those of finite element simulations. Finally we apply the method to the simulation of the moulding flows of Newtonian and non-Newtonian fluids. We obtain stable results for the power-law moulding flow with the highest zero shear rate viscosity being $1.22\times10^4 Pa\cdot s$ and the lowest power-law exponent 0.294, Reynolds number is 3×10^{-4} to 6×10^{-5}, and the highest pressure is in the range $O(10^8)$ to $O(10^{10})Pa$.

INTRODUCTION

Smoothed particle hydrodynamics (SPH) is a mesh free or particle method to simulate large deformation and flow problems in continuum mechanics. It has been attracting a lot of research effort since its invention in 1977 by Gingold & Monaghan (1977) and Lucy (1977) to simulate nonaxisymmetric phenomena in astrophysics. SPH has been widely applied to simulating complex problems in science and engineering. Liu and Liu (2003) gave a detailed description of SPH theory, coding and its applications in their book. Extensive reviews can be found in Monaghan (1992, 2005). Recently, SPH has been applied to simulate non-Newtonian and viscoelastic transient and free surface flows, for example see Shao and Lo (2003), Ellero and Tanner (2005), Fang *et al* (2006), Hosseini *et al* (2007), Rafiee *et al.* (2007) and Rafiee (2008).

Although SPH has been widely used in various branches of science and engineering, its application to polymer processing has not been reported by other researchers. Cleary *et al.* (2002) and Prakash *et al.* (2009) used SPH to simulate metal flow in casting and furnace emptying successfully. Compared to polymeric liquids, liquid metal is Newtonian with lower viscosity and processing pressure. Momentum dominates liquid metal flow. In the SPH momentum equation they used, the interaction force between two particles due to viscosity is always along the line connecting their two centers, i.e. aligned with the interaction force due to pressure. This approximation will be in error for high viscous and low Reynolds number flows. High viscosity puts serious constraints on the length of time step. Morris et al (1997) simulated low Reynolds number incompressible flows using SPH. The value of Reynolds number was as low as 1.25×10^{-2} and the viscosity they used was $10^{-3} Pa\cdot s$, which is the same as water at 20 °C. The time step constraint restricts the further increase of the viscosity for a stable simulation. It has been known that there are two stability problems for the standard SPH: the inconsistency (Bonet and Kulasegaram 2000) and tensile instability (Swegel *et el.* 1995). Inconsistency leads to error in SPH solutions, as a small error in density results in a large fluctuation in pressure. High pressure is usually accompanied with high pressure fluctuation during simulation. The system becomes unstable. The tensile instability is an intrinsic weakness of the kernel function and results in particle clustering.

Mesh methods, such as the finite element method and finite volume method, have been applied to polymer processing. For example, Moldlfow is a popular commercial software for injection moulding and de-mould shrinkage and warpage analysis. Recently, Daun and Li

(2007) used the Characteristic Based Split algorithm in the Arbitrary Lagrangian-Eulerian framework to simulate incompressible non-isothermal non-Newtonian flows in injection moulding. However, there have been no reports, before the publication of our article (Fan *et al.* 2010), on the application of mesh-free methods, especially SPH to injection moulding. In the present chapter, we try to extend the application of SPH to polymer processing. One of the advantages of SPH over mesh methods is that the moving free surfaces can automatically be caught. This makes it much easier to update complex free surfaces during the simulation. But the disadvantages of SPH may frustrate this application. We have to deal with non-Newtonian flows with the viscosity being as high as $O(10^3)$ to $O(10^4)$ *Pa-s* and the pressure about $O(10^8)$ to $O(10^{10})$ *Pa*. This kind of SPH system is highly sensitive to instability. To obtain a stable solution using the standard explicit SPH the time step must be reduced to a very small value according to the diffusive stability constraint. Hence, SPH is thought to be infeasible in simulating polymer processing. If the generation of meshes, especially in three dimensions, could be avoided, then this would be an advantage for injection moulding studies. In the current early stage of development, mesh free methods are slower than meshed methods in simulating injection moulding. However, we believe they need exploration: in the present chapter we want to go a step further to apply SPH to highly viscous isothermal non-Newtonian flows.

This chapter gives a concise review of SPH and then describes its application to the moulding flow of polymer melts. It consists of 8 sections. Sec 2, outlines the SPH kernel estimation and various formations of SPH conservation equations. Secs. 3 and 4 discuss the convergence and stability of SPH formulations and state equations. The various treatments of boundary conditions are summarized in Sec. 5. The description of how to simulate injection moulding flow and some numerical results are followed in Secs. 6 and 7. Finally some conclusions are given.

2. FORMULATIONS

2.1. SPH Kernel Functions

The foundation of a numerical method is its interpolation theory, based on which the governing differential equations and field variables can be evaluated at discrete points in the domain. SPH allows any function to be expressed in terms of its values at a set of particle points using a kernel function.

Consider a continuous function, $f(x)$. Its kernel estimate at position x is defined as

$$\langle f(x)\rangle = \int_V f(x')W(x-x',h)\,dx', \tag{0.1}$$

where V denotes the volume of the flow domain, x' and dx' are the position vector of a point in V and the differential volume element at x', h is the smooth length representing the effective width of the kernel, and W a kernel function with the following properties:

$$\int_V W(\boldsymbol{x}-\boldsymbol{x}',h)d\boldsymbol{x}'=1, \tag{0.2}$$

$$\lim_{h\to 0} W(\boldsymbol{x}-\boldsymbol{x}',h)=\delta(\boldsymbol{x}-\boldsymbol{x}'). \tag{0.3}$$

If $f(\boldsymbol{x})$ is known only at N discrete particles at $\boldsymbol{x}_b\,(b=1,...,N)$, we can approximate $f(\boldsymbol{x}')$ as

$$f(\boldsymbol{x}')=\sum_{b=1}^{N}\delta(\boldsymbol{x}-\boldsymbol{x}')f(\boldsymbol{x}_b)(\Delta V)_b, \tag{0.4}$$

where $\delta(\cdot)$ denotes the Dirac delta function, $(\Delta V)_b$ is the fluid volume with mass m_b associated to the particle *b* at $\boldsymbol{x}_b$ where the fluid density is ρ_b. We have $(\Delta V)_b = m_b/\rho_b$. Substituting Eq. (0.4) into Eq. (0.1), we obtain the kernel estimate of $f(\boldsymbol{x})$ to be

$$\langle f(\boldsymbol{x})\rangle=\sum_{b=1}^{N}\frac{m_b}{\rho_b}f(\boldsymbol{x}_b)W(\boldsymbol{x}-\boldsymbol{x}_b,h). \tag{0.5}$$

It is noted that $\langle f(\boldsymbol{x})\rangle$ is a kernel estimate of $f(\boldsymbol{x})$ and may be viewed as a non-local interpolation of the latter, i.e. $f(\boldsymbol{x})\neq\langle f(\boldsymbol{x})\rangle$.

The kernel function satisfies Eqs. (0.2) and (0.3) usually has a compact support. Hence, there is an influence domain for each particle and only those particles whose influence domains cover the point $\boldsymbol{x}$ contribute to the kernel estimate $\langle f(\boldsymbol{x})\rangle$. The kernels depend on the smooth length, *h*, and the dimensionless distance to centre of the support, $q = x/h$, here $x=|\boldsymbol{x}-\boldsymbol{x}_b|$ is the distance from point $\boldsymbol{x}$ to the kernel centre $\boldsymbol{x}_b$. Some commonly used kernel functions are listed as follows.

1. Gaussian function:

$$W(x,h)=\alpha_D\exp(-q^2), \tag{0.6}$$

   where α_D, is the normalization factor and equal to $1/(\pi h^2)^{d/2}$, with *d* is the number of dimension.

2. Quadratic function:

$$W(x,h)=\begin{cases}\alpha_D\left[\dfrac{3}{16}q^2-\dfrac{3}{4}q+\dfrac{3}{4}\right], & 0\le q\le 2;\\ 0, & q>2,\end{cases} \tag{0.7}$$

with α_D to be $2/\left(\pi h^2\right)$ in 2D and $5/\left(4\pi h^3\right)$ in 3D.

3. Cubic spline:

$$W(x,h)=\alpha_D\begin{cases}1-\frac{3}{2}q^2+\frac{3}{4}q^3, & 0\le q\le 1;\\ \frac{1}{4}(2-q)^3, & 1\le q\le 2;\\ 0, & q>2,\end{cases} \tag{0.8}$$

where α_D is 2/(3h) in 1D, $10/\left(7\pi h^2\right)$ in 2D and $1/\left(\pi h^3\right)$ in 3D.

4. Quintic spline function:

$$W(x,h)=\alpha_D\begin{cases}(3-q)^5-6(2-q)^5+15(1-q)^5, & 0\le q<1;\\ (3-q)^5-6(2-q)^5, & 1\le q<2;\\ (3-q)^5, & 2\le q<3;\\ 0, & q\ge 3,\end{cases} \tag{0.9}$$

Where α_D is $7/\left(478\pi h^2\right)$ in 2D and $1/\left(40\pi h^3\right)$ in 3D.

In the above cases, the kennel functions are symmetrical between two particles, i.e.

$$W\left(\frac{x_{ab}}{h},h\right)=W\left(\frac{x_{ba}}{h},h\right), \tag{0.10}$$

where $x_{ab}:=|\boldsymbol{x}_{ab}|=\sqrt{(\boldsymbol{x}_a-\boldsymbol{x}_b)\cdot(\boldsymbol{x}_a-\boldsymbol{x}_b)}=x_{ba}$. This equation can be written, for short, as

$$W_{ab}=W_{ba}. \tag{0.11}$$

The gradient of the kernel function can be calculated from Eq. (0.10) as follows

$$\nabla_a W\left(\frac{x_{ab}}{h},h\right)=\nabla W\left(\frac{|\boldsymbol{x}-\boldsymbol{x}_b|}{h},h\right)\Bigg|_{\boldsymbol{x}=\boldsymbol{x}_a}=\frac{\partial W(q,h)}{\partial q}\frac{\partial|\boldsymbol{x}-\boldsymbol{x}_b|}{h\partial x_I}\boldsymbol{e}_I\Bigg|_{\boldsymbol{x}=\boldsymbol{x}_a} \tag{0.12}$$

$$=\frac{\partial W(q,h)}{h\partial q}\Bigg|_{q=\frac{x_{ab}}{h}}\hat{\boldsymbol{r}}_{ab},$$

where $\boldsymbol{e}_I$ is the I-th base vector of Cartesian coordinates and the gradient operator is defined as $\nabla := \frac{\partial}{\partial x_I}\boldsymbol{e}_I$ (here the repeated indexes denote the sum over the dimension of space), and $\nabla|\boldsymbol{x}-\boldsymbol{x}_b| = \left.\frac{\partial|\boldsymbol{x}-\boldsymbol{x}_b|}{\partial x_I}\boldsymbol{e}_I\right|_{\boldsymbol{x}=\boldsymbol{x}_a} = \frac{\boldsymbol{x}_{ab}}{x_{ab}} = \hat{\boldsymbol{r}}_{ab}$ is the unit vector pointing from $\boldsymbol{x}_b$ to $\boldsymbol{x}_a$. From Eqs. (0.11) and (0.12), the gradients of the kernel functions between two particles a and b satisfy

$$\nabla_a W_{ab} = -\nabla_b W_{ba}. \tag{0.13}$$

This equation is useful in verifying the momentum and energy conservation of SPH formulations.

2.2. SPH Interpolations

In the SPH literature it is usually assumed that the interpolated value of a function at $\boldsymbol{x}$ is approximated by its kernel estimate expressed in Eq. (0.5), i.e.

$$f(\boldsymbol{x}) \approx \sum_b \frac{m_b}{\rho_b} f_b W\left(|\boldsymbol{x}-\boldsymbol{x}_b|, h\right), \tag{0.14}$$

where the sum goes over all particles whose influence domains cover point $\boldsymbol{x}$ and f_b denotes the value of function $f(\boldsymbol{x})$ at particle $\boldsymbol{x}_b$. If the function is a constant, $f(\boldsymbol{x}) = f(\boldsymbol{x}_b) = const.$, we have

$$\sum_b \frac{m_b}{\rho_b} W(\boldsymbol{x}-\boldsymbol{x}_b) = 1 \tag{0.15}$$

For the density, Eq. (0.14) gives the interpolated density at point $\boldsymbol{x}$ is

$$\rho(\boldsymbol{x}) = \sum_b m_b W(\boldsymbol{x}-\boldsymbol{x}_b). \tag{0.16}$$

This equation implies how the mass of a set of particles is smoothed to produce the interpolated density.

The gradient of $f(\boldsymbol{x})$ at $\boldsymbol{x}$ can be calculated from Eq. (0.14):

$$\nabla f(\boldsymbol{x}) = \sum_b \frac{m_b}{\rho_b} f_b \nabla W(\boldsymbol{x}-\boldsymbol{x}_b). \tag{0.17}$$

If $f(\boldsymbol{x}) = const.$, the above equation should give

$$\sum_b \frac{m_b}{\rho_b} \nabla W(\boldsymbol{x} - \boldsymbol{x}_b, h) = 0. \tag{0.18}$$

In numerical tests, it is found that Eq. (0.18) does not exactly vanish. This means that the gradient of a constant usually is not zero when calculated by using Eq. (0.17). A simple way to ensure a vanishing gradient for the constant is to use the identity (Monahan 2005):

$$\nabla f = \frac{1}{\phi}\left[\nabla(\phi f) - f\nabla\phi\right], \tag{0.19}$$

where ϕ is any differentiable function. According to Eq.(0.17), we can estimate the gradient of the function at $\boldsymbol{x}_a$, $\nabla f(\boldsymbol{x}_a)$ to be

$$\nabla f(\boldsymbol{x}_a) = \frac{1}{\phi_a}\sum_b \frac{m_b}{\rho_b}\phi_b (f_b - f_a)\nabla_a W_{ab}. \tag{0.20}$$

If f is constant, $\nabla f \equiv 0$. Different choices of ϕ give various expressions of derivatives used in the SPH formulation. For example, when $\phi = 1$ we obtain

$$\nabla f(\boldsymbol{x}_a) = \sum_b \frac{m_b}{\rho_b}(f_b - f_a)\nabla_a W_{ab}, \tag{0.21}$$

and when $\phi = \rho$,

$$\nabla f(\boldsymbol{x}_a) = \frac{1}{\rho_a}\sum_b m_b (f_b - f_a)\nabla_a W_{ab}. \tag{0.22}$$

Similarly, for a vector $\boldsymbol{A}$ or symmetric tensor, we have, for $\phi = 1$,

$$\nabla \boldsymbol{A}(\boldsymbol{x}_a) = \sum_b \frac{m_b}{\rho_b}\nabla_a W_{ab}(\boldsymbol{A}_b - \boldsymbol{A}_a), \tag{0.23}$$

$$\nabla \cdot \boldsymbol{A}(\boldsymbol{x}_a) = \sum_b \frac{m_b}{\rho_b}(\boldsymbol{A}_b - \boldsymbol{A}_a)\cdot\nabla_a W_{ab}, \tag{0.24}$$

$$\nabla \times \boldsymbol{A}(\boldsymbol{x}_a) = \sum_b \frac{m_b}{\rho_b}\nabla_a W_{ab} \times (A_b - A_a), \tag{0.25}$$

and, for $\phi = \rho$,

$$\nabla \boldsymbol{A}(\boldsymbol{x}_a) = \frac{1}{\rho_a} \sum_b m_b \nabla_a W_{ab} (\boldsymbol{A}_b - \boldsymbol{A}_a), \tag{0.26}$$

$$\nabla \cdot \boldsymbol{A}(\boldsymbol{x}_a) = \frac{1}{\rho_a} \sum_b m_b (\boldsymbol{A}_b - \boldsymbol{A}_a) \cdot \nabla_a W_{ab}. \tag{0.27}$$

$$\nabla \times \boldsymbol{A}(\boldsymbol{x}_a) = \frac{1}{\rho_a} \sum_b m_b \nabla_a W_{ab} \times (\boldsymbol{A}_b - \boldsymbol{A}_a). \tag{0.28}$$

The above two sets of expressions for the derivatives of scalar and tensor functions, corresponding to $\phi = 1$ and ρ, are equivalent formally. However, when the system involves two or more fluids with large density ratios in contact, the expressions corresponding to $\phi = 1$ are more accurate, because the density information of different fluids is accounted for in the summation (Monaghan 2005)

The above formulae give zero gradients for constants but are not symmetric with respect to indices *a* and *b*. To obtain a symmetrical gradient interpolation, the following identity can be used (Monaghan 1992, Li & Liu 2007):

$$\frac{\nabla f}{\rho} = \frac{f}{\rho^n} \nabla \left(\frac{1}{\rho^{1-n}} \right) + \frac{1}{\rho^{2-n}} \nabla \left(\frac{f}{\rho^{n-1}} \right), \tag{0.29}$$

where n is an integer. When f is a scalar function, the SPH interpolation of Eq. (0.29) is

$$\left(\frac{\nabla f}{\rho} \right)_a = \sum_b m_b \left(\frac{f_b}{\rho_a^n \rho_b^{2-n}} + \frac{f_a}{\rho_a^{2-n} \rho_b^n} \right) \nabla_a W_{ab}. \tag{0.30}$$

Taking $n = 1$, we obtain the symmetrical gradient of the scalar function f at $\boldsymbol{x}_a$ to be

$$\left(\frac{\nabla f}{\rho} \right)_a = \sum_b m_b \frac{(f_a + f_b)}{\rho_a \rho_b} \nabla_a W_{ab}. \tag{0.31}$$

Similarly, for a tensor field $\boldsymbol{A}$, we have

$$\left(\frac{\nabla \boldsymbol{A}}{\rho} \right)_a = \sum_b m_b \nabla_a W_{ab} \frac{(\boldsymbol{A}_a + \boldsymbol{A}_b)}{\rho_a \rho_b}. \tag{0.32}$$

If $\boldsymbol{A}$ is a vector or symmetric tensor, its divergence can be estimated as

$$\left(\frac{\nabla\cdot \boldsymbol{A}}{\rho}\right)_a = \sum_b m_b \frac{\left(\boldsymbol{A}_a + \boldsymbol{A}_b\right)}{\rho_a\rho_b}\cdot\nabla_a W_{ab}. \tag{0.33}$$

Taking $n=2$, we obtain the estimate of $\nabla f/\rho$ at $\boldsymbol{x}_a$:

$$\left(\frac{\nabla f}{\rho}\right)_a = \sum_b m_b \left(\frac{f_a}{\rho_a^2}+\frac{f_b}{\rho_b^2}\right)\nabla_a W_{ab}. \tag{0.34}$$

For the vector or symmetrical tensor field, we have

$$\left(\frac{\nabla \boldsymbol{A}}{\rho}\right)_a = \sum_b m_b \nabla_a W_{ab}\left(\frac{\boldsymbol{A}_a}{\rho_a^2}+\frac{\boldsymbol{A}_b}{\rho_b^2}\right), \tag{0.35}$$

and

$$\left(\frac{\nabla\cdot \boldsymbol{A}}{\rho}\right)_a = \sum_b m_b \left(\frac{\boldsymbol{A}_a}{\rho_a^2}+\frac{\boldsymbol{A}_b}{\rho_b^2}\right)\cdot\nabla_a W_{ab}. \tag{0.36}$$

The second derivatives are involved in momentum and energy conservation equations for Newtonian viscous flow. Though they can be calculated as derivatives of the above expressions for the first derivatives, the results have a number of disadvantages (Monaghan 2005): 1. they are very sensitive to particle disorder; 2. the direction of flux between the given two particles may depends on their separation, because the results involve the second derives of the kernel function, which can change sign when the separation changes; 3. the results will break the conservation laws.

To avoid using the second derivatives of the kernel function, the first derivatives of tensor functions can be approximated by using their finite differences between two particles. For example, when considering $\nabla \boldsymbol{A}$ as a tensor function in Eq. (0.35), we have

$$\left(\nabla\nabla \boldsymbol{A}\right)_a = \rho_a \sum_b m_b \nabla_a W_{ab}\left(\frac{\left(\nabla \boldsymbol{A}\right)_a}{\rho_a^2}+\frac{\left(\nabla \boldsymbol{A}\right)_b}{\rho_b^2}\right). \tag{0.37}$$

The gradients of $\boldsymbol{A}$ at $\boldsymbol{x}_a$ and $\boldsymbol{x}_b$ can be approximated using the finite difference between these two points:

$$\left(\nabla A\right)_a \approx \frac{\left(\boldsymbol{x}_a-\boldsymbol{x}_b\right)}{\left|\boldsymbol{x}_a-\boldsymbol{x}_b\right|}\frac{\left(\boldsymbol{A}_a-\boldsymbol{A}_b\right)}{\left|\boldsymbol{x}_a-\boldsymbol{x}_b\right|} = \frac{\boldsymbol{x}_{ab}\boldsymbol{A}_{ab}}{x_{ab}^2} = \frac{\boldsymbol{x}_{ba}\boldsymbol{A}_{ba}}{x_{ba}^2} \approx \left(\nabla A\right)_b. \tag{0.38}$$

Substituting Eq. (0.38) into Eq. (0.37), we have

$$(\nabla\nabla \boldsymbol{A})_a = 2\rho_a \sum_b m_b \left(\frac{1}{\rho_a^2} + \frac{1}{\rho_b^2}\right) \frac{\nabla_a W_{ab} \boldsymbol{x}_{ab} \boldsymbol{A}_{ab}}{\left(x_{ab}^2 + \xi^2\right)}, \tag{0.39}$$

where ξ is a small parameter to prevent from singularity as x_{ab} approaches zero. If the densities of particles ***a*** and ***b*** are approximated by their mean value, $\frac{1}{2}(\rho_a + \rho_b)$, Eq. (0.39) reduces to

$$(\nabla\nabla \boldsymbol{A})_a \approx 8\rho_a \sum_b m_a \frac{\nabla_a W_{ab} \boldsymbol{x}_{ab} \boldsymbol{A}_{ab}}{(\rho_a + \rho_b)^2 \left(x_{ab}^2 + \xi^2\right)}. \tag{0.40}$$

More elegant estimate approximation of the second derivatives derived by Español and Revenga (2003) and Yildiz *et al.*(2008). Readers are also referred to see Monaghan (2005) for the details of derivation. They obtained the following expression for a scalar function or tensor component f:

$$(\nabla\nabla f)_a = \sum_b \frac{m_b}{\rho_b} \left((d+2)\frac{\boldsymbol{x}_{ab}\boldsymbol{x}_{ab}}{x_{ab}^2} - \boldsymbol{I}\right) \frac{\boldsymbol{x}_{ab} \cdot \nabla_a W_{ab}}{x_{ab}^2} \left(f(\boldsymbol{x}_a) - f(\boldsymbol{x}_b)\right), \tag{0.41}$$

where $\boldsymbol{I}$ is the unit tensor of the second order and $d(=2 \; or \; 3)$ is the dimension of space. The Laplacian can be expressed, in terms of contracting, as

$$\left(\nabla^2 f\right)_a = 2\sum_b \frac{m_b}{\rho_b} \left(f(\boldsymbol{x}_a) - f(\boldsymbol{x}_b)\right) \frac{\boldsymbol{x}_{ab} \cdot \nabla_a W_{ab}}{x_{ab}^2}. \tag{0.42}$$

2.3. SPH Conservation Equations

The conservation equations in fluid mechanics are as follows:

$$\frac{d\boldsymbol{u}}{dt} = -\frac{\nabla p}{\rho} + \frac{1}{\rho} \nabla \cdot \boldsymbol{\tau} + \boldsymbol{b}, \tag{0.43}$$

$$\frac{d\rho}{dt} = -\rho \nabla \cdot \boldsymbol{u}, \tag{0.44}$$

$$\frac{de}{dt} = -\frac{p}{\rho} \nabla \cdot \boldsymbol{u} + \frac{1}{\rho} \boldsymbol{\tau} : \nabla \boldsymbol{u} - \frac{1}{\rho} \nabla \cdot (k \nabla T), \tag{0.45}$$

where $\boldsymbol{u}$ is the velocity vector, ρ the density, p the pressure, $\boldsymbol{\tau}$ the deviatoric stress tensor, $\boldsymbol{b}$ the field force, T the temperature, k the thermal conductivity and e the internal energy. For the generalized Newtonian fluid the deviatoric stress can be expressed to be,

$$\boldsymbol{\tau} = \eta(\dot{\gamma})\dot{\boldsymbol{\gamma}} \quad and \quad \dot{\boldsymbol{\gamma}} = \nabla \boldsymbol{u} + \nabla \boldsymbol{u}^T - \frac{2}{d}\nabla \cdot \boldsymbol{u}\boldsymbol{I} \tag{0.46}$$

here $\dot{\boldsymbol{\gamma}}$ is the strain rate tensor and $\dot{\gamma} = \sqrt{\frac{1}{2}\dot{\boldsymbol{\gamma}} : \dot{\boldsymbol{\gamma}}}$, the mode of the strain rate and $\eta(\dot{\gamma})$ the non-Newtonian viscosity.

Using the SPH estimates for scalar and vectors and their derivatives in Sec. **2.2**, we can discretize the above conservation equations into their SPH formulations. In this chapter, we focus only on isothermal flow and concern the mass and momentum conservations.

There are two kinds of the SPH mass conservation expressions. The first one is the summation density, which is the direct application of the SPH estimate for density, Eq. (0.16) :

$$\rho_a = \sum_b m_b W_{ab} \tag{0.47}$$

Alternatively, applying Eq. (0.24) (or Eq.(0.27)) to the divergence term of the continuity equation, Eq. (0.44), we obtain

$$\frac{d\rho_a}{dt} = \rho_a \sum_b \frac{m_b}{\rho_b} \boldsymbol{u}_{ab} \cdot \nabla_a W_{ab}, \tag{0.48}$$

or

$$\frac{d\rho_a}{dt} = \sum_b m_h \boldsymbol{u}_{ab} \nabla_a W_{ab}, \tag{0.49}$$

where $\boldsymbol{u}_{ab} = \boldsymbol{u}_a - \boldsymbol{u}_b$. These two equations give zero rate of density change for uniform flow.

For the momentum equation, Eq. (0.43), we can apply Eqs. (0.31) and (0.33) or Eqs. (0.34) and (0.36) to obtain two kinds of SPH momentum equations. When Eqs. (0.31) and (0.33) are used, the following SPH momentum equation is obtained:

$$\frac{d\boldsymbol{u}_a}{dt} = \sum_b \frac{m_b}{\rho_a \rho_b} \left[-(p_a + p_b)\boldsymbol{I} + (\boldsymbol{\tau}_a + \boldsymbol{\tau}_b) \right] \cdot \nabla_a W_{ab} + \boldsymbol{b}_a. \tag{0.50}$$

While using Eq. (0.34) and (0.36), we have

$$\frac{d\boldsymbol{u}_a}{dt}=\sum_b m_b\left[-\left(\frac{p_a}{\rho_a^2}+\frac{p_b}{\rho_b^2}\right)\boldsymbol{I}+\left(\frac{\boldsymbol{\tau}_a}{\rho_a^2}+\frac{\boldsymbol{\tau}_b}{\rho_b^2}\right)\right]\cdot\nabla_a W_{ab}+\boldsymbol{b}_a. \tag{0.51}$$

For the generalized Newtonian fluid, the deviatoric stress is expressed in Eq. (0.46). The velocity gradient in the above two equations can be approximated in terms of the finite difference. Similar to Eq.(0.38), we have

$$\nabla\boldsymbol{u}_a=\nabla\boldsymbol{u}_b\approx\frac{\boldsymbol{x}_{ab}\boldsymbol{u}_{ab}}{x_{ab}^2}, \tag{0.52}$$

and

$$\left(\nabla\cdot\boldsymbol{u}\right)_a=\left(\nabla\cdot\boldsymbol{u}\right)_b\approx\frac{\boldsymbol{u}_{ab}\cdot\boldsymbol{x}_{ab}}{x_{ab}^2}. \tag{0.53}$$

The density and viscosity at $\boldsymbol{x}_a$ and $\boldsymbol{x}_b$ in the stress term of Eqs. (0.50) and (0.51) are further approximated by their mean values between particles a and b. Two SPH momentum equations for the generalized Newtonian fluid can be derived. Here, one derived from Eq. (0.51) is written as follows:

$$\frac{d\boldsymbol{u}_a}{dt}=-\sum_b m_b\left[\frac{p_a}{\rho_a^2}+\frac{p_b}{\rho_b^2}+\frac{4(\eta_a+\eta_b)\boldsymbol{x}_{ab}\cdot\boldsymbol{u}_{ab}}{(\rho_a+\rho_b)^2\left(x_{ab}^2+\xi^2\right)}\right]\nabla_a W_{ab}+\sum_b\frac{4m_b(\eta_a+\eta_b)}{(\rho_a+\rho_b)^2\left(x_{ab}^2+\xi^2\right)}\left(\boldsymbol{x}_{ab}\cdot\nabla_a W_{ab}\boldsymbol{u}_{ab}+\boldsymbol{u}_{ab}\cdot\nabla_a W_{ab}\boldsymbol{x}_{ab}\right)+\boldsymbol{b}_a, \tag{0.54}$$

were ξ is the same small parameter as that in Eq. (0.39) to avoid the singular behavior when x_{ab} goes to zero. From Eq.(0.50), we can obtain another SPH momentum equation similar to Eq. (0.54) but with the pressure terms to be $\frac{p_a+p_b}{\rho_a\rho_b}$.

The above SPH formulations are based on the kernel function which satisfies Eq. (0.15). This means that the velocity field of a rigid body translation can be estimated exactly. Hence Galilean invariance can be preserved (Li and Liu 2007). It has also been proved that these SPH equations not only describe the local conservation of fluid properties, such as mass, linear and angular momentums, but also preserve the global conservation of these physical quantities for the corresponding continuum system in absent of external forces (Li and Liu 2007, Bonet and Lok 1999).

The SPH discretization of conservative equations is not unique. For example, we can use either Eq. (0.47) or Eq. (0.48) (or Eq. (0.49)) to describe the mass conservation and either Eq. (0.50) or (0.51) for momentum conservation. In simulation we need only one pair of the discrete mass and momentum equations. There are different combinations which can be chosen. Bonet and Lok (1999) pointed out that the combination is not optional. They found that the SPH momentum equation in the type of Eq. (0.50) is variationally consistent with the continuity equation, i.e. the SPH mass conservation equation in the type of Eq. (0.48) or (0.49); while the momentum equation of Eq. (0.51) with the summation density, Eq. (0.47). If the combination of the momentum and mass conservations is inconsistent variationally, the simulation will result in poor accuracy.

3. Convergence, Stability and Consistency

Consistency and stability are important for a numerical scheme. A convergent numerical method should be consistent and stable. However, the standard SPH method has the problems of inconsistency and instability.

3.1. Tensile Instability

One of the intrinsic weaknesses the SPH has is the tensile instability. This instability was first analyzed by Swegle *et al.* (1995) for a one dimensional SPH equation. This problem was subsequently investigated by several researchers, such as Dyka et al (1995, 1996), Morris (1996), Belyschko et al (2000), Monaghan (2000) and Bonet and Kulasegarm (2001).

To briefly illustrate the tensile instability based on the analysis of Swegle et al (1995) (also see Li and Liu, 2007), we start with the momentum equation without the external field force:

$$\rho \frac{d\boldsymbol{u}}{dt} = \nabla \cdot \boldsymbol{\sigma}, \tag{0.55}$$

where $\boldsymbol{\sigma} = -p\boldsymbol{I} + \boldsymbol{\tau}$ is the total stress tensor. Applying the gradient estimate similar to Eq. (0.17) to the stress tensor, we have

$$\ddot{\boldsymbol{x}}_a = \sum_b \frac{m_b}{\rho_a \rho_b} \boldsymbol{\sigma}_b \cdot \nabla_a W_{ab} = -\sum_b \frac{m_b}{\rho_a \rho_b} \boldsymbol{\sigma}_b \cdot \nabla_b W_{ab}. \tag{0.56}$$

The cubic spline with the support size of *2h*, Eq.(0.8), is chosen as the kernel function. In the one dimensional case, this equation reduces to

$$\ddot{x}_a = -m \left[\frac{\sigma_{a+1}}{\rho_a \rho_{a+1}} \frac{d}{dx_{a+1}} W\left(x_a - x_{a+1}, h\right) + \frac{\sigma_{a-1}}{\rho_a \rho_{a-1}} \frac{d}{dx_{a-1}} W\left(x_a - x_{a-1}, h\right) \right], \tag{0.57}$$

where σ is the unique stress component in 1D. The particles are distributed evenly with the spacing of h and identical mass m. The distance of particles $a+2$ and $a-2$ to particle a is $2h$ at which the gradient of the kernel vanishes and the gradient of the kernel at the centre of the support, $\nabla_a W_{aa}$, is zero, as well. Hence, these particles have no contribution to Eq. (0.57). If we denote $\frac{dW(x_a - x_b)}{dx_b} = \begin{cases} W_b', & x_b > x_a, \\ -W_b', & x_b < x_a. \end{cases}$ and $\rho_a = \rho_b = \rho$, Eq. (0.57) becomes

$$\ddot{x} = -\frac{m}{\rho^2}\left(\sigma_{a+1}W_{a+1}' - \sigma_{a-1}W_{a-1}'\right). \tag{0.58}$$

This means qualitatively

$$\ddot{x} \propto -\Delta(\sigma W') = -\Delta\sigma W' - \sigma\Delta W', \tag{0.59}$$

where Δ denotes the finite difference at x_a.

On the other hand, the momentum equation, Eq.(0.57), reduces to

$$\ddot{x} = \frac{1}{\rho}\frac{\partial\sigma}{\partial x} \approx \frac{\Delta\sigma}{\rho\Delta x} = \frac{\Delta\sigma}{\Delta m}, \tag{0.60}$$

where Δm is the mass of particle a. Hence, we have

$$\left(\frac{1}{\Delta m} + W'\right)\Delta\sigma = -\sigma\Delta W'. \tag{0.61}$$

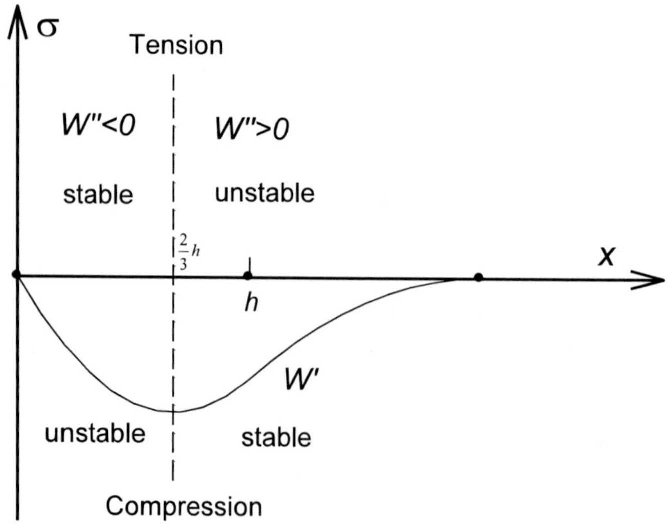

Figure 1. Illustration of tensile instability for the kernel of cubic spline

At each particle, $W' = \frac{d}{dx_a} W_{aa} = 0$, and hence

$$\Delta\sigma \propto -\sigma\Delta W' \propto -\sigma W'' \Delta x, \tag{0.62}$$

where W'' is proportional to the second derivative of the kernel function with respect to its argument.

From Eq. (0.62) we can see that the sufficient condition for the tensile instability is

$$\sigma W'' > 0. \tag{0.63}$$

In this case, when $\Delta x > 0$, $\Delta\sigma < 0$ and when $\Delta x < 0$, $\Delta\sigma > 0$. In extension ($\sigma > 0$), the stress is decreasing as the particle separation increases, while the stress is increasing as the particle separation decreases. In compression ($\sigma < 0$), the increase of the particle separation (less compression) results in the increase of the magnitude of compressive stress, while the decrease of the particle separation (further compress) leads to reduce the magnitude of compressive stress. These mean the system is unstable. However, if $\sigma W'' < 0$, the system is stable, because according to Eq. (0.62) we have that $\Delta\sigma > 0$ as $\Delta x > 0$ and $\Delta\sigma < 0$ as $\Delta x < 0$. Figure 1 shows a schematic illustration of tensile instability for the kernel function of cubic spline, Eq. (0.8). When the separation is less than $\frac{2}{3}h$, $W'' < 0$, tension ($\sigma > 0$) is table since $\sigma W'' < 0$, but compression ($\sigma < 0$) is unstable since $\sigma W'' > 0$. When the separation is larger than $\frac{2}{3}h$, tension is unstable but compression is stable.

Several methods have been proposed to alleviate or eliminate the tensile instability. Dyka *et al.* (1995, 1997) recommended inserting additional stress points into the system. The stress points are only used to calculate the stress. The velocities of the stress points are interpolated from those of neighboring SPH particles. Morris (1996) suggested using high order spline kernels, such as the quintic spline, Eq. (0.9) to suppress the instability. Based on a unified stability analysis of meshless particle methods, Belytschko et al (2000) showed that the tensile instability is, to larger extent, an idiosyncrasy of what we called Eulerian kernels, which are functions of the co-ordinates in the current configuration. In an Eulerian kernel, the stability depends on the stress and the second derivatives of the kernel. This generates the tensile instability. They also showed that when the kernel is a function of the material (Lagrangian) co-ordinates, a Lagrangian Kernel, the tensile instability does not occur. They found that instability due to rank deficiency occurs for both Lagrangian and Eulerian kernels and this instability can be eliminated by stress points but stress points cannot completely stabilize Eulerian kernels. In fluid mechanics, the current configuration is usually chosen as the reference configuration and it is not convenient to use Lagrangian kernels to eliminate the tensile instability. The tensile instability results in a clustering of particles. Monaghan (2000) proposed a simple and intelligent method to eliminate the tensile instability. He used a repulsive force, the so-called artificial pressure, between two close particles to avoid the clustering of particles. A possible form of this force is the Lennard-Jones force. However, to remove the numerical instability in the fluid case, Monaghan used the kennel function to construct a repulsive force, which increases with the particle separation decreases:

$$f_{ab} = \frac{W\left(x_{ab}\right)}{W\left(\overline{\Delta x}\right)}, \tag{0.64}$$

where $\overline{\Delta x}$ denotes the average particle spacing in the neighborhood of particle a, which is proportional to the interpolation length scale h. In the case of elastic body dynamics, Monaghan (2000) introduced the artificial stress, which has been constructed by analogy to the artificial pressure, and effectively eliminated the instability in simulating bouncing rubber rings. We will adopt the artificial pressure in simulating injection moulding flow.

3.2. Consistency and Completeness

Consistency of a discretization of a differential equation is usually applied to finite difference approximations in the study of convergence. The consistency of an approximation is defined by its ability to exactly represent the differential equation in the limit as the maximum distance between neighboring grid points goes to zero. A convergent numerical scheme must be consistent and stable. For unstructured mesh methods and meshless methods, it is difficult to use the consistency in studying convergence. A completeness or reproducing condition is used instead (Belytschko *et el.* 1998). If any polynomial up to order *k* can be represented exactly by an approximation, it has *k*-th order completeness. Assume that an approximation $u^h(x)$ is given by

$$u^h(x) = \sum_b \phi_b(x) u_b, \tag{0.65}$$

where $\phi_b(x)$ are the shape functions and u_b are the nodal values. The *k*-th order completeness (reproducing) requires the following conditions:

$$\begin{gathered} \sum_b \phi_b(x) = 1, \quad \sum_b \phi_b(x) x_b = x, \\ \sum_b \phi_b(x) x_b^2 = x^2, \ \ldots, \ \sum_b \phi_b(x) x_b^k = x^k. \end{gathered} \tag{0.66}$$

In two dimensions, $\boldsymbol{x} = (x, y)$, the linear completeness (reproducing) can be written as

$$\sum_b \phi_b(\boldsymbol{x}) = 1, \ \sum_b \phi_b(\boldsymbol{x}) x_b = x, \ \sum_b \phi_b(\boldsymbol{x}) y_b = y, \tag{0.67}$$

In solving the weak form of a second order differential equation, the approximations are usually required to have up to linear completeness, which is achieved by requiring the derivatives of a polynomial field be correctly reproduced. Differentiation of Eq. (0.67) gives the derivative reproducing conditions for a linear field in two dimensions:

$$\begin{aligned} &\sum_b \phi_{b,x}(\boldsymbol{x}) = 0, \quad &&\sum_b \phi_{b,y}(\boldsymbol{x}) = 0, \\ &\sum_b \phi_{b,x}(\boldsymbol{x}) x_a = 1, \quad &&\sum_b \phi_{b,y}(\boldsymbol{x}) x_b = 0, \\ &\sum_b \phi_{b,x}(\boldsymbol{x}) y_b = 0, \quad &&\sum_b \phi_{b,y}(\boldsymbol{x}) y_b = 1, \end{aligned} \tag{0.68}$$

where $\phi_{b,x}(\boldsymbol{x})$ and $\phi_{b,y}(\boldsymbol{x})$ denote the derivatives of the shape functions centered at point b with respect to the co-ordinates x and y. This can generally be written as

$$\begin{aligned} &\sum_b \phi_{b,i}(\boldsymbol{x}) = 0, \\ &\sum_b \phi_{b,j}(\boldsymbol{x}) x_j = \delta_{ij}, \end{aligned} \tag{0.69}$$

where i and j denote the indexes of the co-ordinates and δ_{ij} is the Kronecker delta, which is equal to 1 when $i = j$ and 0 when $i \neq j$.

In SPH interpolations, the shape function is the product of the kernel function and associated volume of the particle, $\phi_b(\boldsymbol{x}) = W(\boldsymbol{x} - \boldsymbol{x}_b, h)\Delta V_b$. Though the SPH kernel functions satisfy the normalization condition in continuous form, Eq. (0.2) in Sec. 2.1, its discrete counterpart does not necessary satisfy the zero order completeness, the 1-st equation of (0.66), due to the truncation error, i.e. usually

$$\sum_b W(\boldsymbol{x} - \boldsymbol{x}_b, h)\Delta V_b \neq 1. \tag{0.70}$$

When particles are in disordered configuration, the truncation error becomes more serious. For particles near or on the boundary, the problem of particle deficiency makes situation even worse. Hence, the SPH approximation hardly satisfies even the zero order completeness.

The correction of completeness is necessary for the standard SPH approximations. Two types of corrections are usually used: the zero order completeness for the approximate functions and first order completeness for the derivatives of the approximate functions.

A widely used kernel function, which possesses the zero order completeness, is the Shepard function (Shepard 1968):

$$W_a^s(\boldsymbol{x},h)=\frac{W_a(\boldsymbol{x},h)}{\sum_b W_b(\boldsymbol{x},h)\Delta V_b}, \tag{0.71}$$

where $W_a(\boldsymbol{x},h)$ is the kernel function used in the standard SPH approximation. A function, $u(\boldsymbol{x})$, can be estimated

$$u(\boldsymbol{x})=\sum_b \Delta W_b^s(\boldsymbol{x})u_b. \tag{0.72}$$

The gradient of the Shepard function is

$$\nabla W_a^s(x,h)=\frac{\nabla W_a(\boldsymbol{x},h)}{\left(\sum_b \Delta V_b W_b(\boldsymbol{x},h)\right)}-\frac{W_a(\boldsymbol{x},h)\sum_b \Delta V_b \nabla W_b(\boldsymbol{x},h)}{\left(\sum_b \Delta V_b W_b(\boldsymbol{x},h)\right)^2}. \tag{0.73}$$

It is obvious that

$$\sum_a \Delta V_a W_a^s(\boldsymbol{x},h)=1 \tag{0.74}$$

and

$$\sum_a \Delta V_a \nabla W_a^s(\boldsymbol{x},h)=0. \tag{0.75}$$

These mean the Shepard function has zero order completeness. For a constant function, when $u_a=c$ at any particle a, the Shepard function estimates the constant exactly:

$$u^h(\boldsymbol{x})=\sum_a \Delta V_a W_a^s(\boldsymbol{x},h)u_a=\frac{\sum_a \Delta V_a W_a(\boldsymbol{x},h)}{\sum_a \Delta V_b W_b(\boldsymbol{x},h)}c=c. \tag{0.76}$$

Using Eq. (0.75), we estimate the gradient of the constant to be zero, $\nabla u(\boldsymbol{x})=0$, i.e.

$$\sum_b c\Delta V_b \nabla W_b^s(\boldsymbol{x})=0. \tag{0.77}$$

This proves that the Shepard function possesses only zero order completeness.

Higher order completeness can be achieved by using the moving least-square, MLS, approximation (Belytschko *et al.* 1994, Dilts 1999 & 2000) or the reproducing kernel particle method, RKPM, (Liu *et al.* 1995, Cueto-Felgueroso *et al.* 2004). However, the SPH method with the MSL loses conservation and is considerable slower than the standard SPH (Monaghan 2005). For RKPM, after introducing correcting function, the modified kernel function may not be positive anymore. Within the compact support it may becomes negative in some region of the domain (Li and Liu 2007)

Alternatively, for simulating a linear field we can correct the derivatives to satisfy the zero completeness instead of to correct the function to satisfy the first order completeness. Several approaches to derivative corrections have been reported (Belytschko *et al.* 1998, Li and Liu 2007), for example:

1. the symmetrization on derivative approximation (Monaghan 1992);
2. the Johnson and Beissel correction (Johnson and Beissel 1996);
3. the Randles and Libersky renormalization (Randles and Libersky 1996);
4. the Krongauz and Beleytschko corrections (Krongauz and Belytschko 1997) and so on.

To illustrate the derivative correction scheme, we describe the Krongauz-Belytschko method here (Belytschko *et al.* 1998). The corrected derivatives are denoted by $G_{ia}(\boldsymbol{x})$

$$u_{,i}^{h}(\boldsymbol{x})=\sum_{b}G_{ib}(\boldsymbol{x})u_{b},\tag{0.78}$$

where the $G_{ib}(\boldsymbol{x})$ are linear combinations of the derivatives of the Shepard function and is expressed as:

$$G_{ib}(\boldsymbol{x})=\alpha_{ij}(\boldsymbol{x})\nabla_{j}W_{b}^{s}(\boldsymbol{x}) \text{ or } \boldsymbol{G}_{b}(\boldsymbol{x})=\boldsymbol{\alpha}(\boldsymbol{x})\cdot\nabla W_{b}^{s}(\boldsymbol{x}),\tag{0.79}$$

where $\alpha_{ij}(\boldsymbol{x})$ are the coefficients of the linear combination and $\nabla_{j}W_{b}^{S}$ is the j-component of the gradient of the Shepard function. For a linear function, $u_{i}^{h}(\boldsymbol{x})=x_{i}$, we require $u_{i,j}^{h}(\boldsymbol{x})=\delta_{ij}$:

$$u_{i,j}^{h}(\boldsymbol{x})=\sum_{b}\Delta V_{b}G_{b,j}(\boldsymbol{x})x_{bi}=\sum_{b}\alpha_{jk}(\boldsymbol{x})\Delta V_{b}\nabla_{k}W_{b}^{s}(\boldsymbol{x})x_{bi}=\delta_{ij}\tag{0.80}$$

Noting $A_{ik}=\sum_{b}x_{bi}\nabla_{k}W_{b}^{s}(\boldsymbol{x})\Delta V_{b}$, we can write Eq. (0.80) can as

$$\boldsymbol{A}\boldsymbol{\alpha}^{T}=\boldsymbol{I}.\tag{0.81}$$

Hence

$$\boldsymbol{\alpha} = \boldsymbol{A}^{-T} = \left[\sum_b \Delta V_b \nabla W_b^s(\boldsymbol{x}) \boldsymbol{x}_b \right]^{-1} \tag{0.82}$$

Using the zero order completeness of the Shepard function, Eq.(0.77), we can choose local coordinates to reduce the condition number of the matrix $\boldsymbol{A}$:

$$\boldsymbol{\alpha}(\boldsymbol{x}) = \left[\sum_b \Delta V_b \nabla W_b^s(\boldsymbol{x})(\boldsymbol{x} - \boldsymbol{x}_b) \right]^{-1}. \tag{0.83}$$

It is also obvious, due to Eqs. (0.77) and (0.80) that

$$\sum_b \Delta V_b G_{bi}(\boldsymbol{x}) = \alpha_{ik}(\boldsymbol{x}) \sum_b \Delta V_b \nabla_k W_b^s(\boldsymbol{x}) = 0 \tag{0.84}$$

and

$$\sum_b \Delta V_b G_{bi}(\boldsymbol{x}) x_{bj} = \delta_{ij}.$$

The corrected SPH conservation equations can be formulated by using the Shepard function and the derivative corrections. There are two methods that can be used (Belytschko *et al.* 1998):

1. Collocation method, in which the equation is enforced at each particle; this is similar to the kernel approximation in Sec. 2.3. But the function is estimated by the Shepard function, Eq. (0.72), and the gradient by the derivative corrections, Eq. (0.78).
2. Galerkin method, in which the discrete equations are obtained from the weak form of the momentum equation.

The collocation or kernel estimate method is usually used but the total linear and angular momentum conservations are difficult to prove, since unlike the derivatives of the standard SPH kernel the corrected derivative is not anti-symmetric on the particle index:

$$\nabla_{x_a} W_b^s(\boldsymbol{x}_a) \neq -\nabla_{x_b} W_a^s(\boldsymbol{x}_b), \tag{0.85}$$

where $\nabla_{x_a} W_b^s(\boldsymbol{x}_a) = \nabla W_b^s(\boldsymbol{x})|_{\boldsymbol{x}=\boldsymbol{x}_a}$. Belytschko *et al.* (1998) showed that the SPH momentum equations can be obtained using a Petrov-Galerkin method. They used the Shepard function as the test function and the corrected derivatives as the trial functions. The weak form of the momentum equation was integrated with nodal integration. Because the corrected derivatives are not integrable, the Bubnov-Galerkin method, in which the test function is the same as the trial function, results in a discretization which does not converge (Krongauz and Belytschko 1997). The Petrov-Galerkin SPH formulation can conserve the total linear momentum but the total angular momentum conservation can not be proved since the Shepard function has no linear completeness. The MLS (moving least-square) function is integrable and has the zero and first order completeness. Hence, the MLS function can be used as both test and trial function in the Bubnov-Galerkin method to formulate the SHP momentum equations (Belytschko et al 1998).

3.3. Stability Constraints on Time Steps

The explicit integration methods, such as leapfrog, velocity-Verlet, prediction-correction and Runge-Kutta methods, are usually used to integrate SPH equations for fluid flows. The explicit time integration is conditionally stable. A stable time step has to satisfy the following constraints (Lo and Shao 2002):

1. If the Convection Term is Integrated Explicitly, the Time Step Should Satisfy the Convective Stability Constraint, i.e. CFL Condition,

$$\Delta t_c \le \alpha_c \frac{h}{u_{\max}}. \tag{0.86}$$

   where $u_{\max}$ is the maximal velocity of particles and α_c the constant with the order of 0.1.
2. If the diffusion term is integrated explicitly, the time step should satisfy the diffusive stability constraint,

$$\Delta t_d \le \alpha_d \frac{\rho h^2}{\eta_0} \tag{0.87}$$

   where α_d is the constant with the order of 0.1 and η_0 the characteristic viscosity of fluid.
   Hence, for fully explicit integration the stable time step should be

$$\Delta t = \min\left(\Delta t_c, \Delta t_d\right). \tag{0.88}$$

For high viscous flow the viscosity is very large and the particle Reynolds number particle is very small. The diffusive stability constraint becomes determinant and the stable time step may be very small. When the time step is prohibitively small, the simulation would be infeasible. In this case, implicit time integration on the diffusion term has to be used to get rid of the time step limitations.

4. STATE EQUATIONS

In SPH simulations a commonly used state equation for fluid is that of the weakly compressible fluid (WCF) (Monaghan 1994) expressed as

$$p = B\left[\left(\frac{\rho}{\rho_0}\right)^{\gamma} - 1\right] \tag{0.89}$$

where ρ_0 is the reference density at zero pressure and $\gamma = 7$. The value of the pressure coefficient, B, in Eq. (0.89) is usually determined by the speed of sound, c_0:

$$B = \frac{\rho_0 c_0^2}{\gamma} \tag{0.90}$$

The speed of sound depends on the maximum velocity and Mach number limitation, because in the momentum dominated flow the compression is mainly determined by the kinetic energy of fluid. However, in the moulding flow, the fluid velocity is small and the compression is due to the high pressure. In the die, polymer melts exhibit a compressibility of about 5% to 10% under high pressure. The pressure coefficient in Eq. (0.89) would be too small if it is determined based on the maximum velocity. A roughly estimated coefficient *B* for WCF is explained in the Appendix.

5. BOUNDARY CONDITIONS

Solid boundaries are usually constructed by using solid particles in SPH simulations. These particles are fixed or moving with the velocity of moving boundaries and interact with fluid particles so that the fluid particle can not penetrate into the solid and flow satisfies the no slip boundary condition. Since the SPH interpolant is intrinsically non-local, it is more difficult for SPH to enforce the essential boundary condition than mesh methods. Several methods have been used to enforce the boundary conditions on the solid wall surfaces.

5.1. Repulsive Boundary Conditions

The repulsive boundary condition was proposed by Monahhan (1994) and is commonly used by most later researchers. This method simply applies a repulsive force between a boundary particle and a fluid particle when they are close to each other. This force pushes the fluid particle away to prevent it from penetrating the solid boundary. The Lennard-Jones forces are usually used as the repulsive forces (Monaghan 1994):

$$\boldsymbol{f}_{aw} = \frac{D}{r_{aw}}\left[\left(\frac{r_0}{r_{aw}}\right)^{n_1} - \left(\frac{r_0}{r_{aw}}\right)^{n_2}\right]\hat{\boldsymbol{r}}_{aw}, \quad \text{if } r_{aw} \le r_0, \tag{0.91}$$

where $\boldsymbol{f}_{aw}$ is the repulsive force acting on the fluid particle *a* by a wall particle *w*, r_0 the cutoff distance, r_{aw} and $\hat{\boldsymbol{r}}_{aw}$ are the distance and unit vector pointing from the wall particle *w* to the fluid particle *a*, *D* is a coefficient, which usually is the square of the maximum velocity, and $n_1 = 12$ and $n_2 = 4$. Cueto-Felgueroso *et al.* (2004) considered the repulsive force to be always normal to the boundary:

$$\boldsymbol{f}_{aw} = \begin{cases} \dfrac{D}{r_{aw}}\left(\left(\dfrac{r_0}{r_{aw}}\right)^{4} - \left(\dfrac{r_0}{r_{aw}}\right)^{2}\right)\boldsymbol{n}_w, & \boldsymbol{u}_{aw}\cdot\boldsymbol{n}_w < 0; \\ \boldsymbol{0}, & \boldsymbol{u}_{aw}\cdot\boldsymbol{n}_w \ge 0, \end{cases} \tag{0.92}$$

where $\boldsymbol{u}_{aw} = \boldsymbol{u}_a - \boldsymbol{u}_w$, here $\boldsymbol{u}_a$ and $\boldsymbol{u}_w$ are velocities of fluid and wall particles, respectively, $\boldsymbol{n}_w$ is the unit vector at particle *w* which is normal to the wall surface and pointing to the fluid and the cutoff distance r_0 is usually to be $(0.5 \sim 0.7)h$.

This is the simplest way to treat the boundary conditions in SPH. However, the repulsive force is undergoing a large variation when a particle is moving parallel to the boundary and will cause large disturbance to flow near the boundary. Monaghan *et al.*(2004) improved the repulsive boundary condition by assuming:

$$\boldsymbol{f}_{aw} = \frac{m_w}{m_a + m_w} B(x, y)\boldsymbol{n}_w, \tag{0.93}$$

where m_w and m_a are mass of the wall and fluid particles, *y* and *x* are the normal and tangential distances from the wall particle to the fluid particle and *B(x,y)* is a function which rapidly increases as *y* decreases towards zero and the variation with *x* ensures that the force on a particle moving parallel to the wall is constant. This function is further assumed to be the product of $\Gamma(y)\chi(x)$, where the function $\chi(x)$ is defined by

$$\chi(x) = \begin{cases} \left(1 - \dfrac{x}{\Delta p}\right), & \text{if } 0 < x < \Delta p, \\ 0, & \text{otherwise}, \end{cases} \tag{0.94}$$

where Δp is the boundary particle spacing. This function ensures that a fluid particle moving parallel to the wall will always feel the same force because the total force from two neighboring wall particles is constant. The function $\Gamma(y)$ is required to rapidly increase as *y* decreases. Monaghan *et al.* (2004) choose a form related to the gradient of the cubic spline with the argument $q = y/h$:

$$\Gamma(y) = \beta \begin{cases} \frac{2}{3}, & \text{if } 0 < q \le \frac{2}{3}, \\ \left(2q - \frac{3}{2}q^2\right), & \text{if } \frac{2}{3} < q \le 1, \\ \frac{1}{2}(2-q)^2, & \text{if } 1 < q < 2, \\ 0, & \text{otherwise,} \end{cases} \tag{0.95}$$

where $\beta = 0.02c_s^2 / y$, which is an estimate of the maximum force necessary to stop a particle moving at the estimated maximum speed. Monaghan (2005) also showed that these repulsive forces conserve linear and angular momentums for a moving rigid body interacting with a fluid.

5.2. Ghost Particles

The repulsive boundary conditions can effectively prevent the fluid particles from penetrating the solid walls but do not eliminate the problem of particle deficiency near boundaries. Another approach for the boundary treatment is to use the ghost particles proposed by Morris *et al.* (1997). They created ghost particles outside the solid surfaces of the flow domain when generating the particle configuration. During simulation, for each fluid particle, of which the influence domain is truncated by the solid surface, a normal to the boundary surface is drawn and then a tangent line to the boundary can be determined, see Figure 2. The velocities of all ghost particles within the influence domain of particle a are calculated as follows:

$$\boldsymbol{u}_g = -\frac{d_g}{d_a}\boldsymbol{u}_a, \tag{0.96}$$

where $\boldsymbol{u}_g$ and $\boldsymbol{u}_a$ are the velocities of the ghost and fluid particles and d_g and d_a the distance from the ghost and fluid particles to the tangent line through the point which is on the boundary surface and nearest to the fluid particle a, as shown in Figure 2.

The effect of the ghost particle is to enforce the no slip boundary condition on the fixed solid wall surface. When the solid wall is moving at a velocity, $\boldsymbol{u}_w$, the velocity of the ghost particle is calculated to be

$$\boldsymbol{u}_g = \boldsymbol{u}_w + \frac{d_g}{d_a}\left(\boldsymbol{u}_w - \boldsymbol{u}_a\right). \tag{0.97}$$

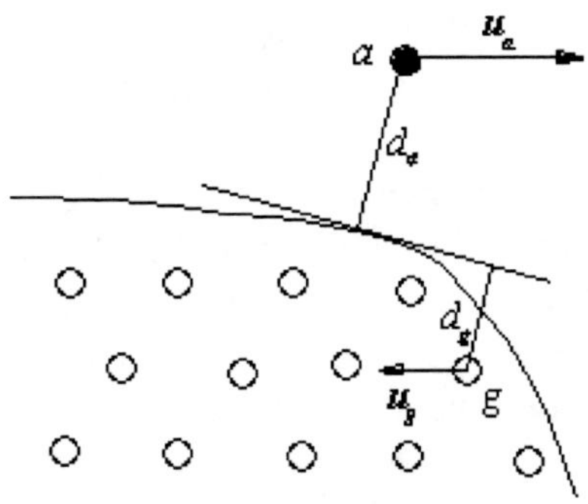

Figure 2. The velocity of the ghost particles (open circles)

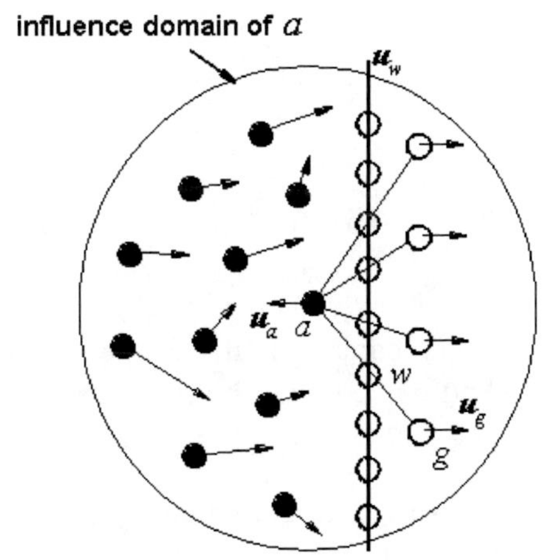

Figure 3. Ghost particles generated via the local point-symmetry

The ghost particles eliminate the particle deficiency near solid boundaries. The configuration of ghost particles is somewhat optional. In the above method (Morris *et al.* 1997), the ghost particles are created in the initial generation of particle configuration and the solid boundary is not represented by particles. If it is preferable to use particles to represent the boundaries, the ghost particles are usually created by mirroring the fluid particles with respect to a plane (or line) which is tangent to the boundary surface (Coagrossi and Landrini, 2003). However, it is emphasized by other researchers that the single boundary tangent method has significant limitations for treating complex boundary shape, because there might be more than one way to construct a normal vector to the boundary. Yildiz *et al.* (2009) proposed a multiple boundary tangent method to overcome the limitation of the single tangent method. For each boundary particle, there are a number of fluid particles within the influence domain of this boundary particle. These fluid particles, called mother particles, are mirrored with respect to the tangent line of this boundary particle. Every boundary particle is associated with its corresponding mirrored particles. A fluid particle near boundary may have several boundary particles in its influence domain and will become a neighbour of all mirrored particles associated with these boundary particles if these mirrored particles and their mother particles are in the influence domain of this fluid particle. The influence domains of the neighbouring boundary particles may overlap. A fluid particle within the overlapped part will have more than one image. The mass of these image particles should be divided by the number of corresponding over-creations as performing SPH summation over ghost particles.

The multiple boundary tangent method can be applied to boundaries with complex geometry but is complicated in creating the mirrored particles. The cell array structure (the Fortran 90 derived data type) has to be used in data processing (Yildiz *et al.* 2009). Ferrari *et al.* (2009) proposed a somewhat simpler method, called the local point-symmetry method, to generate ghost particles. The boundaries are also represented by boundary particles which interact with all adjacent fluid particles. The ghost particles are generated via the local symmetry. An example is show in Figure 3. The fluid particle a has several ghost particles in its influence domain which are symmetrical to a with respect to corresponding boundary particles. The ghost particle g has following associated properties:

$$\boldsymbol{x}_g = 2\boldsymbol{x}_w - \boldsymbol{x}_a, \quad \boldsymbol{u}_g = 2\boldsymbol{u}_w - \boldsymbol{u}_a, \quad \rho_g = \rho_a, \quad m_g = m_a. \qquad (0.98)$$

5.3. Imposing the Boundary Condition

This method was proposed by Randles and Libersky (1996). There are two problems involved in imposing the boundary condition in SPH: a) definition and location of the boundary surface and b) a method of applying the boundary condition to the variable and governing equations. We can denote $N_{(a)}$ to be the set of neighboring particles of a particle a (including the particle a itself). If particle a is a boundary particle, all particles in $N_{(a)}$ are divided into three subsets, as shown in Figure 4:

$I_{(a)}$: all interior particles which are neighbors of a,

$B_{(a)}$:all boundary particles which are neighbors of a.

$E_{(a)}$:all missing exterior particles which would be neighbors of a.

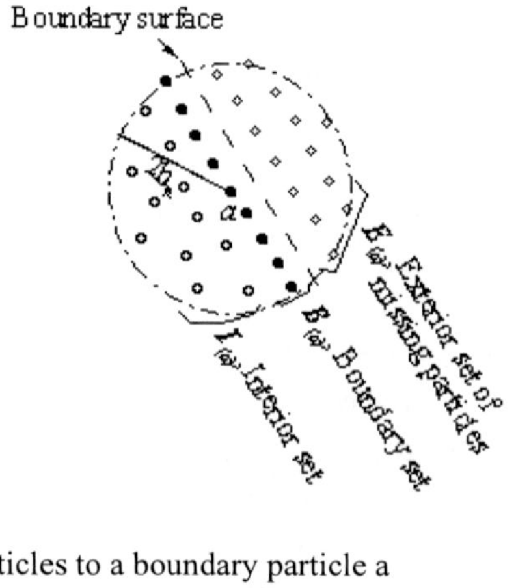

Figure 4. Neighboring sets of particles to a boundary particle a

The boundary surface is located nominally $h_a/2$ away outside the boundary particle *a*.

The particle deficiency has been removed because the addition of the exterior set $E_{(a)}$. If the truncation error due to the interior particle disorder is negligible, the kernel estimate should have zero order completeness, i.e. Eq. (0.15) should be valid:

$$\sum_{b\in I_{(a)}} \frac{m_b}{\rho_b} W_{ab} + \sum_{b\in B_{(a)}} \frac{m_b}{\rho_b} W_{ab} + \frac{m_w}{\rho_w} \sum_{b\in E_{(a)}} W_{ab} = 1 \tag{0.99}$$

where it has been assumed that all exterior particles in set $E_{(a)}$ have the prescribed mass and density m_w and ρ_w. The density estimate at $\boldsymbol{x} = \boldsymbol{x}_a$ is

$$\rho_a = \sum_{b\in I_{(a)}} m_b W_{ab} + \sum_{b\in B_{(a)}} m_b W_{ab} + m_w \sum_{b\in E_{(a)}} W_{ab}. \tag{0.100}$$

If further assume that all particles in the boundary set $B_{(a)}$ also have the prescribed mass, m_w and density ρ_w, the density at $\boldsymbol{x} = \boldsymbol{x}_a$ can be obtain from Eqs. (0.99) and (0.100) :

$$\rho_a = \frac{\sum_{b\in I_{(a)}} m_b W_{ab}}{\sum_{b\in I_{(a)}} \frac{m_b}{\rho_b} W_{ab}} = \sum_{b\in I_{(a)}} m_b W_{ab}^s. \tag{0.101}$$

The boundary condition for a general scalar field $f(\boldsymbol{x})$ can be enforced in the similar way. We assume again all particle in the exterior set $E_{(a)}$ and boundary set $B_{(a)}$ have the same mass and density and $f(\boldsymbol{x})$ has the prescribed value at all particles in $E_{(a)}$, i.e. $f_b = f_w$ as $b \in E_{(a)}$. The kernel estimate gives

$$f_a = \sum_{b\in I_{(a)}} f_b \frac{m_b}{\rho_b} W_{ab} + f_a \sum_{b\in B_{(a)}} \frac{m_b}{\rho_b} W_{ab} + \left(f_w \frac{m_a}{\rho_a} \sum_{b\in E_{(a)}} W_{ab} \right) \tag{0.102}$$

Multiplying Eq. (0.99) with f_w and subtracting it from Eq. (0.102), we obtain the following boundary correction:

$$f_a = f_{bc} + \frac{\sum\limits_{b \in I_{(a)}} (f_b - f_w) \frac{m_b}{\rho_b} W_{ab}}{\left(1 - \sum\limits_{b \in B_{(a)}} \frac{m_b}{\rho_b} W_{ab} \right)}. \tag{0.103}$$

This method allows the field variable on the boundary particle to be influenced by both the interior field values and imposed boundary value. The particle set $E_{(a)}$ is imaginary and has no influence on the results. The boundary particle *a* is a pseudo-boundary particle with a buffer of material between it and the actual material boundary. The boundary value is imposed at the position beyond the particle a for $h_a / 2$. This boundary condition treatment is suitable for impact and contact problems.

5.4. Direct Forcing of Rigid-Fluid Coupling

Becker *et al.* (2009) proposed a novel boundary treatment to directly force one- or two-way rigid-fluid coupling for both velocity and position, based on a predictor-corrector scheme. When the fluid and rigid body are both represented by particles, the enforcing various boundary conditions is reduced to control the relative velocities and positions for the fluid and rigid particle collision and contacting problem.

If we detect a collision between a fluid particle at $\boldsymbol{x}_f$ with the velocity $\boldsymbol{u}_f$, mass m and a rigid body particle at $\boldsymbol{x}_p$, we can calculate the contact point to be $\boldsymbol{x}_{cp} = \boldsymbol{x}_p + r_p \boldsymbol{n}$, where $\boldsymbol{n}$ is the normal vector, $\left(\boldsymbol{x}_f - \boldsymbol{x}_p\right) / \left|\boldsymbol{x}_f - \boldsymbol{x}_p\right|$, at $\boldsymbol{x}_{cp}$, as shown in Figure 5.

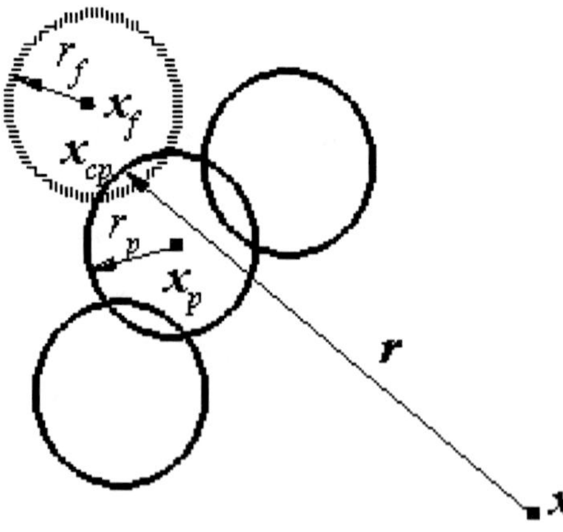

Figure 5. Rigid particle and fluid particle in contact. Here $\boldsymbol{r}$ denotes the position vector of the contact point from the centre of the rigid body and $r_f \left(= h_f / 2\right)$ and $r_p \left(= h_p / 2\right)$ are the radii of fluid and rigid body particles

The rigid-body velocity at the contact point is, in the predict stage,

$$\boldsymbol{u}_{cp}^{*}(t+\Delta t)=\boldsymbol{u}_{c}^{*}(t+\Delta t)+\boldsymbol{\omega}_{c}^{*}(t+\Delta t)\times\boldsymbol{r}^{*}(t+\Delta t), \quad (0.104)$$

where $\boldsymbol{r}^{*}(t+\Delta t)=\boldsymbol{x}_{cp}^{*}(t+\Delta t)-\boldsymbol{x}_{c}^{*}(t+\Delta t)$ is the relative position of the contact point with respect to the mass centre of the rigid body, the superscript * denotes the predicted values at $t+\Delta t$, and $\boldsymbol{u}_{c}^{*}(t+\Delta t)$ and $\boldsymbol{\omega}_{c}^{*}(t+\Delta t)$ are the linear and angular velocity of the rigid body. In the correct stage, we need to impose a boundary condition on the relative velocity $\boldsymbol{u}_{r}=\boldsymbol{u}_{f}-\boldsymbol{u}_{cp}$ in the following form, for $\boldsymbol{u}_{r}^{*}(t+\Delta t)\cdot\boldsymbol{n}<0$,

$$\boldsymbol{u}_{r}(t+\Delta t):=\boldsymbol{u}_{f}(t+\Delta t)-\boldsymbol{u}_{cp}(t+\Delta t)=\zeta\left[\boldsymbol{u}_{r}^{*}(t+\Delta t)\right]_{t}-\delta\left[\boldsymbol{u}_{r}(t)\right]_{n}, \quad (0.105)$$

where the normal velocity at the previous time step is $\left[\boldsymbol{u}_{r}(t)\right]_{n}=\left(\boldsymbol{u}_{r}(t)\cdot\boldsymbol{n}\right)\boldsymbol{n}$ and the uncontrolled tangential velocity at the present time step $\left[\boldsymbol{u}_{r}^{*}(t+\Delta t)\right]_{t}=\boldsymbol{u}_{r}^{*}(t+\Delta t)-\left[\boldsymbol{u}_{r}^{*}(t+\Delta t)\right]_{n}$, ζ and δ are two parameters to control the slip and collision between the fluid and rigid-body particles. When $\zeta=1$ the relative tangential velocity without any damping. This corresponds to free slipping. When $\zeta=0$, there is no slip between the fluid and rigid-body particles. $\delta=1$ corresponds to a perfectly elastic collision, while $\delta=0$ results in the purely inelastic collision. When $\boldsymbol{u}_{r}^{*}(t+\Delta t)\cdot\boldsymbol{n}>0$, the fluid and rigid-body particles are moving away from each other and the normal component of the relative velocity is unchanged:

$$\boldsymbol{u}_{r}(t+\Delta t)=\zeta\left[\boldsymbol{u}_{r}^{*}(t+\Delta t)\right]_{t}+\left[\boldsymbol{u}_{r}^{*}(t+\Delta t)\right]_{n}. \quad (0.106)$$

Next we explain how to enforce the velocity constraints for a single fluid particle in contact with the rigid body. If the contact point of the fluid particle a and the rigid body is $\boldsymbol{x}_{cp,a}^{*}(t+\Delta t)$ and $\boldsymbol{r}_{a}^{*}(t+\Delta t)$ is the relative position vector of this contact point with respect to the centre of the rigid body, we need to determine the force, $\boldsymbol{F}_{a}$, between the fluid particle and the rigid body, which is required for the relative velocity constraint, Eq. (0.105). The velocity of particle a under the action of $\boldsymbol{F}_{a}$ at the correct stage is

$$\boldsymbol{u}_{a}(t+\Delta t)=\boldsymbol{u}_{a}^{*}(t+\Delta t)+\frac{\Delta t}{m_{a}}\boldsymbol{F}_{a}, \quad (0.107)$$

where m_{a} is the mass of the fluid particle a. And the velocity of the contact point $\boldsymbol{x}_{cp,a}^{*}$ is

$$\boldsymbol{u}_{cp,a}(t+\Delta t)=\boldsymbol{u}^{*}_{cp,a}(t+\Delta t)-\frac{\Delta t}{m_c}\boldsymbol{F}_a+\Delta t\left(r_a^{*2}(t+\Delta t)\boldsymbol{I}-\boldsymbol{r}_a^{*}(t+\Delta t)\boldsymbol{r}_a^{*}(t+\Delta t)\right)\cdot\boldsymbol{J}^{-1}\cdot\boldsymbol{F}_a, \tag{0.108}$$

where $\boldsymbol{J}$ is the inertia moment tensor of the rigid body with respect to its mass centre, m_c the mass of the rigid-body. The relative velocity at the contact point, $\boldsymbol{u}^{*}_{r,a}(t+\Delta t)=\boldsymbol{u}^{*}_{a}(t+\Delta t)-\boldsymbol{u}^{*}_{cp}(t+\Delta t)$, should satisfy the constraint, Eq. (0.105). Form Eqs. (0.107), (0.108) and (0.105), we obtain the force acting to the fluid particle as

$$\boldsymbol{F}_a=\frac{1}{\Delta t}\left[\left(\frac{1}{m_a}+\frac{1}{m_c}\right)\boldsymbol{I}+\left(r_a^{*2}\boldsymbol{I}-\boldsymbol{r}_a^{*}\boldsymbol{r}_a^{*}\right)\cdot\boldsymbol{J}^{-1}\right]^{-1}\hat{\boldsymbol{u}}_a, \tag{0.109}$$

where the parameter, $t+\Delta t$, of $\boldsymbol{r}_a^{*}$ has been omitted for short and $\hat{\boldsymbol{u}}_a$ is defined as $\hat{\boldsymbol{u}}_a:=\zeta\left[\boldsymbol{u}^{*}_{r,a}(t+\Delta t)\right]_t-\delta\left[\boldsymbol{u}_{r,a}(t)\right]_n-\boldsymbol{u}^{*}_{r,a}(t+\Delta t)$. The velocity of the fluid particle can then be calculated using Eq. (0.107). This method can be generated to the case of a single rigid body in contact with multi fluid particles, see Becker *et al.* (2009) for details.

In updating the position of fluid particles, the non-penetration condition with respect to the boundary is required. This condition is enforced in a separate sub-step using an additional control jump of the fluid particle, $\boldsymbol{j}_a$, in the normal direction, so that the distance, r_f, from the centre of the fluid particle a to the contact point, $\boldsymbol{x}_{cp,a}$, remains unchanged:

$$\left[\boldsymbol{x}_a^{**}(t+\Delta t)+\Delta t\boldsymbol{j}_a-\boldsymbol{x}_{cp,a}(t+\Delta t)\right]\cdot\boldsymbol{n}=r_f, \tag{0.110}$$

where both $\boldsymbol{x}_{cp,a}$ and the predicted future fluid particle positions are calculated using the modified velocities from the subsequent step. The jump velocity is calculated to be

$$\boldsymbol{j}_a=\frac{1}{\Delta t}\left[\left(\boldsymbol{x}_{cp,a}(t+\Delta t)-\boldsymbol{x}_a^{**}(t+\Delta t)\right)\cdot\boldsymbol{n}+r_f\right]\boldsymbol{n}. \tag{0.111}$$

This correction of the position occurs in the normal direction and will not influence the slip but leads a higher compression of the fluid at the boundary layer. However the higher density will rapidly balanced in subsequent time steps (Becker *et al.* 2009).

5.5. Blending Function

In Galerkin methods, the test functions must vanish on the essential boundaries. However, the test function in the SPH Petrov-Galerkin formulation does not vanish there. Belytschko *et al.*(1998) introduce a layer of linear finite elements along the essential boundaries and used a blending function between the finite elements on the boundaries and the meshfree nodes in the interior. The shape functions in the blending region are given by a linear combination of the finite element shape functions and meshfree shape functions according to the weight of a blending function. The blending function vanishes on the essential boundaries and the shape function reduces to the finite element shape function; and it is one in the meshfree region and the shape function is equal to the meshfree one. This method has a difficulty that the derivatives of the finite element shape functions are discontinuous at nodes and a special treatment is required; see Belytschko *et a*l. (1998).

6. APPLIICATION TO INJECTION MOULDING FLOW

SPH has been applied to various branches of science and engineering. However, its application to polymer processing had not been reported before the publication of the paper by the authors (Fan *et al.* 2010). In the present chapter we explain how to apply SPH to simulate the injection molding for polymer melts and present some preliminary results of this application.

6.1. Implicit SPH Formulations with the Artificial Pressure

Firstly we have to choose a suitable SPH formulation as the basic equations for our simulation. There mainly are two kinds of SPH formulation: the standard SPH formulation and the corrected SPH formulation. The momentum equation of the corrected SPH is more complicated than that of the standard SPH and its angular momentum conservation is unproved. Hence, as the first attempt, we chose the standard SPH momentum equation with the finite difference approximation in the viscous terms, i.e. Eq. (0.54), and its variationally consistent mass conservation, Eq. (0.16), as the conservation equations in the simulation. The serious problems for these formulations are inconsistency and instability, as mentioned in Sec. 3. Inconsistency not only affects the accuracy of the results but also causes instability due to numerical error. Instead of using the corrected SPH formulation, we can use the Shepard function as a filter to reduce the numerical error due to the inconsistency. Randles and Libersky (1996) generated the conservative smoothing approach to the multi dimension case. For a solution U at particle a, the spatial smoothing function takes the form:

$$S_{cs}\left(U_a,\alpha_{cs}\right)=U_a+\alpha_{cs}\left[\frac{\sum\limits_{b\neq a}\frac{m_b}{\rho_b}U_bW_{ab}}{\sum\limits_{b\neq a}\frac{m_b}{\rho_b}W_{ab}}-U_a\right], \quad (0.112)$$

with $\alpha_{cs}=0\sim 0.5$. The filters can smooth the $2h$-wavelength oscillations and be applied to the solution every 30 time steps or so.

To effectively stabilize the SPH system, we adopted the artificial pressure, proposed by Monaghan (2000) to remove the tensile instability. Here we use a type of Lennard-Jones force as the artificial pressure, $\tilde{p}_{ab}$, between two particles *a* and *b*:

$$\tilde{p}_{ab} = \frac{\varepsilon}{r_{ab}}\left[\lambda^2\left(\frac{h_{ab}}{r_{ab}}\right)^4 - \left(\frac{h_{ab}}{r_{ab}}\right)^2\right] = \tilde{p}_{ba}, \tag{0.113}$$

were ε is a parameter controlling the strength of the artificial pressure, r_{ab} the distance between these two particles and h_{ab} is neutral distance which can be the average particle spacing.

The tensile instability in SPH causes a clustering of SPH particles and results in the failure of the numerical procedure. Swegle *et al.*(1995) showed that the sufficient condition for tensile instability involves only the sign of the product of the total stress times the second derivative of the kernel function, W'', as expressed in Eq. (0.63). If we only consider the artificial pressure, $\tilde{p}$, of which the positive direction is opposite to that of the stress, the sufficient condition for $\tilde{p}$ to stabilize the simulation is that

$$W''\tilde{p} < 0. \tag{0.114}$$

For the cubic spline function, Eq. (0.8), we know that $W'' < 0$ when $r < \frac{2h}{3}$ and $W'' > 0$ when $r > \frac{2h}{3}$. It is obvious from Eq. (0.113) that this condition can be satisfied if $\lambda = \frac{2}{3}$, since $\tilde{p}_{ab} > 0$ when $r_{ab} < \frac{2h_{ab}}{3}$ and $\tilde{p}_{ab} < 0$ when $r_{ab} > \frac{2h_{ab}}{3}$. However, in the simulations we set $\lambda = 1$ to minimize the effect of the artificial pressure on the fluid volume. The strength of the artificial pressure is adjusted using the parameter ε. For Poiseuille flow, the artificial pressure is not necessary, and we can set $\varepsilon = 0$. In simulating the moulding flow with $\eta_0 = 1.22\times10^4\,Pa\cdot s$, we set $\varepsilon = 0.1$, which makes the artificial pressure large enough to avoid particle clustering. Adding artificial pressure to the hydrostatic pressure in Eq. (0.54), we obtain a SPH momentum equation with the artificial pressure:

$$\frac{d\boldsymbol{u}_a}{dt} = -\sum_b m_b\left[\frac{p_a}{\rho_a^2} + \frac{p_b}{\rho_b^2} + \tilde{p}_{ab}\left(\frac{1}{\rho_a^2} + \frac{1}{\rho_b^2}\right) + \frac{4(\eta_a+\eta_b)\boldsymbol{x}_{ab}\cdot\boldsymbol{u}_{ab}}{(\rho_a+\rho_b)^2\left(x_{ab}^2+\xi^2\right)}\right]\nabla_a W_{ab}$$
$$+\sum_b \frac{4m_b(\eta_a+\eta_b)}{(\rho_a+\rho_b)^2\left(x_{ab}^2+\xi^2\right)}\left(\boldsymbol{x}_{ab}\cdot\nabla_a W_{ab}\boldsymbol{u}_{ab} + \boldsymbol{u}_{ab}\cdot\nabla_a W_{ab}\boldsymbol{x}_{ab}\right) + \boldsymbol{b}_a, \tag{0.115}$$

This equation can be solved using an explicit time integration scheme. However, as mentioned in Sec.3.5, the time step should satisfy two constraints: the convective and diffusive stability constraints, Eqs. (0.86) and (0.87). When the viscosity is very high, Δt_d becomes very small and the time step would be prohibitively small. For example, if $h = 10^{-4} m$, $\rho = 10^3 kg/m^{-3}$, $\eta_0 = 10^3 Pa \cdot s$, and $\alpha_d = 0.1$, we obtain $\Delta t \le 10^{-9} s$ according to the diffusive constraint, Eq. (0.87). Hence, Morris *et al.* (1997) took the viscosity to be as low as $10^{-3} Pa \cdot s$ in their simulation of low Reynolds number flow $\left(\mathrm{Re} = 1.25 \times 10^{-2}\right)$. Now we are simulating low Reynolds number flows of polymeric liquids with viscosity of $O\left(10^3\right)$ to $O\left(10^4\right)$ $Pa \cdot s$. The explicit time integration would be very inefficient in this case. We only can use it to simulate the flow with low viscosity. For high viscosity flow, The implicit time integration must be used. Eq. (0.115) can be discretized, at time t_{n+1}, as follows,

$$\frac{\boldsymbol{u}_a^{n+1}}{\Delta t} - \sum_b \frac{4m_b\left(\eta_a + \eta_b\right)}{\left(\rho_a + \rho_b\right)^2\left(x_{ab}^2 + \xi^2\right)}\Big[\left(\boldsymbol{u}_a^{n+1} - \boldsymbol{u}_b^{n+1}\right)\boldsymbol{x}_{ab} \cdot \nabla_a W_{ab} + \boldsymbol{x}_{ab}\left(\boldsymbol{u}_a^{n+1} - \boldsymbol{u}_b^{n+1}\right) \cdot \nabla_a W_{ab}$$
$$- \left(\boldsymbol{u}_a^{n+1} - \boldsymbol{u}_b^{n+1}\right) \cdot \boldsymbol{x}_{ab} \nabla_a W_{ab}\Big] = \frac{\boldsymbol{u}_b^n}{\Delta t} - \left\{\sum_b m_b\left[\frac{p_a}{\rho_a^2} + \frac{p_b}{\rho_b^2} + \tilde{p}_{ab}\left(\frac{1}{\rho_a^2} + \frac{1}{\rho_b^2}\right)\right]\nabla_a W_{ab}\right\}^{n+1/2} + \boldsymbol{b}_a^{n+1/2}. \quad (0.116)$$

Its variationally consistent mass conservation is Eq. (0.47), that is:

$$\rho_a = \sum_b m_b W_{ab}, \quad (0.117)$$

A fully implicit scheme should be unconditionally stable and the time step constraints can be removed. However, Eq. (0.116) is not fully implicit. In this form, the viscosity η, density ρ, $\boldsymbol{x}_{ab}$ and $\nabla_a W_{ab}$ on the left hand side are all unknown at the time step of $n+1$. This is a nonlinear algebraic equation for particle velocities and densities. Its linearization can be done using the Newton-Raphson method but this requires complicated algebra. To simplify the numerical treatment, the particle viscosity is calculated in terms of the velocity values at time step n and the particle density is replaced by its value at step $n+\frac{1}{2}$. This treatment of linearization leads Eq. (0.116) to become a semi-implicit version but its convergence is not as sensitive to the initial value as the Newton-Raphson method. Moreover, in this algorithm, the convection term in Eq.(0.116) is handled explicitly, therefore the convective stability constraint for the time step based on the maximum flow velocity , Eq. (0.86), should still hold.

The matrix of Eq. (0.116) is symmetrical and it can be solved in terms of an iterative technique, such as the Conjugate Gradient (CG) method. The Generalized Minimal Residual (GMRES) is an iterative solver for symmetry and non-symmetric equations. We found that it resulted in a stable convergence of iterative solutions for this equation. Hence, we use GMRES in the simulation. The iterative solvers are much more efficient than the direct solvers for large systems and do not require us to generate and store the system matrix. Hence, they are suitable for the present SPH simulations.

6.2. State Equations for Polymer Melts

The weakly compressible fluid (WCF), Eq. (0.89), has been widely used in simulating incompressible liquid flows with and without free surfaces. We can also use this state equation in simulating the polymer melts flow with high viscosity. However, as mentioned in Sec. 4, polymer melts are compressible under high pressure. The pressure coefficient B should be determined in a way other than that for simulating incompressible fluids. A method to assess the coefficient B is outlined in Appendix.

There have been some PVT models (Rodgers 1993) available for polymer processing. Some of them can be adopted as the state equations in the SPH simulation of polymeric liquids. Here, we choose the Tait model and Sun, Song and Yan's (SSY) model as examples.

The Tait model is commonly used in polymer processing. In the isothermal condition, the Tait equation reduces to

$$\frac{\rho_0}{\rho} = 1 - c\ln\left(1 + \frac{p}{B(T)}\right), \tag{0.118}$$

where ρ_0 is the density at the reference temperature at which $p = 0$, and $c = 0.0894$. The pressure can be expressed as, from Eq.,

$$p = B(T)\left\{\exp\left[\frac{1}{c}\left(1 - \frac{\rho_0}{\rho}\right)\right] - 1\right\}. \tag{0.119}$$

In the Tait model, there is only one adjustable parameter, B, which can be determined by fitting to experimental data.

Sun *et al.* (1992) proposed an empirical isothermal state equation for polymer solids and liquids:

$$p = \frac{B(T)}{n-m}\left[\left(\frac{\rho}{\rho_0}\right)^{n+1} - \left(\frac{\rho}{\rho_0}\right)^{m+1}\right], \tag{0.120}$$

where $n = 6.14$ and $m = 1.16$, B is adjustable parameter, as well. The model is reduced to WCF, Eq. (0.89), if $n = 6$ and $m = -1$. We determined parameter B in Eqs. (0.119) and (0.120) using the PVT data of Moldflow material testing for Borealis polypropylenes (Moldflow 2004). The values of B at four different temperatures for the above two models are listed in Table 1. The curves of two models fitting to experimental data with parameters in

Table 1 are plotted in Figures 6 and 7, in which the open circles are the experimental data and solid lines are the model predictions. It is seen from the figures that the two models are both accurate for polymer melts.

Table 1. Pressure coefficient, B

Temperature (^{0}C)	B (Pa), Sun *et al*	B (Pa), Tait
166.8	0.937467E+09	0.784484E+08
183.7	0.873428E+09	0.722741E+08
198.4	0.816189E+09	0.678670E+08
217.7	0.716068E+09	0.591723E+08

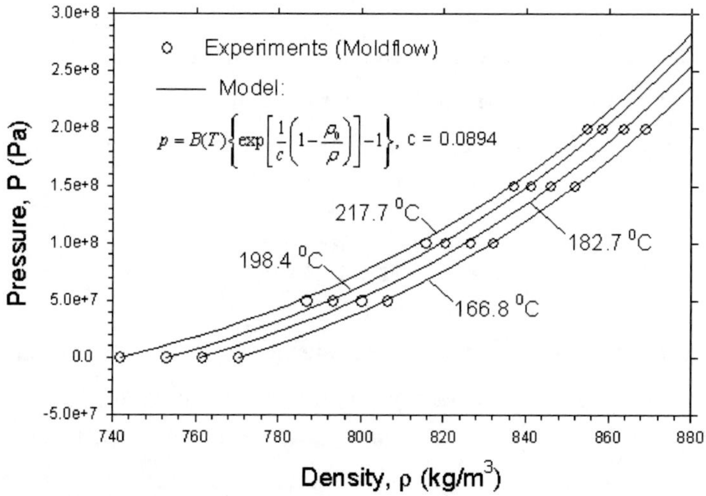

Figure 6. Prediction of the Tait equation, Eq.(0.119)

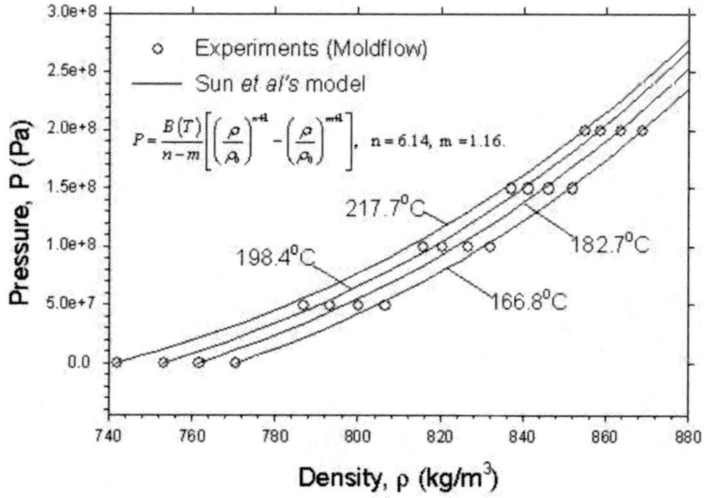

Figure 7. Predictions of Sun *et al's* model, Eq.(0.120)

6.3. Viscosity of Polymer Melts

Polymer melts are viscoelastic fluids but are usually treated as generalized non-Newtonian fluids in shear-dominated polymer processing such as injection moulding. The viscosities of Borealis polypropylenes at three different temperatures have been measured in Moldflow material testing (Moldflow 2004) and are shown in the curves of Figure 8. These curves can be fitted to by the power-law or Cross model. In the present work, we use the power-law model:

$$\eta(\dot{\gamma}) = \begin{cases} m\dot{\gamma}_0^{n-1}, & \dot{\gamma} \le \dot{\gamma}_0 \\ m\dot{\gamma}^{n-1}, & \dot{\gamma} > \dot{\gamma}_0 \end{cases} \tag{0.121}$$

where m and n are the power-law parameters, $\dot{\gamma}_0$ is the critical shear rate. If shear rate is lower than this critical value the viscosity is constant, that is the zero-shear-rate viscosity, $\eta_0 = m\dot{\gamma}_0^{n-1}$. Fitting Eq.(0.121) to the curve of 200 ^{0}C in Figure 8 with assuming $\dot{\gamma}_0 = 10s^{-1}$, we determine that

$$n \approx 0.294, \quad m \approx 6.2\times10^3 Pa\cdot s^n \quad \text{and} \quad \eta_0 \approx 1.22\times10^3 \mathrm{Pa\cdot s} \tag{0.122}$$

To test the present numerical method, we prefer to set the zero shear rate viscosity to be $1.22\times10^4 Pa\cdot s$, which is 10 times as high as the measured figure in the material testing. We will investigate the effect of the power-law index, n, on the flow behaviour but keeping the zero viscosity unchanged. For convenience, we set $\gamma_0 = 1.0S^{-1}$ and keep $m = 1.22\times10^4 Pa\cdot s^n$ as n is varying.

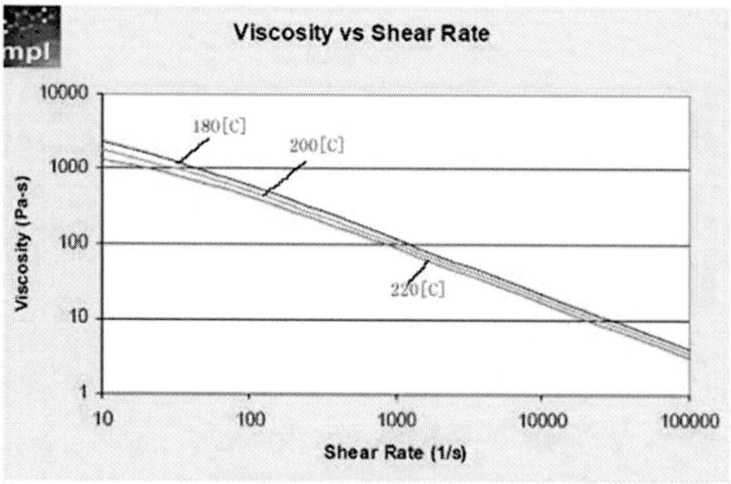

Figure 8. Viscosity versus shear rate at various temperatures for Borealis polypropylenes

There are other viscosity models, such as the Carreau-Yasuda model (Bird *et al.* 1987), Cross model (Cross 1965) etc., which can be employed in the present SPH simulation. However, we choose the power-law model first, because an analytical solution for the slit flow is available for the power-law model.

The Borealis polypropylene is a semi-crystalline polymer. Following resent work of Tanner and Qi (2005) on crystallization rheology, it would be possible to incorporate crystallization effects into the constitutive model if a full processing including both molten flow and subsequent solidification of the polymer is to be simulated. This chapter is limited to isothermal molten flows at a high temperature and therefore takes no account of crystallization effects.

6.4. Boundary Condition Treatment

In Sec. 5, we described several methods to impose the boundary condition on the solid surfaces. Most of these methods are mainly used in the explicit SPH. To simulate the flow of polymer melts, the explicit SPH is infeasible. The implicit SPH has to be used. In the implicit scheme, it is difficult to use ghost particles (Sec. 5.2) and directly forcing of rigid-fluid coupling (Sec. 5.4). The method of imposing the boundary condition (Sec. 5.3) (Randles and Libersky 1996) may be used but is more complicated than the repulsive force method (Sec. 5.1). In the present simulation, we use the latter to implement the boundary condition. The repulsive force method is simple to implement and flexible to cope with complex geometry of boundaries. It can effectively prevent the fluid particles from penetrating the solid walls but causes an extra perturbation to the flow near boundaries. In the attempt of simulating the injection moulding, we prefer to use this simple and flexible method to enforce the boundary conditions. We use one layer of solid particles to represent the solid wall and add the repulsive force between a solid particle and neighbouring fluid particles to prevent them from penetrating through the solid boundary. A type of Lennard-Jones forces is used as the repulsive force, $\boldsymbol{f}_w$, as a fluid particle approaches the wall particle, $\boldsymbol{u}_{aw} \cdot \boldsymbol{n}_w < 0$:

$$\boldsymbol{f}_{aw} = \frac{D}{r_{aw}}\left(\left(\frac{r_0}{r_{aw}}\right)^{12} - \left(\frac{r_0}{r_{aw}}\right)^{4}\right)\boldsymbol{n}_w, \qquad (0.123)$$

where r_{aw} is the distance between a pair of wall and fluid particles, r_0 the critical distance, which is a half of the mean spacing of the pair of particles, when $r \geq r_0$ or $\boldsymbol{u}_{aw} \cdot \boldsymbol{n}_w \geq 0$, $\boldsymbol{f}_w = \boldsymbol{0}$, $\boldsymbol{n}_w$ the normal unit vector of the wall; and D the coefficient. Eq. (0.123) is similar to Eq. (0.92) used by Cueto-Felgueroso *et al.* (2004) but we use larger exponents in simulating the moulding flow to prevent the high pressure in the die from driving fluid particles to penetrate the wall. For moulding flow with $\eta_0 = 1.22 \times 10^4 Pa \cdot s$, we set $D = 700$; but for Poiseuille flow D can be zero. The no slip condition is approximately enforced by setting the velocities of all wall particles to be equal to the known wall velocity in solving Eq. (0.116).

6.5. Solution Procedure

The solution procedure can be outlined as follows. The initial velocity is usually set to be zero and the initial particle configuration and density are all known. Starting from the solution at step (n), we obtain the solution at step (n+1) as follows:

1. Advance particle configuration for ½ step:

$$\mathbf{x}_a^{n+1/2} = \mathbf{x}_a^n + \frac{1}{2}\Delta t\mathbf{u}_a^n; \tag{0.124}$$

2. Calculate $\rho^{n+1/2}$ in terms of Eq. (0.117);
3. Calculate the pressure and artificial pressure at $n+\frac{1}{2}$, obtain the right hand side of Eq. (0.116); if particle a is in the influence domain of a wall particle, the repulsive force is calculated using Eq. (0.123) and added to the right hand side;
4. Calculate the viscosity of particles using Eq. (0.121) and the velocity obtained at step n;
5. Solve Eq. (0.116) for u_a^{n+1}, using the iterative method, GMRES;
6. Update the particle configuration:

$$\mathbf{x}_a^{n+1} = \mathbf{x}_a^n + \frac{1}{2}\Delta t\left(\mathbf{u}_a^{n+1} + \mathbf{u}_a^n\right). \tag{0.125}$$

7. Some Numerical Results

7.1. Simple Flows

The present method is firstly applied to simulating simple flows, such as transient slit flow and shear flow to test how well the implicit scheme works.

Eq. (0.116) is also valid for Newtonian fluid provided the viscosity is constant. Hence, we first simulate the transient slit flow of a Newtonian fluid, for which the series solutions are available (Morris 1997, Liu and Liu 2003). Totally 800 fluid particles and 40 wall particles are used. There are 40 particles distributed along the width of the slit, which is 10^{-3}m. The density and viscosity of fluid are $1000 kg/m^3$ and 10^{-3} $Pa \cdot s$, respectively. The Reynolds number is 2.5×10^{-2} when the flow is fully developed. The time step is 10^{-4} s. Figure 9 shows the comparison of the simulated velocity profiles in a transient slit flow against the series solutions. Parameter h_0 in the figure is the initial spacing of particle centres. It is seen that the maximum error of the maximum steady velocity is about 2.75%.

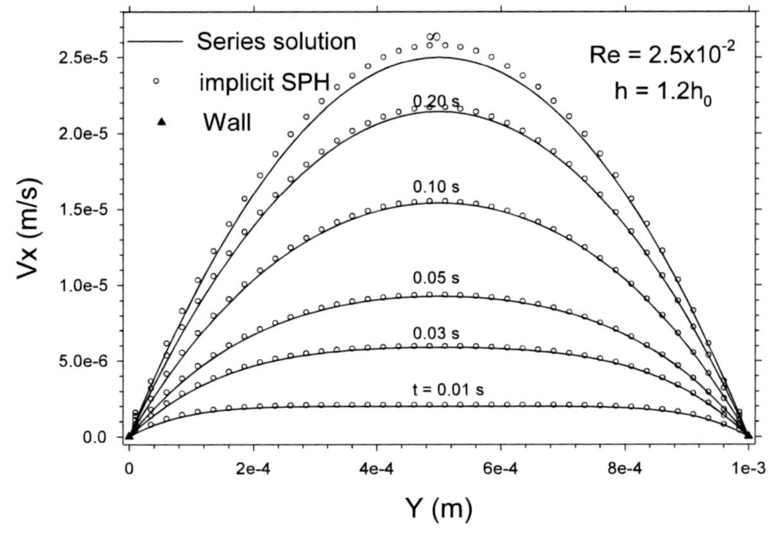

Figure 9. Velocity profiles for transient slit flow, a comparison of the implicit SPH results against the series solutions

We also simulated the transient shear flow of the Newtonian fluid. The particle configuration and fluid parameters are set the same as those for the transient slit flow. The shear flow is driven by the moving upper wall with the velocity of 0.25×10^{-4} m/s. The Reynolds number for fully developed shear flow is 2.5×10^{-2}, as before. Figure 10 shows that the simulated velocity profiles in the transient shear flow agree excellently with those of the series solutions (Morris 1997, Liu and Liu 2003).

The above two simulation tests indicate that the implicit scheme has been correctly implemented for Newtonian flow. The next test is to apply the present scheme to the slit flow of a power-law fluid. The particle configuration is the same as that for the Newtonian case. The density and zero shear rate viscosity of the fluid are 1000 kg/m^3 and $m=1.22\times10^4 Pa\cdot s^n$. The power-law exponent is 0.294. The Reynolds number based on the zero shear rate viscosity, $\eta_0 = 1.22\times10^4 Pa\cdot s$, is 0.934×10^{-5}. The analytical solution of steady slit flow for a power-law fluid is known as

$$u_x = u_{\max}\left[1-\frac{|y-H|^{\frac{1}{n}+1}}{H}\right], \text{ with } u_{\max} = \frac{nH^{\frac{1}{n}+1}}{n+1}\left(\frac{\rho b}{m}\right), \tag{0.126}$$

where x and y are the coordinates along and across the slit, H is the half width of the slit, b the field force driving the slit flow.

The comparison of the numerical result with the analytic solution is shown in Figure 11. It is found that the maximum error in the maximum steady velocity is about 0.2% for the implicit SPH.

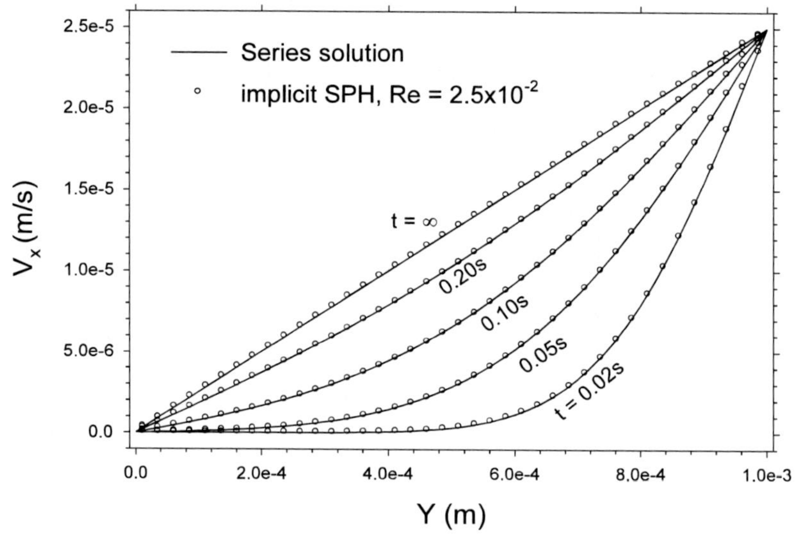

Figure 10. Velocity profiles for transient shear flow, a comparison of the implicit SPH results against the series solutions

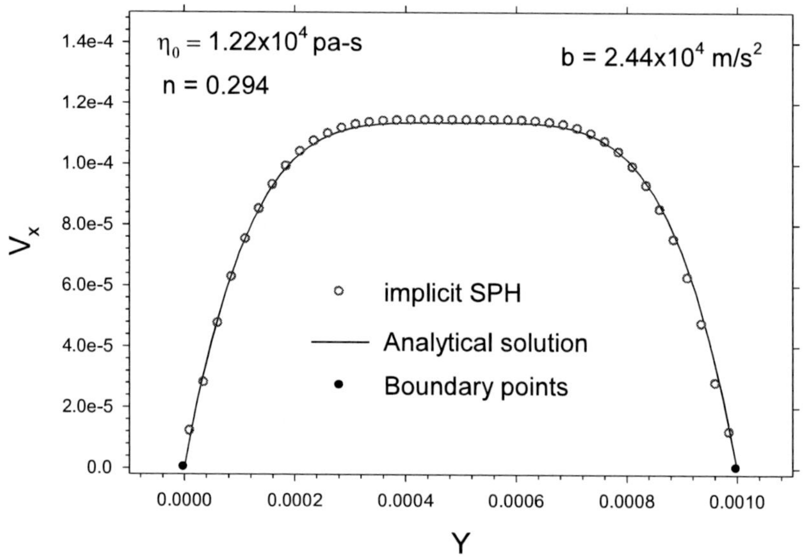

Figure 11. Velocity profiles for the steady slit flow of power-law fluid.A comparison of the implicit SPH result against the analytic one

The above results indicated that the error of Poiseuille flow is bigger than that of other two flows. This may be due to the finite difference approximation in calculating the velocity gradient. From Eq. (0.52) we can see that this approximation implies $\boldsymbol{\tau}_a = \boldsymbol{\tau}_b$. For Poiseuille flow, the stress is linear across the slit while it is constant for shear flow. Hence, this approximation is accurate for shear flow but not so good for Poiseuille flow. The slit flow of the power-law fluid is in between of the two. The stress in a large central part of slit is near zero. In this part the approximation works well.

7.2. Fountain Flow

Fountain flow is a phenomenon encountered in the advancing front of injection moulding flow, where the fluid particles decelerate as they approach the advancing front and spill over toward the walls. In traditional numerical method for injection moulding simulations based on the so-called "Hele-Shaw approximation", the fountain flow was neglected, and special approximate treatments have to be introduced to account for effects of fountain flow on temperature and orientation (Crochet *et al* 1994).

Mavridis *et al.* (1986) performed a numerical simulation on the isothermal fountain flow using the finite element method and calculated correctly the front position for the first time. Their work showed that there are no qualitative differences between Newtonian and shear thinning fluids, of which the free surface shapes are all surprisingly close to a semi-circle. Their calculations are fully in agreement with experimental observations and other approximate analysis. Recently, Bogaerds *et al.*(2004) simulated the fountain flow of a viscoelastic fluid using the extended pom-pom model (Verbeeten *et al.* 2001) and showed that the flow front is not affected appreciably by the elasticity of the fluid and it does not deviate from a semi-circle for a wide range of flow rate in a planar domain. It can be concluded that the fountain flow is a purely hydrodynamic phenomenon and thus the flow patterns are almost independent of the fluid model used. To reveal numerically subtle differences existing for different fluid models, Mitsoulis (2010) used the Herschel-Bulkley viscoplastic model (Bird *et al.* 1982) to simulate the fountain flow. This model reduces to the Bingham, power-law or Newtonian model with a suitable choice of parameters. He employed polar grids in the front flow region to more accurately determine the front location. He found that for highly non-Newtonian fluids the flow front is not a semi-circle. In the case of power-law fluids, shear-thinning substantially increased the front location compared with that of the Newtonian fluid, while shear-thickening does not affect it appreciably. He obtained that the dimensionless centerline front location, χ, increase from 0.887 to 1.409 when the power-law exponent, *n*, decreases from 2.0 to 0.25. Here the definition of the dimensionless centerline front location is $\chi = \frac{\Delta x_{cl}}{H}$ with $\Delta x_{cl} = x_{cl} - x_w$, where x_{cl} is the centerline front position, x_w the wall position and *H* the channel half width. We choose the cases of $n = 1.0$ and 0.5 as the bench mark for the SPH simulation. From Table 4 of his paper (Mitsoulis 2010) it is listed that $\chi = 0.895$ as $n = 1.0$ and $\chi = 1.007$ as $n = 0.5$.

The initial configuration for the SPH simulation of the fountain flow is shown in Figure 12. The size of the simulated domain is $5m \times 1m$. Two parallel walls are $y = 0$ and $y = 1m$ with the velocity of -0.1m/s moving from right to left. A fixed wall parallel to the y-axis is set at $x = 0.03333m$. A periodic boundary condition is applied in the x-direction for moving wall particles. The total number of particles is 4654 including 362 wall particles and 4292 fluid particles. The fluid particles are distributed in the region from $x = 0.0633m$ to $0.45m$. The initial spacing between particles is 0.0315m. There are about 29 particles distributed over each column along the channel width. In Figure 12 only about 1/3 particles are displayed to make the symbols of particles clearly separated.

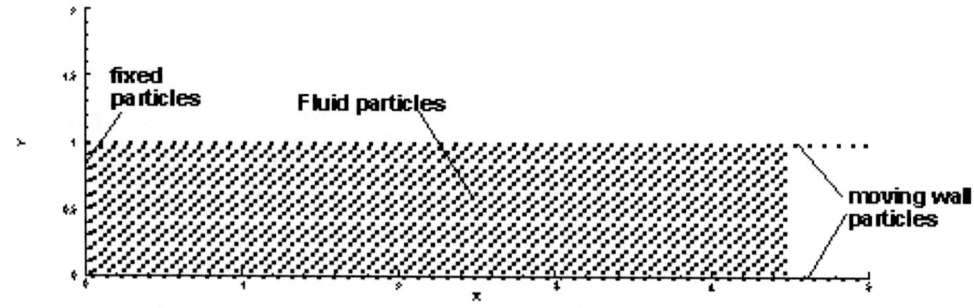

Figure 12. The initial configuration for the fountain flow simulation

The parameters used in simulating the fountain flow are as follows. The density of the fluid is $1000kg/m^3$. The viscosity is $1.22\times10^3 Pa\cdot s$ for the Newtonian fluid. The power-law parameters are that $m = 1.22\times10^3 Pa\cdot s^n$, $n = 0.5$ and $\dot{\gamma}_0 = 1.0s^{-1}$. These result in the zero-shear-rate viscosity being $1.22\times10^3 Pa\cdot s$. The coefficient of the repulsive wall force, *D*, in Eq. (0.91) is 0.01 and the coefficient of the artificial pressure, ε, in Eq. (0.113) is 0.1. The time step is taken to be 10^{-3} s. The coefficient of the weakly compressible fluid model, Eq. (0.89), is $25\times10^4 Pa$.

The fluid particles are driven by the two moving walls but the left fixed wall blocks the flow passage. This is equivalent to setting a uniform velocity profile on the entrance boundary, while in the finite element simulation the fully developed velocity profile is usually specified there. The effect of the entrance boundary condition on the flow front is decreasing as the channel length increases. However, the particle number and thus the computation requirements are increased accordingly. In the present simulation we only use a shorter channel with the aspect ratio of 4.5 for a qualitative comparison with the finite element simulations. This specification of the boundary condition will result in some numerical error. A total of 16×10^4 time steps are simulated for each case and the filters are used after every 30 time steps to alleviate the particle deficiency near boundaries and reduce short wave disturbances. The Shepard function, Eq. (0.71) , and spatial smoothing function, Eq. (0.112), with $\alpha_{cs} = 0.5$ are used as the filters for the density and velocity, respectively. The steady front shape is achieved after $8.5\times10^3 s$ ($t = 85s$) or so.

The simulated flow patterns of the front part are shown in Figure 13 for the Newtonian case and in Figure 14 for the shear thinning case with $n = 0.5$. From these figures, it is seen that the shape of the fountain flow free surface is close to a semi-circle no matter whether the fluid is shear thinning or not but the shear thinning obviously increases the front location of the free surface. We can roughly measure the dimension centerline location for $n = 1$ to be about 0.9 from Figure 13, and for $n = 0.5$ the free surface almost coincides with the semi-circle (see Figure 14). Hence, we can see that the results of the present SPH simulation are in agreement with those of the finite element simulation and that the SPH method simulates the fountain flow naturally, while mesh methods need some special treatment for free surfaces (Mavridis *et al.* 1986, Mitsoulis 2010) - the location and shape of the free surface boundary are not known in advance and the mesh has to be updated iteratively during the solution procedure.

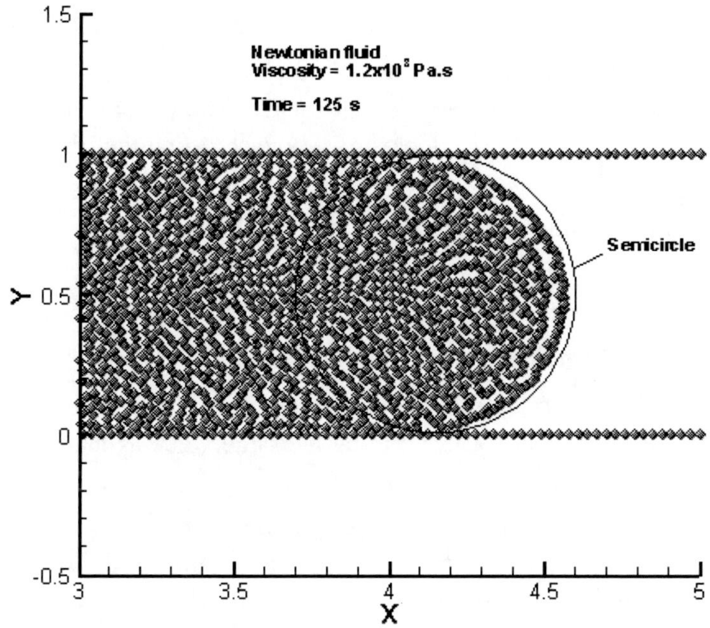

Figure 13. Newtonian fountain flow pattern at $t = 125s$

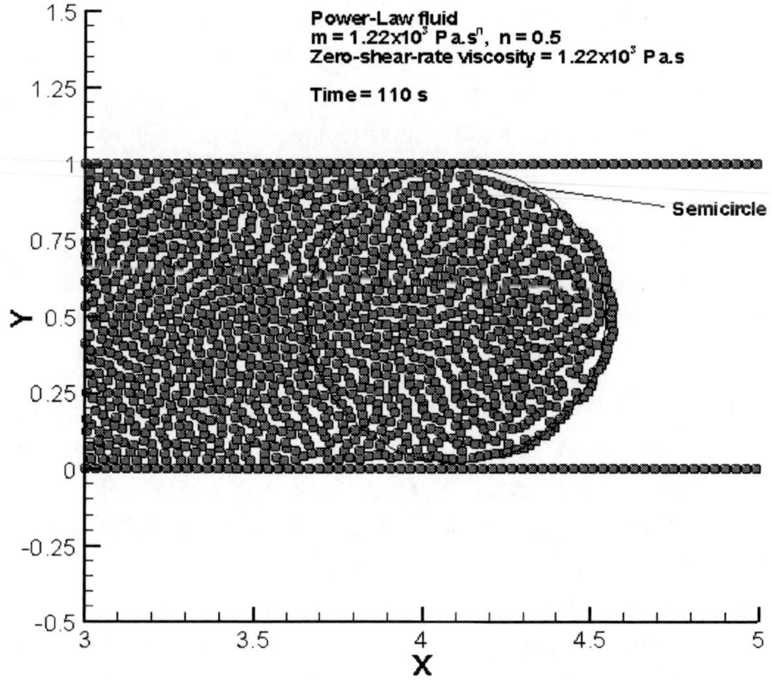

Figure 14. Power-law fountain flow pattern with $n = 0.5$ at $t = 110s$

7.3. Moulding Flow of Polymer Liquids

Now we are in position to apply the present method to simulate the moulding flow of a power-law fluid. Totally 5987 particles are used, including 1136 wall particles. There is a gravity force (0, -9.81 m/s^2) acting on every fluid particle. The flow is driven by a piston, modelled using a moving wall (see Figures 19 to 22). The parameters of fluid are set as follows. The density is 753.012 kg/m^3; for more seriously testing the program, the zero shear rate viscosity is taken to be $1.22\times10^4 Pa\cdot s$, which is 10 times the value measured in Moldflow material testing (Moldflow 2004); the power-law parameters are $m=1.22\times10^4 Pa\cdot s^n$ and $n=$ 0.8 and 0.294 for two cases, respectively. The velocity of the piston is set to be 0.5*m/s* and 0.1*m/s*, the width of the sleeve of the die is 0.01*m* and that of the gate is 0.002*m*. The Reynolds number based on the zero shear rate viscosity is 0.31×10^{-3} and 0.6×10^{-4}, respectively. The weakly compressible fluid model used as the state equation in which the pressure coefficient, B, is set to be the same value as that for Newtonian case, i.e. 0.165×10^{10} *Pa*, estimated using Eq. (A.6) with the correction coefficient of 3.0, see Appendix. The degree of compression of fluid during moulding is shown in Figure 15 for the piston velocity of 0.5*m/s*. It can be seen that the fluid is significantly less compressed when the power-law exponent decreases from 0.8 to 0.294, because of the increased shear thinning. The maximum compression occurs at $t = 0.036s$, when 1.49% of compression is found for $n =$ 0.8 but only 0.05% for $n = 0.294$. Hence, for the power-law fluid, the pressure coefficient can be reduced. An estimation of B can be done roughly using Eq. (A.7) for the power-law fluid. Alternatively, fitting the WCF equation to the experimental data of the Moldflow material testing at $T = 198.4^0C$ (Moldflow 2004), we get the pressure coefficient to be only 0.133249×10^9, which is 8% of that for Newtonian fluid. The fitting of the WCF model to the experimental data is worse than that of the state equations for polymers, such as the Tait and SSY models. Reducing the pressure coefficient will increase the stability of the system and make the simulation more stable.

The density distribution along flow direction at $t = 0.036s$ is shown in Figure 16 for $n = 0.8$ and in Figure 17 for $n = 0.294$. The density of the fluid is decreasing significantly while the fluid moves along the sleeve to the exit of the gate, at $X=0$, when $n = 0.8$; but slightly when $n = 0.294$.

Three state equations have been used for the power-law fluid with $\eta_0 = 1.22\times10^4 Pa\cdot s$ and $n=0.294$. Their pressure coefficients are determined by fitting to the Moldflow experimental PVT data for $T = 198.4\ ^0C$ (Moldflow 2004). The values of the pressure coefficient for the SSY and Tait models are listed in Table 1. For WCF state equation it is $0.13325\times10^9 Pa$. The degrees of compression during the moulding are shown in Figure 18 for these models. It is seen that three models gave similar results, when their pressure coefficients were determined in terms of same experimental data, but the WCF model gives a somewhat smaller degree of compression, because it does not fit the experimental data as well as the two other models. The maximum compression for this power-law fluid is less than 0.7%.

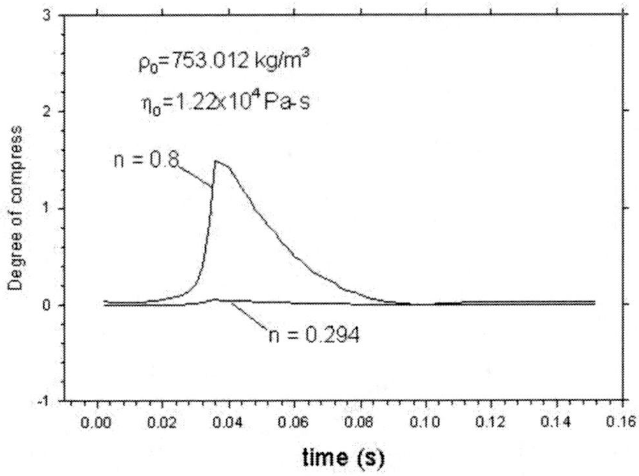

Figure 15. Average compression of power-law fluid during molding using WCF model with same pressure coefficient as that for Newtonian fluid

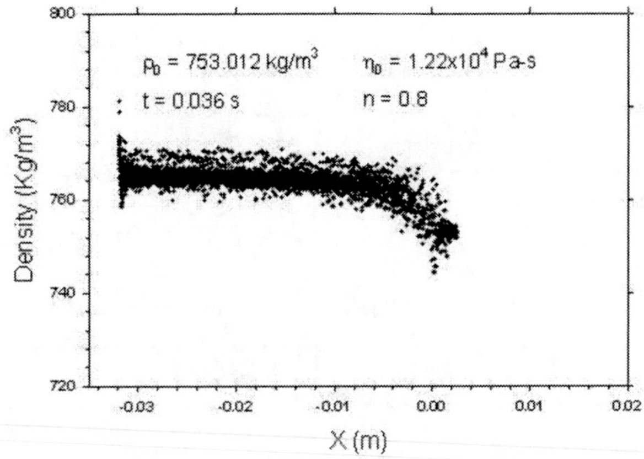

Figure 16. Density distribution along flow direction at t = 0.036s for n = 0.8

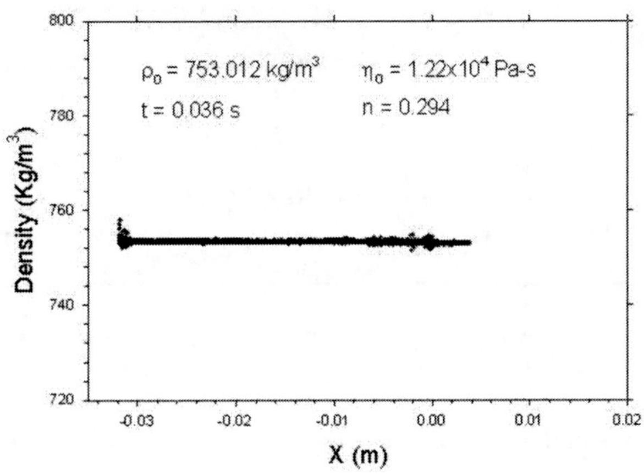

Figure 17. Density distribution along flow direction at t = 0.036 for n = 0.294

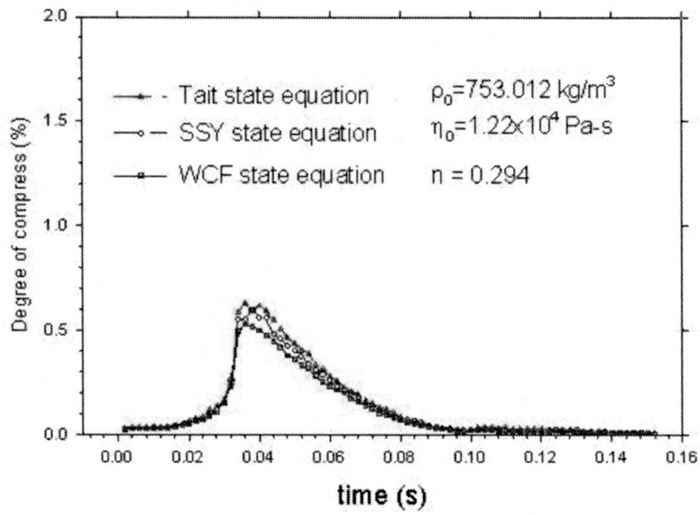

Figure 18. Average density during the moulding for three state equations

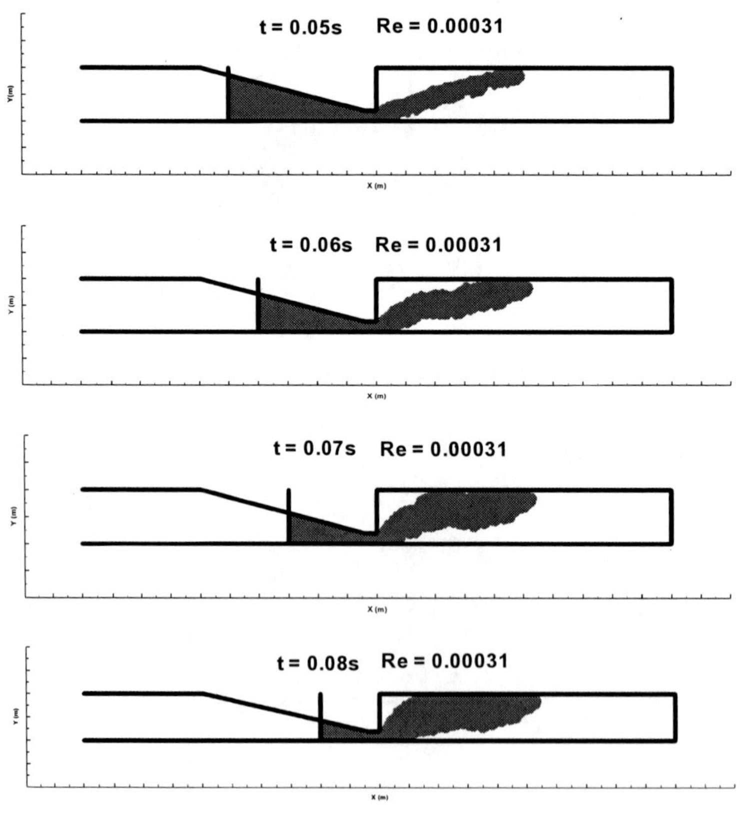

Figure 19. Flow patterns of moulding process for Pa-s, n = 0.8 and $B=0.165\times10^{10}$
(WCF) at t = 0.05 to 0.08s

The flow patterns of moulding flow for two different values of *n*, using the WCF model, are shown in Figures 19 to 20. We can see the effect of the shear thinning on the flow patterns during moulding from these Figures. Some of simulated flow patterns, using the SSY and Tait models are shown in Figures 21 to 22. We can see that if the state equations fit the same experimental results, the flow patterns obtained by using different state equations are very similar. The Reynolds number for the simulations shown in Figures 19 to 22 is 0.31×10^{-3}. We find that the simulation is still very stable for $Re = 0.6\times10^{-4}$ when reducing the velocity of the piston to 0.1m/s.

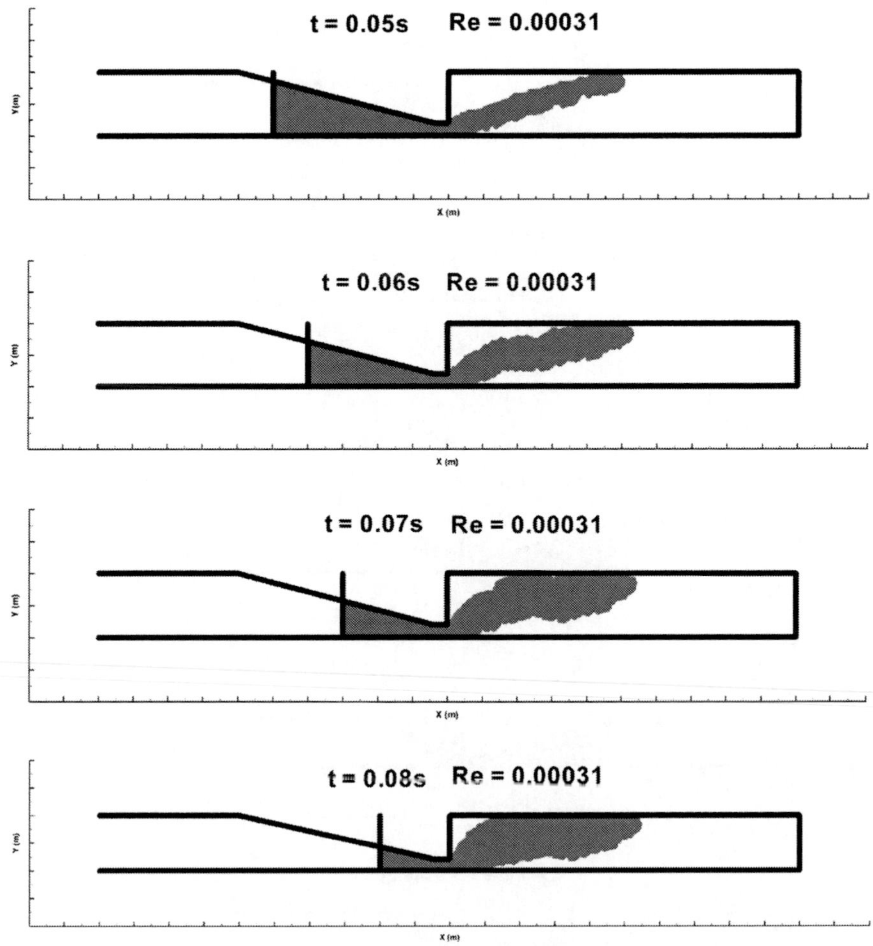

Figure 20. Flow patterns of moulding process for $\eta_0 = 1.22\times10^4$ Pa-s, n = 0.294 and $B = 0.165\times10^{10}$ (WCF) at t = 0.05 to 0.08s

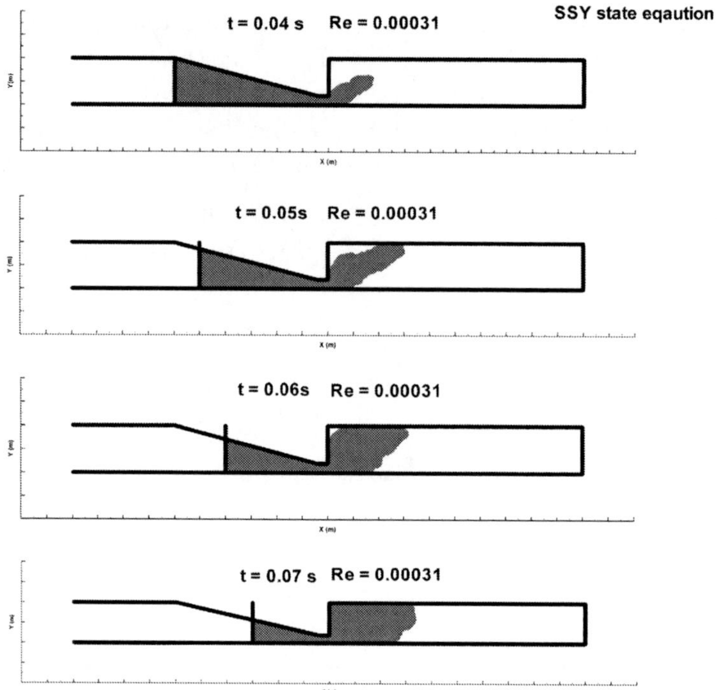

Figure 21. Flow patterns of moulding process for Pa-s and n = 0.294 at t = 0.04 to 0.07s, using the Tait model

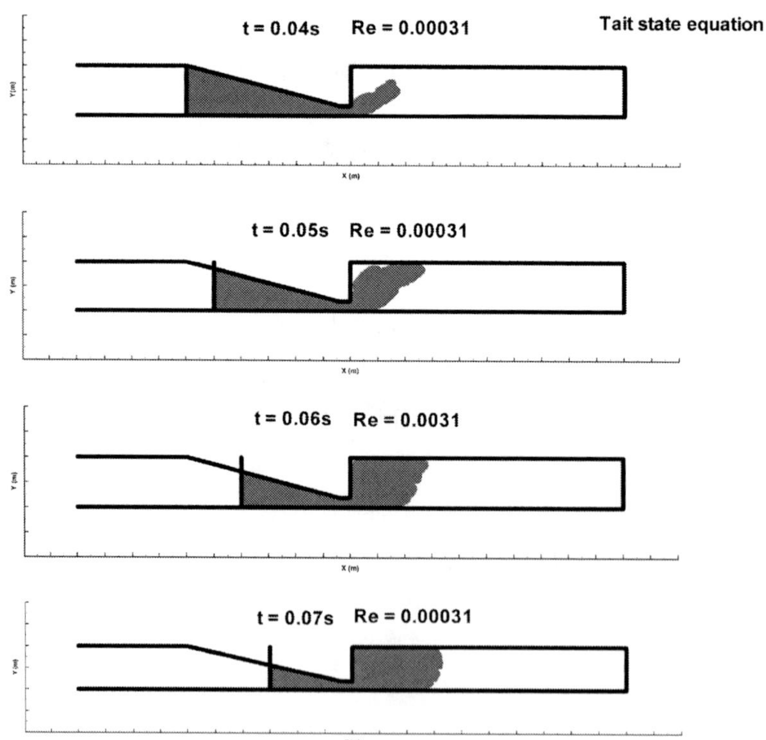

Figure 22. Flow patterns of moulding process for Pa-s and n = 0.294 at t = 0.04 to 0.07s, using the SSY model

CONCLUSION

The following conclusions can be drawn:

1. In this chapter, we proposed a SPH method to simulate high viscous non-Newtonian flows with high pressure. The features of this method are:
   a. The compressibility due to high pressure can be described by employing the state equations commonly used in polymer processing. Realistic viscosity models for the polymer melts can be adopted in the simulation.
   b. The standard SPH systems with such high viscosity and pressure are highly unstable. We used the artificial pressure (Monaghan 2000) to stabilize the simulation. The artificial pressure is a simple and effective way to avoid the tensile instability.
   c. The viscous term in the SPH momentum equation is integrated implicitly in simulating high viscous flows. The implicit scheme removes the diffusive stability constraint on time step so that a reasonably large time step can be used. The implicit scheme requires solving the system equation for the velocities of all particles. The computation would be very expensive for the SPH system with a large number of particles. An iterative solver which does not need to store the system matrix has been used to make the simulation more robust.

   Hence, the present method can successfully simulate the isothermal moulding flow of polymer fluids.
2. At the current stage of developing, the SPH can not compete in speed with the mesh methods, such as the finite element method etc., in simulating polymer processing. The SPH method is more computation demanding than the finite element method. However, SPH is completely mesh free and automatically catches the free surface in simulating fountain flow and injection moulding, while the special treatment is required for the finite element method, see Duan et al. (2007), Mavridis et al. (1986) and Mitsoulis (2010).
3. Imposing the boundary condition on the solid surfaces is an important problem for SPH simulation. Several treatments on boundary conditions are outlined in Sec. 5. Most methods were developed for the explicit SPH. In this Chapter we used a repulsive force to prevent fluid particles from penetrating the wall and specify the wall velocities to the solid particles during iterative solutions. Since the SPH interpolant is intrinsic non-local, this method is not accurate. The no-slip boundary condition can only satisfied approximately. How to enforce the boundary condition for the implicit SPH is a subject worthy of further investigation.
4. The computation is more demanding for the power-law fluid than for the Newtonian fluid. For Newtonian moulding flow the time step can be 2×10^{-6} s and a total of 8×10^{4} steps are required to simulate the whole moulding process. But for the power-law flow the time step should be reduced with decrease of n and, as a result, the total time steps increase. For example, when the zero viscosity is 1.22×10^{4} $Pa\cdot s$, when n = 0.8 and the time step is 10^{-6}s, a total of 16×10^{4} steps and 2.9

hours of CPU time are used to simulate the whole process on the workstation; when n = 0.5 and time step is 10^{-7}s, a total of 16×10^5 steps and 19.56 hours of CPU time are used; and when n = 0.294, 27.44 hours of CPU time are used using the same time step as for n = 0.5. The increase of CPU time is due to more iterations are required in the Krylov space at every time step.

5. The Tait model and Sun et al's are more realistic state equation for polymer fluids and suitable to SPH simulation but require more computations. The total CPU time used is 43.08 hours on the workstation for a job using Sun et al's model and 43.26 hours using the Tait model.
6. All simulations reported in this chapter are two dimensional. Injection moulding flow can usually be simplified to two dimensional flows. However, the codes of the present method can be applied to three dimensional flows with minor modification but the computational demand will increase drastically. A parallel computation is required for further simulation of three dimensional flows.
7. SPH has been applied to heat conduction (Cleary and Monaghan 1999) and semiconductor crystal growth (Rook et al 2007). The possibility to extend the SPH to non-isothermal injection moulding needs further exploration.

APPENDIX

The value of the pressure coefficient, B, in Eq. (0.89) is usually determined by the sound velocity, which depends on the maximum velocity and Mach number limitation, because in the momentum dominated flow the compression is determined by the kinetic energy of fluid. However, in the moulding flow, the velocity of fluid is small and the compression of the fluid is due to the high pressure. The pressure coefficient would be too small if it is determined based on the maximum velocity.

Alternatively, we can roughly estimate the pressure coefficient as follows. Assuming the convergent channel is composed of an infinite number of infinite short slits and the flux through each slit is identical. We assume the volume flux, Q, of Poiseuille flow for the slit at x:

$$Q = \frac{1}{12\mu}\left(-\frac{dp}{dx}\right)Y^3(x) \tag{A.1}$$

where μ is viscosity and $Y(x)$ the slit width at x. The pressure drop for the infinitely short slit is given as

$$-dp = 12\mu Q\frac{dx}{Y^3(x)} \tag{A.2}$$

For the channel with length L the pressure drop is $-\int_0^L dp = p_0 - p_L$. Hence,

$$p_0 - p_L = 12\chi\mu Q \int_0^L \frac{dx}{Y^3(x)} \tag{A.3}$$

where Y as a function of x is known for the die of moulds and χ (=O(1)) is a correction coefficient because the flow is not the Poiseuille flow and local pressure head losses have not been accounted for in Eq. (A.1). For the moulding flow simulated in the present work, we have

$$Y(x) = \begin{cases} 0.01, & 0 \le x < 0.0028 \\ 0.01 - 0.286(x - 0.0028), & 0.0028 \le x < 0.0308 \\ 0.002, & 0.0308 \le x < 0.0328 \end{cases} \tag{A.4}$$

When $\mu = 10^4 \; Pa \cdot s$ and the velocity of the piston is 0.5m/s, we get, when $P_L = 0$,

$$p_0 = 0.2237 \times 10^9 \chi \tag{A.5}$$

If 5% of compressibility is allowed

$$B = \frac{p_0}{1.05^7 - 1} = 0.55 \times 10^9 \chi \tag{A.6}$$

The numerical test shows that $\chi = 3.0$ can be used. Hence $B = 0.165 \times 10^{10}$.
For the power-law fluid , we obtain similarly:

$$p_0 - p_L = \frac{2^{n+1}(2n+1)^n \eta_0 Q^n \chi}{n^n} \int_0^L \frac{dx}{Y^{2n+1}(x)}. \tag{A.7}$$

Acknowledgments

This study was supported by the Polymer Cooperative Research Centre, Melbourne. We are grateful for this support.

References

Becker, M; Tessendorf, H; Teschner, M. *IEEE Trans. Visua. Comp. Graphics*, 15, 2009, 493-503.

Belytschko, T; Lu, YY; Gu, L. Int. *J. Numer. Meth. Engng*, 1994, 37, 229-256.

Belytschko, T; Krongauz. Y; Dolbow, J; Gerlach, C. *Int. J. Numer. Meth. Engng*, 1998, 43, 785-819.

Belytschko, T; Guo, Y; Liu, WK; Xiao, SP. Int. *J. Numer. Meth. Engng*., 2000, 48, 1359-1400.

Bird, RB; Dai, GC; Yarusso, B. *J. Rev. Chem. Engig*, 1982, 1, 1-70.

Bird, RB; Armstrong, RC; Hassager, O. Dynamics of polymeric liquids, Volume 1: *Fluid mechanics*, John Wiley and Sons, 1987.

Bogaerds, ACB; Hulsen, MA; Peters, GW.M; Baaijens, FPT. *J. Rheol*., 2004, 48, 765-785.

Bonet, J; Lok, TSL. *Comput. Methods Appl. Mech. Engrg*, 1999, 180, 97-115.

Bonet, J; Kulasegaram, S. *Int. J. Numer. Meth. Engng*, 2000, 47, 1189-1214.

Bonet, J; Kulasegaram, S. Int. *J. Mumer. Meth. Engng*, 2001, 52, 1203-1220.

Cleary, P; Monaghan, JJ. *J. Comput. Phys*., 1999, 148, 227-264.

Cleary, P; Ha, J; Alguine, V; Nguyen T. *Appl. Math. Model*., 2002, 26, 171-190. Colagrossi, A; Landrini , M. *J. Comput. Phys*., 2003, 191, 448-475.

Crochet, MJ; Dupret, F; Verleye, V. *In Flow and Rheology in Polymer Composites Manufacturing;* Advani, S.G. Ed; *Elsevier*, Amsterdam. 1994. 434-436, and 442-444.

Cross, MM. *J. Colloid Sci*., 1965, 20, 417-437.

Cueto-Felgueroso, L; Colominas, I; Mosqueira, G; Navarrina, F; Casteleiro, M. *Int. J. Numer. Meth. in Engng*, 2004, 60, 1475-1512.

Dilts GA. *Int. J. Numer. Meth. Engng, Part I*, 1999, 44, 1115-1155. Part II, 2000, 48, 225-239.

Duan, Q; Li., X. *Compt. Methods Appl. Mech. Engrg*, 2007, 196, 4911.

Dyka, CT; Ingel, RP. *Computer and Structures*, 1995, 57, 573-780.

Dyka, CT; Randles, PW; ingel, RP. *Int. J. Numer. Meth. Engng*, 1997, 40, 2325-2341.

Ellero M; Tanner RI. *J. Non-Newt. Fluid Mech*., 2005, 132, 61-72.

Español, P. Revenga, M., *Phys. Rev. E*, 2003, 67.026705.

Fan, XJ; Tanner, RI; Zheng, R. *J. Non-Newt. Fluid. Mech*., 2010, 165, 219-226.

Fang, JN., Owens, RG., Tacher, L; Parriaux, A; J. *Non-Newt. Fluid Mech.,* 2006,

Ferrari, A; Dumbser, M; Toro, EF; Armanini, A. *Comp. Fluids*, 2009, 38, 1203-1217.

Gingold, RA; Monaghan, JJ. Mon. Not. R. *Astron. Soc*., 1977, 181, 375-389.

Hosseini, SM; Manzari, MT; Hannani, SK. *Int. J. Num. Method Heat & Fluid Flow*, 2007, 17, 715-735.

Johnson, GR; Beissel, SR. *Int. J. Numer. Meth. Engng*., 1996, 39, 2725-2741.

Krongauz, Y; Belytschko, T. *Appl. Mech. and Eng*. 1997, 146, 371-386.

Li, SF; Liu, WK. *Meshfree Particle Methods*, ISBN 978-3-540-22256-9, Springer-Verlag Berlin Heidelberg New York. 2007.

Liu, GR; Liu, MB. Smoothed particle hydrodynamics, a meshfree particle method, *World Scientific*, Singapore. 2003.

Liu, WK; Jun, S; Li, S; Adee, J; Beleytscko, T. *Int. J. Numer. Meth. Engng*, 1995, 38, 1655-1680.

Lo, EYM; Shao, S. *Appl. Ocean Research*, 2002, 24, 275-286.

Lucy, LB. *Astron. J.* 1977, 82, 1013-1024.

Mavridis, H; Hrymak, N. Vlachopoulos, *J. Polym. Engig Sci*., 1986, 26, 449-454.

Mitsoulis, EJ. *Non-Newt. Fluid Mech*., 2010, 165, 45-55.

Moldflow Pty Ltd, Material testing report, MAT2067, HD601CF, 2004.

Monaghan, J. *J. Annu. Rev. Astron. Astrophys.*, 1992, 30, 543-574.
Monaghan, JJ. *J. Comput. Phys.*, 1994, 110, 339-406.
Monaghan, JJ. *J. Comp. Phys.*, 2000, 159, 290-311.
Monaghan, JJ; Kos, A; Issa, N. *J. Waterway Port. Coastal Ocean Engng*, 2004, 129, 250-249.
Monaghan, J. *J., Rep. Prog. Phys.*, 2005, 68 ,1703-1759.
Morris, JP. Publ. Astron. Soc. *Aust.*, 1996, 13, 97-102.
Morris, JP; Fox, PJ; Zhu, Y. *J. Comp. Phycs.*, 1997, 136, 214-226.
Prakash, M; Cleary, P; Grandfield, J. *J. Mater. Proc. Tech.*, 2009, 209, 3396-3407.
Rafiee, A; Manzari, MT; Hosseini, M. Int. *J. Non-linear Mech.*, 2007, 42, 1210-1223.
Rafiee, A. *ANZIAN J.*, 2008, 49, 411-422.
Randles, PW; Libersky, LD. *Comput. Methods Appl. Mech. Eng.*, 1996, 139. 375-408.
Rodgers, PA. *J. Appl. Polym. Sci.*, 1993, 48, 1061-1080.
Rook, R; Yildiz, M; Dost, S. *Numer. Heat Trans., part B*, 2007, 51, 1-23.
Shao, S; Lo, EYM. Advan. *Water Resources*, 2003, 26, 787-800.
Shephard, D. A two dimensional function for irregularly spaced data, in: *ACM National Conference*, 1968, 517-524.
Swegle, JW; Hicks, DL; Attaway, SW. *J. Comp. Phys.*, 1995, 116, 123-134.
Sun, ZH; Song, M; Yan, ZT. *Polymer*, 1992, 33, 328-331. 139, 68-84.
Tanner, RI; Qi, F. *J. Non-Newt. Fluid Mech.*, 2005. 127, 131-141.
Verbeeten, WMH; Peters, GWH; Baaijens, FPT. *J. Rheol.*, 2001, 45, 823-844.
Yildiz, M., Rook, RA., Suleman, A., Int. *J. Numer. Meth. Engng.*, 2009, 77, 1416-1438.

In: Injection Molding: Process, Design, and Applications ISBN: 978-1-61761-307-4
Editor: Phoebe H. Kauffer

Chapter 4

MELT/MELT WELDLINE IN INJECTION MOLDING

Hong Wu
Polymer Research Institute of Sichuan University
The State Key Laboratory of Polymer Materials Engineering, Chengdu, China

INTRODUCTION

As one of the most widely employed methods, injection molding can be characterized by high productivity, high degree of automation and good dimensional stability. However, molding complicated parts, multigated mold cavities and cavities containing inserts may generate serious difficulties in terms of mold filling and final production especially. In fact, molding of such parts usually produces weldline once the melt fronts have joined either by impingement flow or around an insert [1, 2]. Weldlines also form when jetting occurs along the flow path where the flow is suddenly accelerated (e.g. near the gate) [3].

Weldlines can be categorized as cold and hot weldline [4, 5]. In previous one, two melt fronts meet head to head and additional flow could not take place after combination. This type leads lowest strength. Hot weldlines (stream lines or melt lines) form when two or more melt streams recombine to each other such as breaking up and rejoining of flow around pins. An additional flow after recombination occurs.

1. Formation Mechanisms and Effect Factors

The formation of a weldline is regarded as one of the most undesirable phenomena in injection molding process, since it results in poor appearance as well as poor mechanical properties. It is necessary to have a clear knowledge the formation mechanisms and effect factors of weldline in order to eliminate or weaken the negative effect on properties.

The weakness of the weldline for homopolymer systems arise from the poor intermolecular diffusion at the weldlines, the molecular orientation parallel to the interface induced by the fountain flow and the existence of V-notches at the surface around the weldline caused by the entrapped air which is forced to the wall. The properties of a weldline will depend on whether the polymer chains had enough time to diffuse across the interface to form a strong bond [5-8].

Weldlines are particularly problematic in reinforce plastics because, unlike molecular orientation in homopolymer systems, the flow induced filler orientation does not relax. For filled and reinforced polymers, loss in weldline strength is found to be related with aspect ratio of the fillers. A serious reduction in strength is observed for polymers reinforced with fibers (glass) or platelets (talc) having high aspect ratio. However, spherical additives (glass beads or calcium carbonate) lead almost identical strength for both specimens with and without weldline [8-10]. On the other hand, the weldline strength depends on the mold shape and the filler concentration. Effect of other variables such as distance of flow, insert diameter or lateral expansion on the weldline strength is not significant [11].

In polymeric blends, the phase morphology is a critical factor affecting weldline characteristics. The characteristics of the weldline morphology are quite different from other regions. The concentration of a minor phase, the addition of compatibilizers, and the viscosity ratio of dispersed phase and matrix also influence the morphology of the weldline region [12-24].

2. Theory and Models for the Prediction of Weldline Strength

Attempts to quantify the weldline strength have been presented. Models based on molecular diffusion have been used for homopolymers [25-27] and for immiscible polymer blends [14]. Kim and Suh [25] assumed that the entire load is "beard" by the bounded area of the specimen. This idea is expressed in the following equation [25],

$$\sigma_w = \sigma_b \left(\frac{A_0 - A_N}{A_0} \right) \tag{1}$$

where σ_w and σ_b are the tensile strength of the injection molded part with and without weldline respectively, A_0 is the initial cross-sectional area of the interface, A_N is the non-bonded area, and $A_0 - A_N$ is the bonded area at specific contact time.

They [25] further supposed that the time variation rate of the non-bounded area is given by Fick's Law of diffusion:

$$\frac{dA_N}{dt} = C \frac{D\Delta G}{kT} \tag{2}$$

where C is the constant, D is diffusion coefficient, and k is the Boltzmann constant, T is the absolute temperature, ΔG is the free-energy difference, and written as:

$$\Delta G = -2\gamma A_N - T\Delta S_m \tag{3}$$

where γ is the surface tension, and ΔS_m is the entropy of mixing determined from the Flory-Huggins theory of lattice [28].

With an expression of the entropy of mixing given by the Flory-Huggins lattice theory they obtained an expression of the new bounded area as a function of the contact time and other parameters appearing in Eq. (1-3) From Eq. (1) they then obtained an expression for σ_w as a function of the same set of parameters.

$$\Delta S_m = -kTA_N\,\delta x\delta\, n_0 \ln\left(\tfrac{1}{2}\right) \tag{4}$$

where δx and δn_0 are the diffusion thickness and the number of lattice per unit volume respectively.

Combining equations 2, 3 and 4, $\frac{dA_N}{dt}$ may be expressed as:

$$\frac{1}{A_N}dA_N = -\frac{CD}{kT}\left(2\gamma - kT\delta x\delta n_0 \ln\left(\frac{1}{2}\right)\right)dt \tag{5}$$

In the model prediction for amorphous polymers proposed by Kim and Suh [25], δx is considered as a constant. Hence integrating equation 5, the model prediction for amorphous polymers is given by [25]:

$$\frac{\sigma_w}{\sigma_b} = 1 - \exp\left\{-\frac{CD}{kT}\left(-2\gamma + kT\delta x\delta n_0 \ln\left(\frac{1}{2}\right)\right)t\right\} \tag{6}$$

Here we consider δx as a function of time, $\delta x = 2(D_t)^{1/2}$, we rewrite the model as follows: For glassy and amorphous polymers:

$$\frac{\sigma_w}{\sigma_b} = 1 - \exp\left\{-\frac{CD}{kT}\left[2\gamma - \frac{4}{3}kTD^{1/2}\,\delta n \ln\left(\frac{1}{2}\right)t^{1/2}\right]t\right\} \tag{7}$$

Hamada et al [27] proposed a useful parameter, weldline index φ to estimate weldline property of injection molding.

$$\phi = \int_{-h}^{h}\int_{t}\frac{\partial t}{\tau}\partial y \tag{8}$$

Where 2h is thickness, t is time, τ is viscoelastic relaxation time of polymer melt which was measured using cone plate type rheometer. The relationship of τ with mechanical property of injection moldings with a weldline was discussed for PS and PC. Polymer melt whose τ was relatively short can easily enhance the entangement across the weldline interface and weldline strength. Polymers, which have large weldline index basically, tend to have high weldline strength. Weldline index itself considers not only material property, but also molding conditions such as temperature, pressure etc.

Pecorini and Seo [26] assume that sufficient interpenetrates has occurred when the chains have diffused one radius of gyration obtained the following equation for the healing time:

$$t_h = R_g^2 / 2D_s \tag{9}$$

where using the following equation for the self-diffusion coefficient D_s

$$D_s = \left(\frac{\rho RT}{270}\right)\left(\frac{Mc}{M}\right)^2\left(\frac{R_g^2}{M}\right)\left(\frac{1}{\eta_{Mc,T}}\right) \tag{10}$$

Where ρ is the melt density, T is the processing temperature, R is the gas constant, Mc is the critical molecular weight, M is the absolute molecular weight, and $\eta_{Mc,T}$ is the zero-shear viscosity at the critical molecular weight and processing temperature. With a detailed rheological characterization, they obtained the healing time as a function of temperature. The theoretical results compared well to weldline Izod experimental data.

Mekhilef et al. [14] extended the approach of Kim and Suh [25] to the case of immiscible polymer blends. Contribution to the weldline strength from AA and BB self-diffusion was taken into account using Kim and Suh [25] approach. While AB diffusion was also taken into account using an expression for the free energy comprising the entropy of mixing of two different components and a contribution of the free surface of each constituent.

The model prediction for immiscible polymer blends is expressed as (11):

$$\begin{aligned}\frac{\sigma_w}{\sigma_b} &= 1-\phi_A^2\exp\left\{\frac{C_A D_A}{kT}\left(2\gamma_A+kT\delta x_A\delta n_A\ln\left(\frac{1}{2}\right)\right)t\right\} \\ &-\phi_B^2\exp\left\{\frac{C_B D_B}{kT}\left(2\gamma_B+kT\delta x_B\delta n_B\ln\left(\frac{1}{2}\right)\right)t\right\} \\ &-\left(1-\phi_A^2-\phi_B^2\right)\exp\left\{-\frac{C_{AB}D_{AB}}{kT}\left[\gamma_A+\gamma_B-RT\delta_{AB}\left(\frac{\rho_A}{M_A}\phi_A'\ln\phi_A'+\frac{\rho_B}{M_B}\phi_B'\ln\phi_B'+\chi_{AB}\phi_A'\phi_B'\right)\right]t\right\}\end{aligned} \tag{11}$$

Comparison of model predictions with experimental results obtained with a polycarbonate (PC) and high density polyethylene (HDPE) blends showed that the model predicts fairly well the experimental data for a concentration and temperature utilized. Discrepancy between model predictions and experimental data for certain composition was attribuated to the fact that their approach, unlike the one of Kim and Suh[25], does not take into account molecular orientation which may take place at the flow front due to the so-called "fountain flow", nor does take into account the effect of temperature gradients that might be generated during the non-isothermal filling. Their set of experimental results did not allow to explore a wide range of filling temperature and hence to study the effect of temperature on the weldline strength. Indeed the processing condition required by their specific blend did not allow for a good evaluation of the effect of minor phase (HDPE) to the weldline strength as the lowest injection temperature and the contact time used for their experiments always result in a coagulate healing of the HDPE phase.

For amorphous polymer blends, we also consider δx as a function of time, $\delta x=2(D_t)^{1/2}$, a modified Mekhilef et al. model prediction for weldline strength is given by:

$$\begin{aligned}\frac{\sigma_w}{\sigma_b} &= 1-\phi_A^2 \exp\left\{-\frac{C_A D_A}{kT}\left[2\gamma_A-\frac{4}{3}kTD_A^{1/2}\delta n_A \ln\left(\frac{1}{2}\right)t^{1/2}\right]t\right\}\\ &-\phi_B^2 \exp\left\{-\frac{C_B D_B}{kT}\left[2\gamma_B-\frac{4}{3}kTD_B^{1/2}\delta n_B \ln\left(\tfrac{1}{2}\right)t^{1/2}\right]t\right\}\\ &-\left(1-\phi_A^2-\phi_B^2\right)\exp\left\{-\frac{C_{AB} D_{AB}}{kT}\left[\gamma_A+\gamma_B-\frac{4}{3}RTD_{AB}^{1/2}\delta n_{AB} t^{1/2}\left(\frac{\rho_A}{M_A}\phi'_A \ln\phi'_A+\frac{\rho_B}{M_B}\phi'_B \ln\phi'_B+\chi_{AB}\phi'_A\phi'_B\right)\right]t\right\}\end{aligned} \quad (12)$$

3. Eliminate or Weaken the Weldline

3.1. Materials

Weldline strength greatly replies on the properties of materials [29-32]. For unfilled polymers, the lowest weldline strength was obtained for brittle amorphous polymers and rod-like polymers like PS, PMMA, SAN, LCP, Acetal etc., higher values are usually obtained for ductile amorphous or crystalline polymers such as PP, PE, PBT, PET, PVC and PC. At the same time, the weldline strength decreases with increasing molecular weight of materials.

For semicrystalline polymers, the situation is even more complicated. Indeed, in addition to the above mentioned parameters, the degree of crystallinity, the size and the distribution of the crystallites in the weldline region have also important effects on the weldline strength [33-35]. The effect of processing parameters (mold and melt temperatures, cooling rate, and injection pressure) is however found to be similar to the observed for their amorphous counterpart [36-40]. The more pronounced shrinkage of the semicrystalline polymers as compared to amorphous ones is found to have an important effect on the V-notch which forms a sharp ridge upon cooling [41-42]. The presence of inhomogeneities (abraded particles, impurities, mold release agent, phenolic antioxidant additive) is also found to affect weldline strength through nucleation of crystallites [5, 41-42].

In the case of immiscible polymer blends weldlines are found to be much weaker than for homopolymers. The situation is even more complicated by the presence of preformed structural differences and minor phase concentration gradients in the weld region [12-24]. The addition of compatibilizers and the viscosity ratio of dispersed phase and matrix are the most important effect on the morphology and property of the weldline.

3.1.1. Compatibilizer

The injection molded specimens without weldline exhibited traditional skin-core morphology in which dispersed domains at sub skin layers are elongated toward the flow direction and those at core are not deformed. Elongated domains were reduced by increasing the amount of compatibilizer. But in the presence of weldline and absence of compatibilizer, elongated domains were perpendicular to flow direction at weldline region. Increasing amount of compatibilizer leaded to fine and isotropic morphology [21]. Similar observation was reported that the blend with the compatibilizer reduced thickness of the skin layer, decreased width of weldline region and decreased anisotropy of the dispersed domains at weldline region compared with a specimen without the compatibilizer, and tensile stress of the blend increased with the addition of the compatibilizer [22]. Taking inorganic filler as compatibilizer, a new approach [43] was researched to increase weldline strength of incompatible polymer blend composites by using talc as inorganic filler blending with two immiscible organic phases (PP and PA6). The effects of filler levels as well as compatibilizer on weldline strength of each polymeric phase and blend were investigated. The results show that talc filler which is selectively wetted by dispersed PA phase dramatically reduces the elongated domain size in the weline region and causes to much more homogeneous microstructure, so that it is beneficial in increasing weldline strength via increasing the viscosity of the PA dispersed phase. It is also clear that the difference between the strength of the PP and PA blend composites without and with weldline is not very significant especially at higher filler loading degrees. Compared to uncompatibilized blend composites, compatibilized ones represent much higher weldline strength for all levels of talc loading.

The effect of weldline on the morphology and mechanical properties of polystyrene and polyamide-6 blends with various amounts of poly(styrene-co-maleic anhydride) (SMA) as compatibilizer was investigated, as well as the relationship between the morphology of weldline and fracture mechanism were investigated [44].

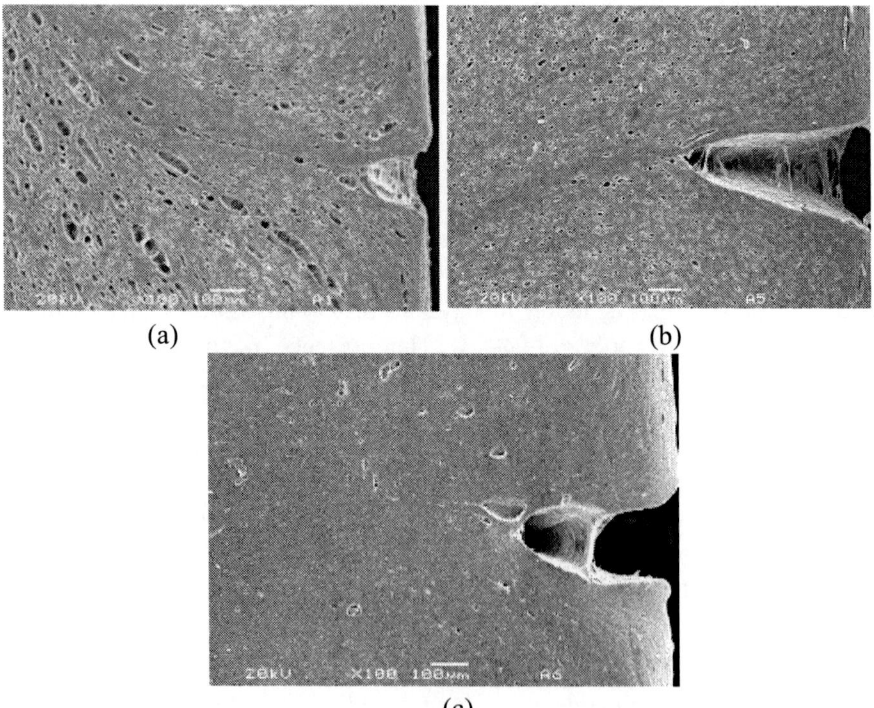

Figure 1. SEM micrographs of weldline for blends containing various contents of compatibilizer SMA; (a) SMA-0; (b) SMA-2.5; (c) SMA-7.5

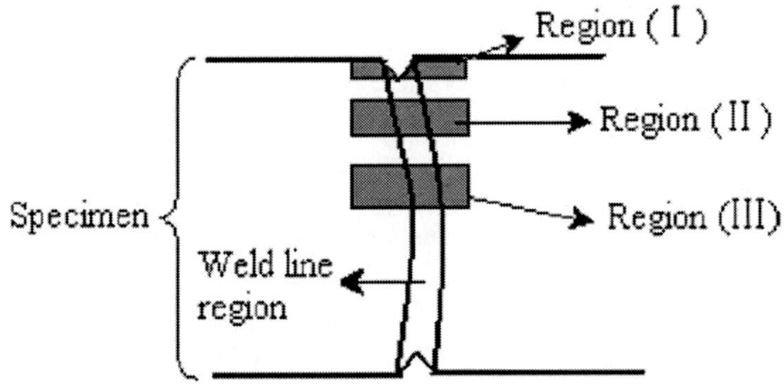

Figure 2. Schematic diagram of the section of specimen with weldline classified three regions

Figure 1 gives the SEM micrographs of weldline for PS/PA blend containing various contents of compatibilizer SMA. All blends with weldline have a V-notch at the weldline where two fronts meet together. And the weldline is visible in Figure 1 (a, b) because the dispersed phase morphology near the weldline is different from the weldline morphology. But the weldline in Figure 1 (c) could not be distinguished from the bulk. In order to show the difference of weldline morphology, three regions(I), (II)and (III) for weldline specimens were classified as shown in Figure 2: reign (I) near the cavity, called skin; reign (III) near the center of specimens, called core; reign (II) between reign (I) and (III), called sub-skin.

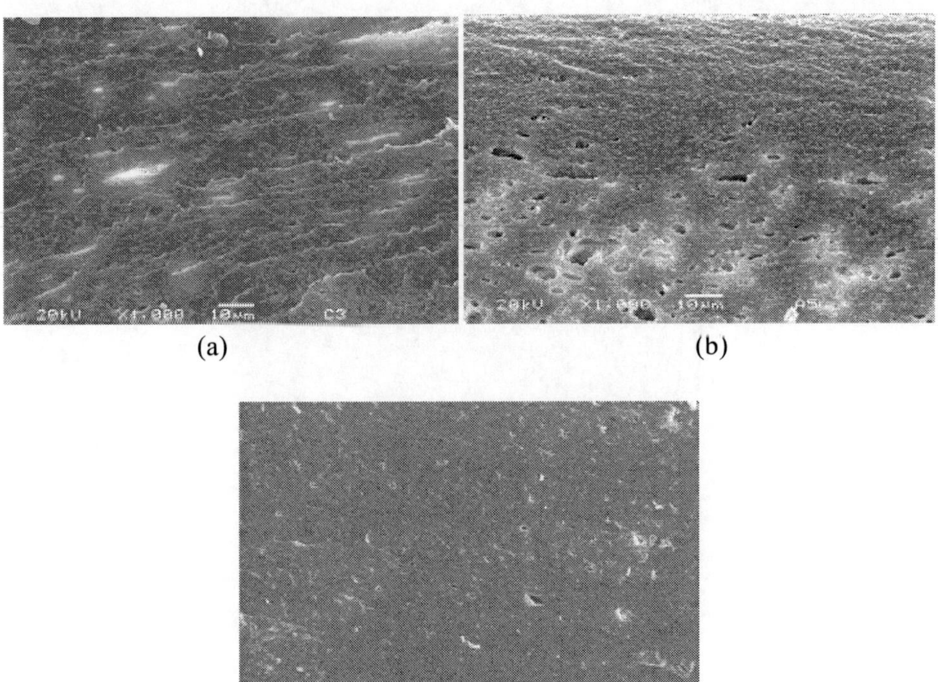

(a) (b)

(c)

Figure 3. SEM micrographs of skin for blends containing various contents of compatibilizer SMA; (a) SMA-0; (b) SMA-2.5; (c) SMA-7.5

The magnified SEM micrographs of three different regions were revealed in Figure 3,4,5 respectively. As shown in Figure 3, the dispersed phases are barely observed at skin in all blends. But at the sub-skin the weldline morphology of specimens is different. As shown in Figure 4 (a, b), the elongated dispersed domains near the weldline are aligned toward the perpendicular direction. The reason is that during injection molding, dispersed domains near the weldline undergo elongation flow. It is also observed that the dispersed domains size with spherical shapes inside the weldline zone is smaller than that outside the weldline zone. These smaller domains are formed by the breakup of highly elongated dispersed domains during injection process. As compared with that of the blend SMA-0 in the absence of SMA compatibilizer, the size and elongation of dispersed domains in the blends with SMA, inside or outside the weldline zone, get smaller. For SMA-7.5 the amount of compatibilizer is higher than that of SMA-2.5, therefore the size of dispersed domains is so small that the dispersed domains cannot be deformed even under the elongation flow. This is why for SMA-7.5 the dispersed domains outside the weldline zone are spherical shapes, which is the same as that inside weldline, as shown in Figure 4 (c). As shown in Figure 5, at the core the dispersed domains have just entered the elongation flow field and have not been deformed to the same extent as the dispersed domains at the sub-skin. So the slightly elongated particles of SMA-0 and SMA-2.5 at the core are probably induced into bulk by the elongation flow at the gate. But for SMA-7.5 the particles were so small that could keep spherical at the core.

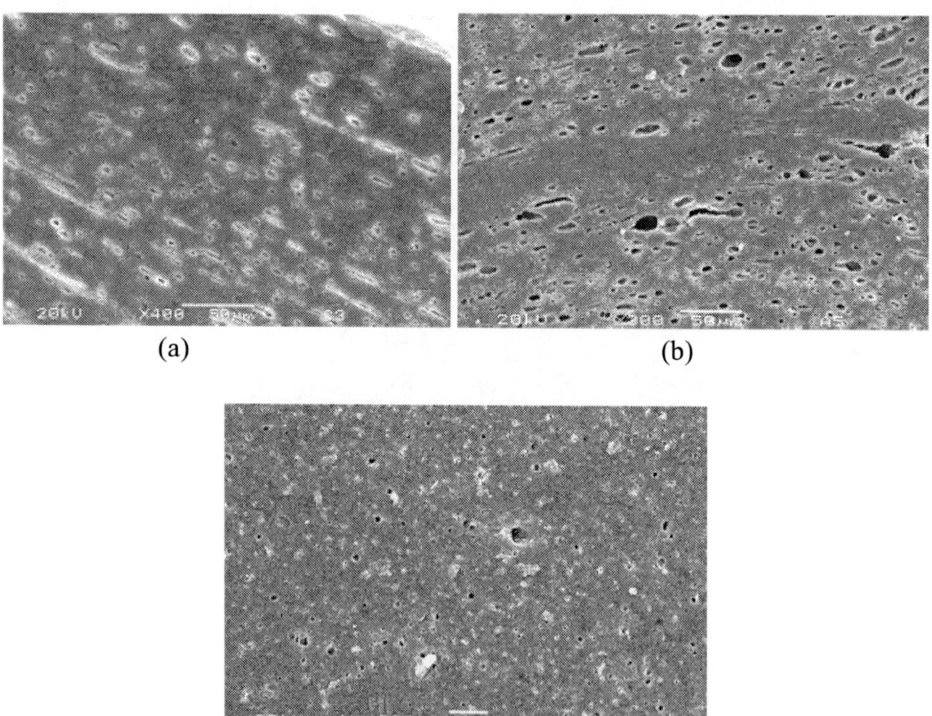

(c)

Figure 4. SEM micrographs of sub-skin for blends containing various contents of compatibilizer SMA; (a) SMA-0; (b) SMA-2.5; (c) SMA-7.5

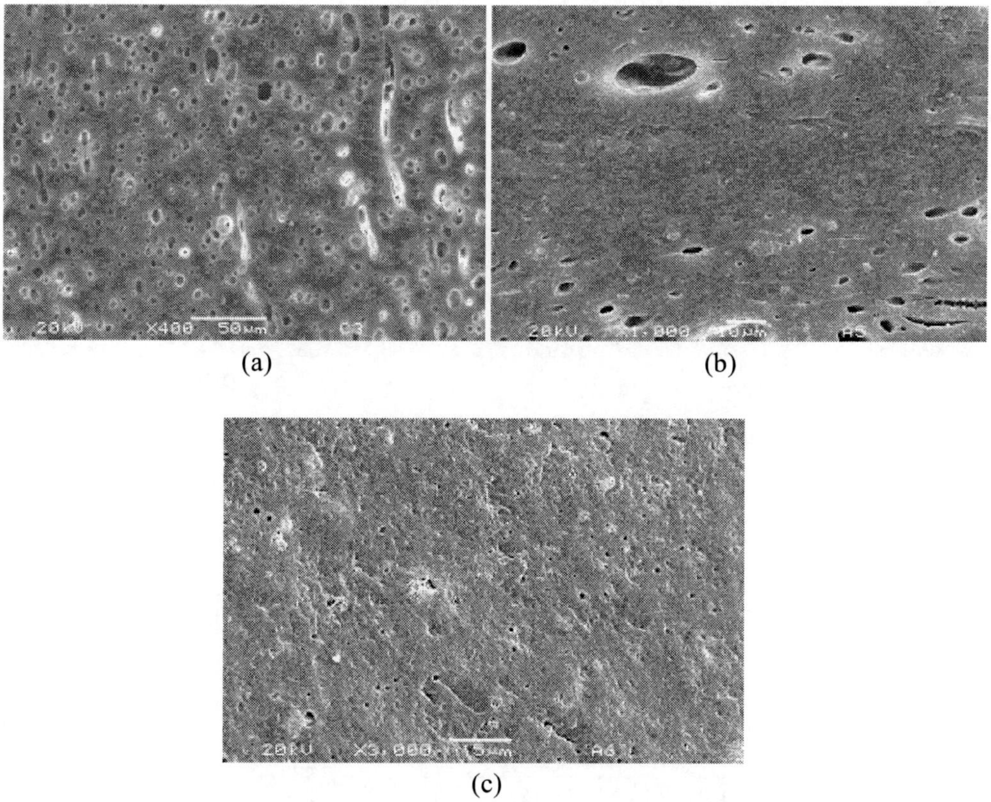

Figure 5. SEM micrographs of core for blends containing various contents of compatibilizer SMA; (a) SMA-0; (b) SMA-2.5; (c) SMA-7.5

From the above discussion, we can observe that for SMA-0 and SMA-2.5 the morphology of dispersed phase near the weldline is different from that of SMA-7.5. In order to study the fracture mechanism, the morphology of tensile fractured sample SMA-0 with weldline was investigated. As shown in Figure 6 (a), the fractured site is near the weldline zone where the dispersed domains are elongated parallel to the weldline. And it is also observed that the tensile fractured surface is full of elongation-dispersed domains, which is according with the morphology near the weldline, Figure 6 (b). So for SMA-0 the fracture mechanism is that the elongated dispersed domains may act as the stress concentrator, causing the fracture of sample. The same morphology in weldline zone was observed in SMA-2.5 and SMA-0, so their fracture mechanism is the same, which is induced by the stress concentration caused by the high oriented dispersed domains. But for SMA-7.5 the morphology of dispersed phase in the weldline with the spherical shapes is the same with that near the weldline. It is hard to confirm that the fractured site is inside or outside the weldline. For SMA-7.5 the morphology of dispersed phase inside or outside the weldline is almost the same, and then the stress concentration must be the same. In weldline zone the insufficient diffusion of molecular chains across the polymer-polymer interface caused the weakness, which didn't exist near the weldline zone. So the fractured site must be inside weldline. So its fracture is induced by insufficient diffusion of molecular chains across the interface in the weldline zone.

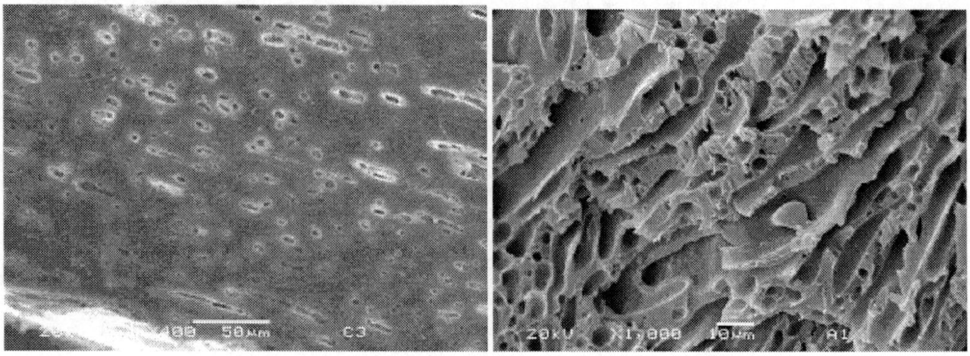

(a) (b)

Figure 6. SEM micrographs of sub-skin for SMA-0; (a) cut open section; (b) fractured surface

Table 1. Tensile properties and Izod impact strength (IS) of blend and neat PS without and with weldline

Blend type	Composition (wt%)	σ_b (MPa)	ε (%)	IS (KJ/m^2)	F_{kl}
Without weldline					
Neat PS	PS(100)	42.0	2.0	8.1	
SMA-0	PS(70)-PA(30)	30.1	1.9	6.3	
SMA-2.5	PS(70)-PA(27.5)-SMA(2.5)	31.5	2.6	7.1	
SMA-7.5	PS(70)-PA(22.5)-SMA(7.5)	33.4	1.5	7.2	
With weldline					
Neat PS	PS(100)	32.1	1.4	4.1	0.76
SMA-0	PS(70)-PA(30)	9.2	0.8	1.6	0.31
SMA-2.5	PS(70)-PA(27.5)-SMA(2.5)	10.4	0.8	0.9	0.33
SMA-7.5	PS(70)-PA(22.5)-SMA(7.5)	23.7	1.1	2.3	0.71

F_{kl}= Tensile stress with weldline/ Tensile stress without weldline

The data listed in Table 1 show that the tensile stress (σ_b) and impact strength of the compatibilized PS/PA blends without weldline were improved a few over their uncompatibilized blends due to their good adhesion strength between the matrix and the dispersed domain. While for the blends with weldline the tensile stress and the impact strength are greatly decreased, compared with the corresponding blends without the weldline, which is due to the morphology difference of dispersed phase in the weldline. It can be seen that the morphology of the dispersed domains in the weldline plays very important role in the mechanical properties of the blends. For SMA-7.5 the tensile stress with weldline is higher than that of SMA-0 and SMA-2.5. Fully elongated domains appear on the tensile fractured surface of SMA-0 (Figure 6b), indicating that the fractured site is near the weldline, where the dispersed domains are elongated parallel to the weldline. This causes a dramatic decrease of the tensile stress of injected SMA-0 and SMA-2.5 parts with weldline. While for SMA-7.5 the fractured site is inside the weldline and the dispersed domains in weldline are spherical, which is the same as those near weldline. This indicates that addition of enough content of compatibilizer SMA greatly promotes the molecular diffusion of polymer blends across the weldline, causing a great increase of weldline strength as compared with that of SMA-0 and

SMA-2.5. The data listed in Table 1show that the weldline strength of injection molded SMA-7.5 parts is two times as high as that of SMA-0 and SMA-2.5.

The data listed in Table 1 also show that for blends with weldline the impact strength greatly decreases compared with blends without weldline. And the impact strength of blends is lower than that of neat polymer, indicating that the presence of weldline decreases the mechanical properties of blends to a greater extent than that of neat polymer. Even the good phase dispersity for SMA-7.5 in weldline was achieved, its impact strength is not also greatly improved, indicating that the impact strength of the blends has weaker dependence on the morphology of dispersed phase as compared with their tensile stress. Interestingly, for SMA-2.5 the impact strength is lower than that of SMA-0, which may be due to the higher depth of V-notch in SMA-2.5 than in SMA-0 (Figure 2).

So the results show that SMA greatly improves the compatibility of PS/PA-6 blends and their weldline morphology of injection molded blends parts. PA-6 particles near the weldline are highly oriented parallel to the weldline for the blends without and with low content of SMA, while PA-6 domains at weldline show spherical particles. For the blend with high content of SMA, its morphology shows fine and isotropic morphology inside and outside weldline zone. The difference of morphology in weldline causes the distinction of fracture mechanism. The content of compatibilizer SMA has greatly influenced in the tensile stress. For the specimens with high anisotropic morphology in weldline the tensile stress is much lower than that of the specimens with isotropic morphology, which imply that the morphology at weldline is an important factor affecting the tensile stress. However, the morphology of blends with weldline has only slight effect on impact stress, even for the blend with fine isotropic morphology in weldline.

3.1.2. Viscosity ratio

The viscosity ratio of the dispersed phase over a matrix is a very important parameter for control of weldline morphology of an immiscible polymer blend. The dispersed particles with low viscosity become more deformed and are easier to break up because of the shear or elongational stress. Failure at weldline changes from brittle to ductile nature by changing the disperse phase viscosity with molecular weight, since increase in viscosity ratio of blend leads to decrease in elongated disperse domains which is the reason of weakness in strength at weldline [20]. The similar tendency was reported in many composite systems [12, 19-20].

Guo et al [45] studied the weldline morphology-properties relationship of polystyrene/poly(methylmethacrylate) (PS/PMMA). The results showed that the viscosity ratio of dispersed phase over matrix was a very important parameter for control of weldline morphology of the immiscible polymer blend. In the PS/PMMA (30/70) blend, the viscosity of the dispersed PS phase was lower than that of the PMMA matrix, the PS phase was absent at the weldline, and PS particles were highly oriented parallel to the weldline. In the PS/PMMA (70/30) blend, the PMMA dispersed domains at the core of weldline were spherically shaped, which was the same as bulk. And the difference of morphology at weldline caused tensile stress of PS/PMMA (30/70) much lower than that of PS/PMMA (70/30) blend, even though the tensile stress of PS/PMMA(30/70)without weldline was much more higher than that of PS/PMMA(70/30).

On the other hand, a model based on melt diffusion and Flory-Huggins free energy theory for predicting the weldline strength of injection molded amorphous polymers and polymer blends parts were modified by considering the diffusion thickness in the interface as a function of contact time. The modified model for weldline strength prediction of homopolymers and polymer blends were respectively used to evaluate the weldline strength of PS and PMMA, and that of PS/PMMA blends.

Guo et al [45] use equation 7 for predicting weldline strength of PS and PMMA, and equation 12 for predicting the weldline strength of PS/PMMA blends.

Self diffusion can be calculated by following equation [46].

$$D\eta = \left({}^{A\rho kT}\!/_{36} \right) {}^{R^2}\!/_{M} \tag{13}$$

where A is Avagodro's number, ${}^{R^2}\!/_{M}$ is essentially a constant for any bulk polymer. If η is known, it is possible to predict D. The viscosities of PS, PMMA and PS/PMMA are shown in Table. 2. The viscosity at other temperatures can easily calculated by Arrhenius equation. Fox and Allen [47] has reported that the constant ${}^{R^2}\!/_{M}$ for PS and PMMA is 4.68×10^{-17} and 3.72×10^{-17} respectively.

The mutual diffusion coefficient D_{AB} is given by following expression [48]

$$D_{AB} = \phi(1-\phi)\Lambda_0 \frac{N_e}{N}\left(\frac{1}{N\phi} + \frac{1}{N(1-\phi)} + 2\left|\chi_{AB}\right| \right) kT \tag{14}$$

where ϕ is volume fraction of dispersed phase, Λ_0 is the mobility coefficient, N_e is the degree of polymerization for entanglement and N is the degree of polymerization. Usually for $\Lambda_{0A} \neq \Lambda_{0B}$ and $N_A \neq N_B$, the diffusion coefficient is given by

$$D_{AB} = D_0 \frac{\phi(1-\phi)}{\phi + k(1-\phi)} \tag{15}$$

where

$$D_0 = 2\left|\chi_{AB}\right| kT\Lambda_{0B} \quad k = \frac{\Lambda_{0B}}{\Lambda_{0A}} \quad N\angle N \tag{16}$$

Table 2. Zero shear viscosities and flow activation energy of PS/PMMA blends

PS/PMMA	100/0	80/20	70/30	30/70	20/80	0/100
180 °C	9.512E5	1.089E6	1.371E6	3.442E6	4.017E6	3.406E6
200 °C	1.738E5	2.099E5	2.425E5	5.855E5	6.545E5	5.637E5
220 °C	7.107E4	6.109E4	6.803E4	1.544E5	1.496E5	1.332E5
Ea, kJ/mol	121.0	134.0	139.7	144.4	153.0	150.7

$$D_0 = 2\left|\chi_{AB}\right| kT\Lambda_{0B}\frac{N_e}{N_B} \quad k = \frac{\Lambda_{0B}}{\Lambda_{0A}}\frac{N_A}{N_B} \quad N \angle N_e \tag{17}$$

According to Rouse theory [28], the friction ξ can be related to the viscosity:

$$\xi = \frac{1}{\Lambda_0} = \frac{36 M_0}{\rho R^2 N_A}\eta_{0cr} \tag{18}$$

where M_0 is molecular mass of a monomer unit, η_{0cr} is a critical viscosity which is given by:

$$\eta_{0cr} = \eta_0 \left(\frac{M_{cr}}{M_w}\right)^{3.4} \text{ if } M_w > M_{cr}$$

$$\eta_{0cr} = \eta_0 \left(\frac{M_{cr}}{M_w}\right) \text{ if } M_w \angle M_{cr} \tag{19}$$

where η_0 is zero shear viscosity, M_{cr} is a critical molecular weight. $M_{cr} = 2M_e$, $M_e = \rho \frac{RT}{G^0}$, G^0 is the plateau modulus. η_0 and G^0 can be measured by oscillatory rheometer. η_0 is listed in Table 1. In literature [49], G^0 is equal to 4.5×10^6 dyne/cm^2 for PMMA, 180 °C, and 2.0×10^6 dyne/cm^2 for PS, 190 °C. In our measurement, G^0 is taken G' corresponding to when tanδ is minimum at plateau region [50], G^0 is equal to 1.7×10^5 N/m^2 for PS, 180 ^{0}C, and 4.45×10^5 N/m^2 for PMMA, 180 °C which coincides with that reported in literature. Introducing the friction coefficients in the diffusion equation, we can calculate D_{AB}.

The surface tension from literature is listed in Table 3 [50].

Figure 7 and 8 show the prediction of the weldline strength as a function of contact time at different melt temperatures for pure PS and PMMA respectively. The bonding area at the weldline increases with the increase of contact time and temperature. The total bonding can be achieved at 280°C and 300°C within 10 s for PS and PMMA respectively. However, when the melt temperature drops close the glass temperature, incomplete bonding is occurred. Since the diffusion of molecular chains at temperature close glass temperature of the materials is weaken.

Table. 3. Surface Tension (mN/m) of PMMA and PS

	20 °C	150 °C	200°C	-dγ/dT, mN/m°C
PS	40.7	31.4	27.8	0.072
PMMA	42.7	31.0	26.5	0.090

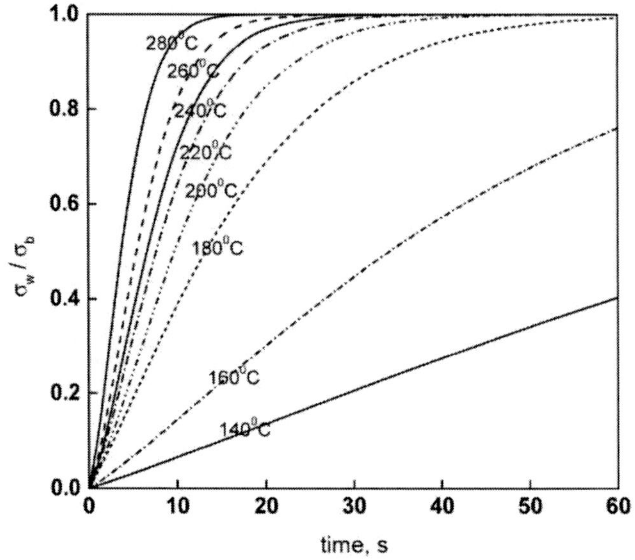

Figure 7. Model predictions for weld-line strength of PS as a function of temperature and contact time

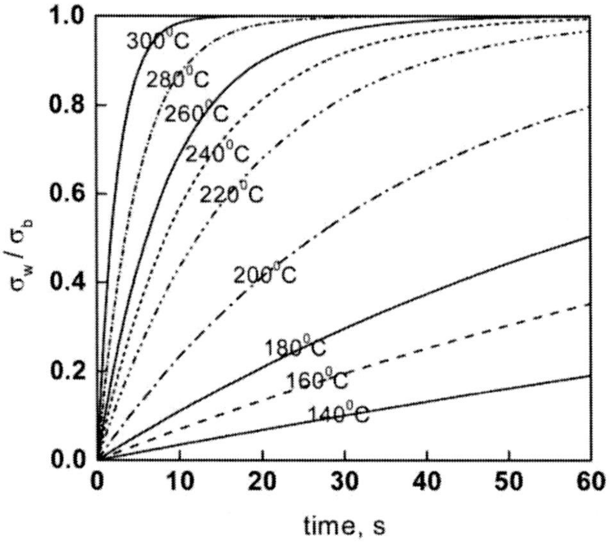

Figure 8. Model predictions for weld-line strength of PMMA as a function of temperature and contact time

For polymer blends, the bonding area at the interface of PS/PMMA (80/20, 70/30, 20/80) increases with the rise of contact time and melt temperature (Figure 9). However, the bonding area for PS/PMMA blends (80/20 and 70/30) increases faster with contact time and temperature than for PMMA/PS blends for PMMA matrix does. In this temperature range T>160°C, both phases contribute to the adhesion of the melt fronts. This slow increase is also awarded to the variation of the diffusion coefficient D of PMMA and D_{AB} for PMMA matrix blend are lower than D of PS, and D_{AB} for PS matrix blend respectively.

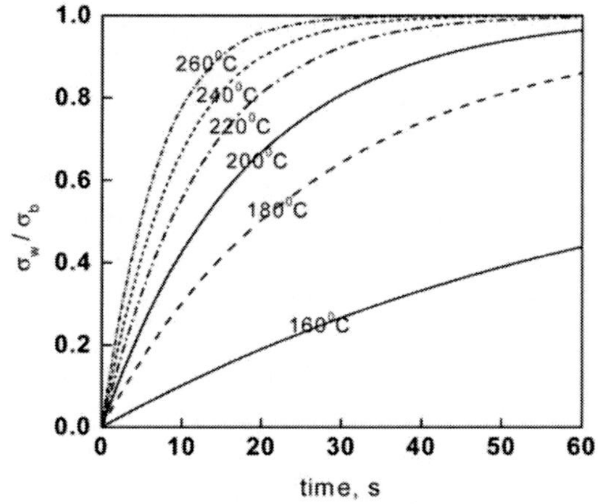

Figure 9. Model predictions for weld-line strength of PS/PMMA (80/20) as a function of temperature and contact time

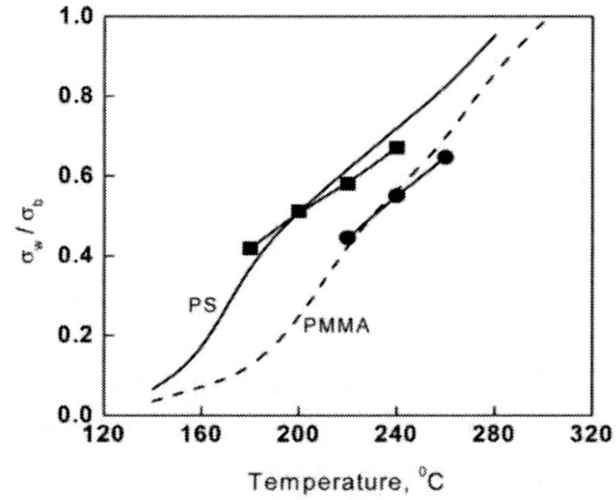

Figure 10. Comparison of model predictions for weld-line strength of PS and PMMA with their corresponding experimental data

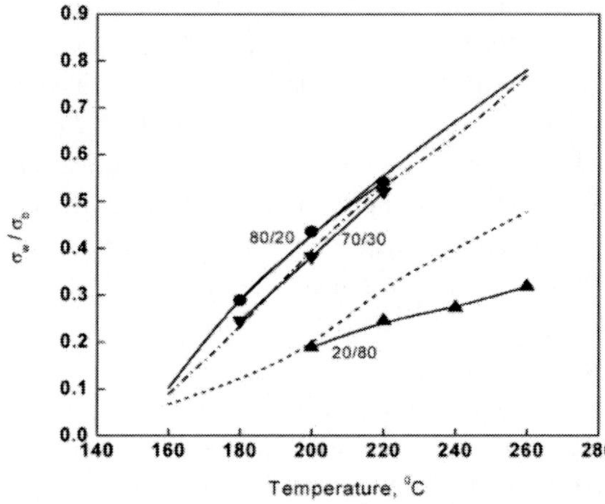

Figure 11. Comparison of model prediction for weld-line strength of PS/PMMA (80/20, 70/30, 20/80) blends with experimental data

As shown in Figure 10 and 11, for PS, PMMA, PS/PMMA (80/20), PS/PMMA (70/30), the predicted σ_w / σ_b is found in good agreement with experimental results (Figures 12-14). However, for PMMA/PS (80/20), PMMA/PS (70/30), the predicted σ_w / σ_b is much higher than the experimental results. Since, in our model for polymer blends, we consider three kinds of diffusion: PMMA - PMMA, PS - PS, PMMA - PS which all contribute to weldline strength. But in fact, there is only one kind of diffusion, matrix PMMA self-diffusion for PMMA/PS (80/20, 70/30). The dispersed PS phase highly oriented along the weldline inhibits the diffusion of PMMA on weldline. Then, we can conclude that the weldline strength for amorphous homopolymers and their blends, and where the viscosity of the matrix lower than that of dispersed phase, can be modeled by Fick's diffusion Law. However, the use of the model taking into account just the coefficient diffusion, as well as the operating parameters, e.g., melt temperature and contact time overestimate the orientation and size of dispersed phase at weldline. The weldline strength is not only attributed to diffusion phenomenon, as well the bonding area. Therefore, it is important to consider the orientation effect at weldline. This might that for PMMA/PS blend, the predicted σ_w / σ_b goes away from experimental results (Figure 11).

3.2. Processing parameters

A number of studies have focused on the effects of processing parameters like melt temperature, holding pressure, holding time, injection velocity and mold temperature on weldline structure and properties of thermoplastics [4, 6, 8, 25, 29, 36, 37, 40, 51-58]. The results showed that weldline strength could be greatly improved by selection of suitable injection molding processing parameters. Melt temperature, injection velocity, holding pressure, holding time and mold temperature will all affect weldline strength to various degrees depending on materials. In general, high melt temperature, high hold pressure and high mold temperature can produce high weldline strength. The melt temperature is the most important parameter for influencing weldline strength. It is obvious that the processing parameters beneficial to molecular diffusion across weldline can improve weldline strength.

For polymer blends, with the effects of processing conditions, the morphology of the dispersed domain and the interaction between the dispersed phase and the matrix should be considered. With a higher injection flow rate and at a lower injection temperature, the force acting on the dispersed phase became larger and the particles became more deformed and were easier to break up. The lager injection rate results in a larger size of the dispersed particles caused by phase coalescence [15, 17, 19]. It is well known that the rheological properties of the blend components and the compatibilization, which might be influenced by the injection processing parameters, are intrinsic factors that control the morphology of the dispersed phase at the weldline [1, 14, 15, 17, 20, 59-61].

3.3. Innovative techniques

In recent years, some novel techniques for increasing the weldline strength were reported. Nichaeli and Galuschka [62] used a push-pull injection molding process to increase the weldline strength. The equipment is made up of two injection units which are separately controlled. The cavity is filled by both units via two separate gates. Similarly, Malloy et al. [63] used a multi live feed injection molding process to drive polymer flow across the weldline, resulting in the increase of the weldline strength. Brian K. Pour et al. [64] reported a mold modification to induce transient flow across the weldline. This mold modification is consisted of the addition of a relief tab located near the weldline. As the molten plastic resin is injected into the mold cavity, the material flows past the opening of the relief tab and forms a thin skin over that opening. Once the pressure increases in the mold cavity, the skin is ruptured and the material is forced to flow into the relief tab, which causes the molten polymer to flow across the weldline. In addition, Garder and Cross [65] were able to generate weldline diffusion by oscillating ejector pin in the vicinity of the weldline. Another approach reported by Gardner and Malloy [66] consists of an in-mold moving boundary system that promotes local mixing in the weldline region during mold filling. The process utilizes one or more cam-operated reciprocating pins to promote lateral displacement of the melt during mold filling. D. J. Dooley [67] used a flow diverter, which was machined in the tensile bar mold to split the flow, to achieve flow across the weldline. The basic idea behind the development of those novel techniques is how to influence the weld line topography to improve weld line strength.

Ultrasonic oscillations were induced to injection molding to improve molecular diffusion across weld line so as to enhance the weld line strength. The mechanism of ultrasonic improvement of weld line strength of PS and PS/HDPE blend was studied. The effect of ultrasonic oscillations induced modes on weld line strength and morphology was also studied.

The injection mold used in this work was a special ultrasonic oscillations mold described in Figure 12. The maximum output power and frequency of the ultrasonic probe are 360W and 15KHz respectively. The oscillations of ultrasonic probe in the fluid can generate the cavitation. Cavitation in the ultrasonic field implies nucleation growth and subsequent claps of bubbles or cavities, resulting in violent shock waves a high pressure of ~1000 bar. The violent shock waves can cause the oscillations of mold through fluid medium translated. During injection molding, the ultrasonic oscillations induced modes are as follows: modes I: Ultrasonic oscillations were induced into mold at the whole process of injection molding; modes II: Ultrasonic oscillations were induced into mold after injection mold filling.

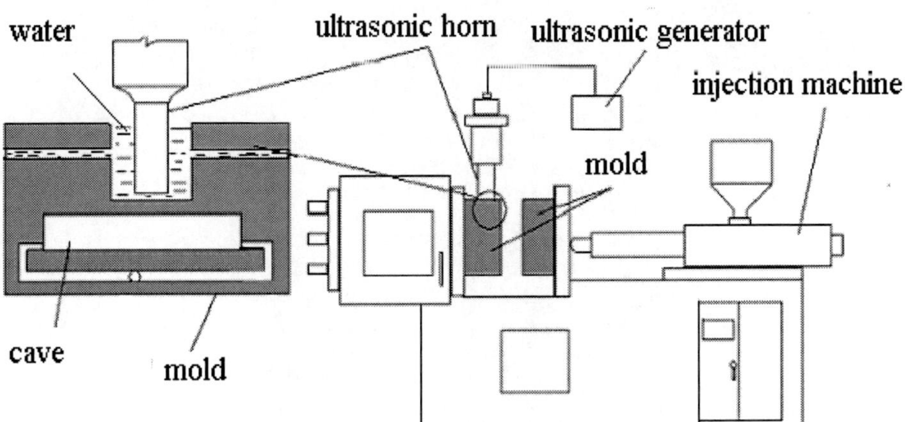

Figure 12. Schematic diagram of injection and mold, equipped with ultrasonic oscillation system

Neat Polystyrene

As shown in Figure 13, melt temperature of PS and ultrasonic oscillations induced modes have great influence on weld line strength of PS. The increase of melt temperature of PS is benefited to the increase of weld line strength of PS. The presence of ultrasonic oscillations can enhance the weld line strength of PS. Compared with mode, the induced ultrasonic oscillations as mode II can greatly enhance the weld line strength of PS.

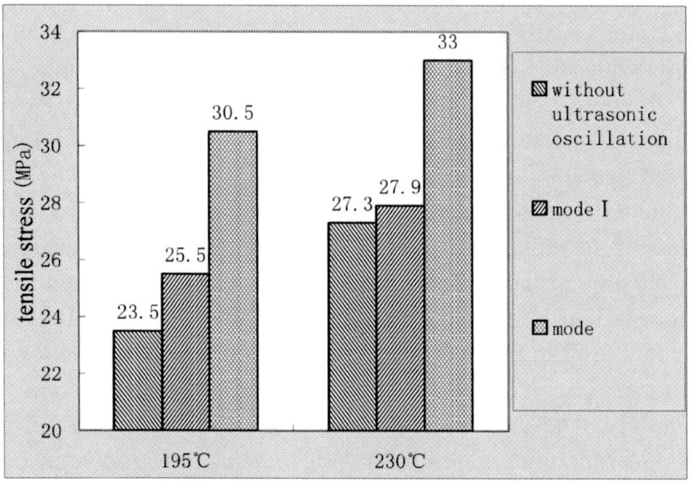

Figure 13. Tensile stress of PS with weld line under different process conditions

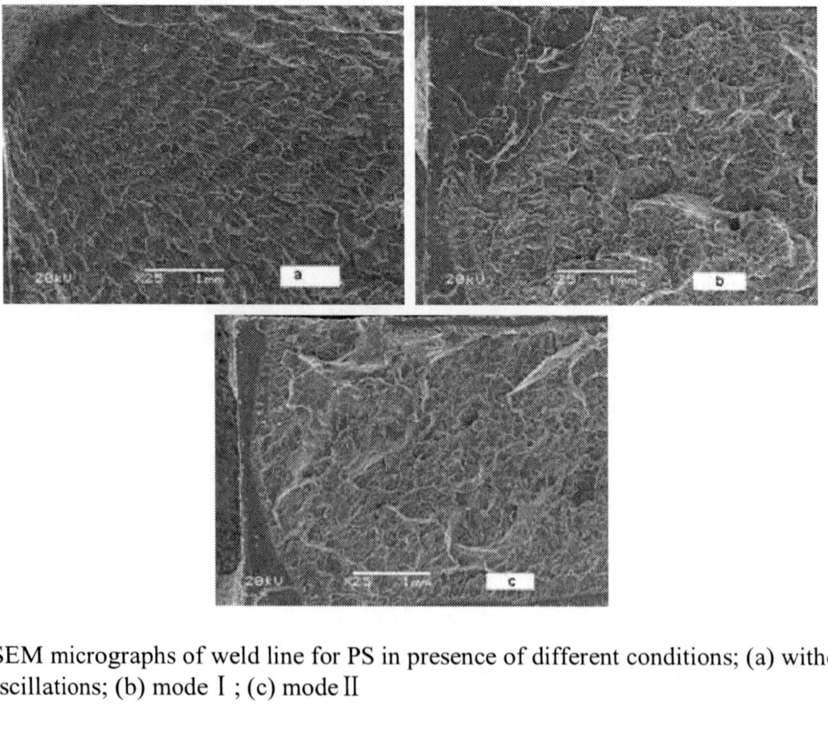

Figure 14. SEM micrographs of weld line for PS in presence of different conditions; (a) without ultrasonic oscillations; (b) mode I ; (c) mode II

In order to study the mechanism for ultrasonic improvement of weld line strength of PS, the tensile fractured surface of injection molded PS parts with weld line were observed through SEM. As shown in Figure 14, compared with the fracture surface of the specimens in the absence of ultrasonic oscillations, the presence of ultrasonic oscillation causes a rougher elongation fractured surface. In order to show the difference of weld line morphology, two regions (Ⅰ) and (Ⅱ) for weld line specimens were classified: region (Ⅰ) near the cavity, called skin; region (Ⅱ) near the center of specimens, called core, (Figure 15, and magnified views were taken, (Figure 16, 17). As shown in Figure 16, in the presence of ultrasonic oscillations the elongation fractured surface at the core is rougher than that of specimens in the absence of ultrasonic oscillations, which implies that the ultrasonic oscillations promote molecular diffusion across the interface in the weld line region, resulting in the improvement of weld line strength of PS. As shown in Figure 17, at the skin the fracture surface is smoother than that at the core, which implies that in weld line region the intermolecular diffusion at the skin is weaker than that at the core due to the shorter time diffusion of the melt molecules at the skin. It can also be observed that the induced ultrasonic oscillations cause a broader skin. The reason may be that in weld line region intermolecular diffusion at the skin is poor, resulting in the poor bond strength of melts at interface. Although the ultrasonic oscillations can promote molecular diffusion across the interface in the weld line region, the bond strength of melts at the skin is too poor to bear the oscillations, resulting in the separation of melts at interface, which causes a wider skin than that in the absence of ultrasonic oscillations. It can also be observed that the induced ultrasonic oscillations as modeⅡ cause the decrease of the width of the skin, compared with mode Ⅰ. The difference between mode Ⅰ and modeⅡ is that for mode Ⅰ the melts at the skin bear the ultrasonic oscillations before the melt fronts meet. But for modeⅡ, the melt fronts meet before the ultrasonic oscillations are induced, which means that the molecular chains at weld line can get primary diffusion before ultrasonic oscillations are induced, resulting in the stronger bond strength of melt at the skin to resist the ultrasonic oscillations than that of mode Ⅰ. That is why the induced ultrasonic oscillations as modeⅡ can decrease the width of the skin and accordingly enhance the weld line strength of PS, Compared with mode Ⅰ. The increase of melt temperature of PS is beneficial to the increase of weld line strength of PS ascribed to have the longer time to promote molecular diffusion across the interface at high melt temperature.

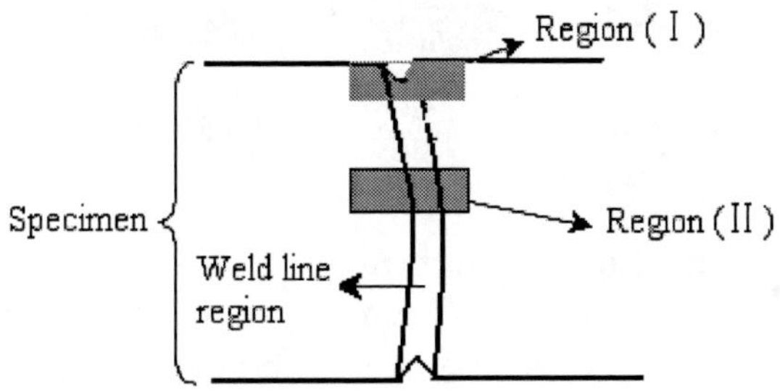

Figure 15. Schematic diagram of the section of specimen with weld line classified two regions.

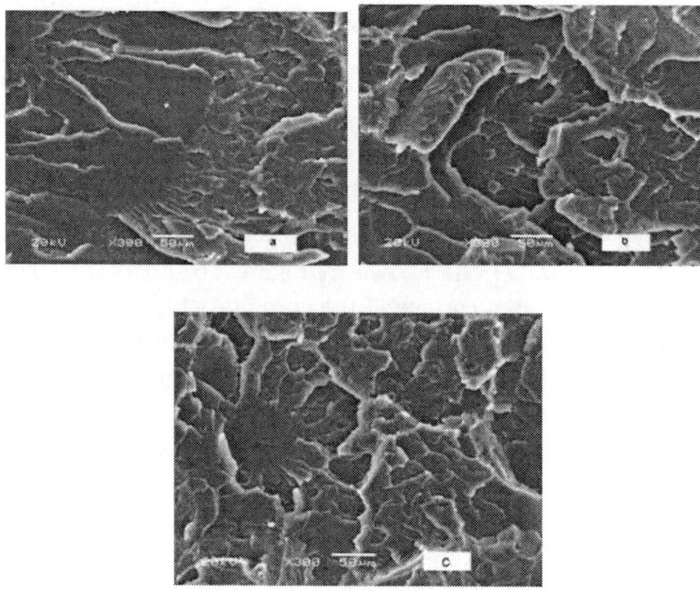

Figure 16. SEM micrographs of core for PS in presence of different conditions; (a) without ultrasonic oscillations; (b) mode I ; (c) mode II

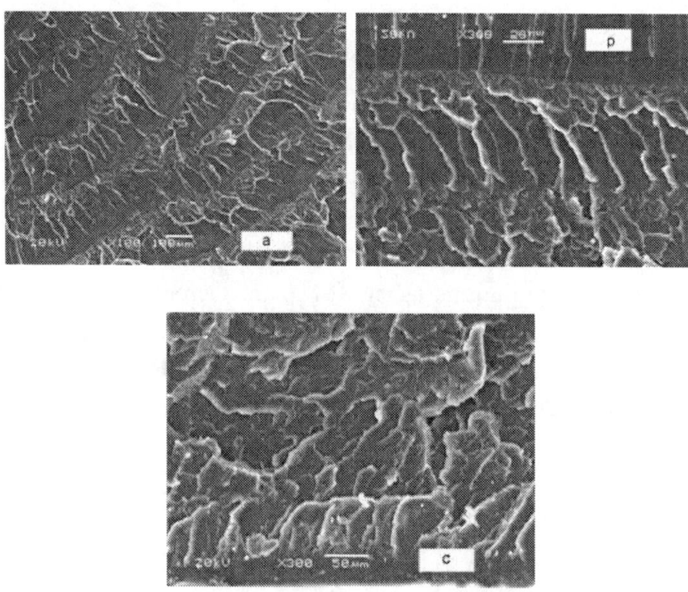

Figure 17. SEM micrographs of sub-skin for PS in presence of different conditions; (a) without ultrasonic oscillations; (b) mode I ; (c) mode II

So the mechanism for ultrasonic improvement of weld line strength of PS can be primary concluded as follows: ultrasonic oscillations can improve the molecular diffusion across weld line at the core and make against the fusion of melt at the skin.

In order to identify that ultrasonic oscillations can improve the molecular diffusion across weld line, testing of molecular diffusion in the presence of ultrasonic oscillations was taken. As shown in Figure 18, for the sample in the absence of ultrasonic oscillations the length of the black segment is almost the same as before. But in the presence of ultrasonic oscillations the black segment is longer obviously than that of in the absence of ultrasonic oscillations. It is obvious that the presence of ultrasonic oscillations promotes the molecular diffusion, causing the blend between PS and PS blended with black pigment, resulting in the elongation of the black segment.

Although the test of molecular diffusion proved that ultrasonic oscillations could promote the molecular diffusion, the test was taken at high mold temperature (180°C) and long time induced ultrasonic oscillations (10min). It was well known that at the process of injection molding the mold temperature is low and the cooling time of the melt in the mold is short. Then the time that the melts bear the ultrasonic oscillations is far less than 10min. So the test of molecular diffusion can't prove whether the presence of ultrasonic oscillations can promote the molecular diffusion or not at the process of injection molding. In order to investigate the effect of ultrasonic oscillations on molecular diffusion at the process of injection molding, positron annihilation lifetime and Infrared dichroism techniques were taken.

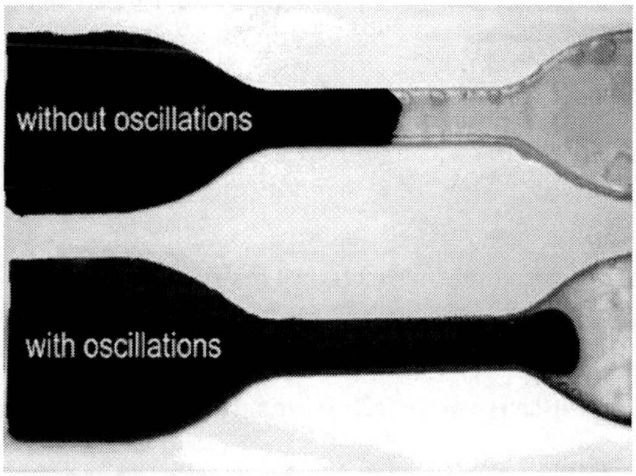

Figure 18. Effect of ultrasonic oscillations on inter-diffusion of PS.

Table. 4. The effect of ultrasonic oscillations on free volume of PS.

Sample	The longest lifetime τ_3 (ps)	Content of τ_3 (%)
Without ultrasonic oscillations		
Skin	2003.6	36.4
Core	2007.1	28.8
With ultrasonic oscillations		
Skin	1997.7	38.2
Core	1989.7	38.5

Positron annihilation lifetime technique is utilized as an effective probe to estimate the free volume size and content since the lifetime (τ_3 , in ~ nm) and intensity of σ-Ps are proved to be well correlated with the average free volume radius and content of free volumes. As shown in Table 4, although in the presence of ultrasonic oscillations the lifetime is not increased the intensity of σ -Ps of the skin and core is increased, indicating that induced ultrasonic oscillations promote the movement of molecular at the skin and core at the process of injection molding, resulting in the increase of content of free volumes.

For the single gate injection molded parts, molecular chain is orientated along flow direction. If ultrasonic oscillations could promote the molecular movement at the process of injection molding, then in the presence of ultrasonic oscillations the orientation of molecular chain is decreased because the direction of induced ultrasonic oscillations is perpendicularity to the direction of orientation of molecular chain. Infrared dichroism is utilized as an effective technique to estimate the orientation of molecular chain of PS. the Dichroic radio measurements allow one to calculate the second-order moment of the orientation function according to the relation:

$$\langle p_2(\cos\theta)\rangle_{av} = \frac{1}{2}\left(3\langle\cos^2\theta\rangle_{av} - 1\right) = \frac{R-1}{R+2}\frac{R_0+2}{R_0-1}$$

with $R_0 = 2\cos t^2\alpha$ where α is the angle between the dipole moment vector of the considered vibration and the chain axis and θ is the angle between the chain axis and absorption bands to measure orientation in PS has been reported. [68] The 1028cm^{-1} and 2850cm^{-1} absorption bands were used. As shown in Table 5, no matter at the surface or at the core (0.2 or 1mm), in presence of ultrasonic oscillations the orientation function is decreased, compared with the sample without ultrasonic oscillations, indicating that ultrasonic oscillations can promote the movement of chain, resulting in the decrease of orientation of molecular chains along the flow direction.

Table 5. The effect of ultrasonic oscillations of orientation fuction $\langle p_2 \cos\theta\rangle$

Sample code	Wave numler(cm^{-1})	α (^{0})	$A_\perp$	$A_{//}$	R($A_{//}$/ $A_\perp$)	$\langle p_2 \cos\theta\rangle$
W-0	1028	32	0.356	0.361	1.014	0.037
	2850	34	0.59	0.588	0.997	-0.01
U-0	1028	32	0.398	0.393	0.987	-0.033
	2850	34	0.585	0.604	1.032	0.096
W-0.2	1028	32	0.372	0.385	1.035	0.09
	2850	34	0.55	0.572	1.04	0.119
U-0.2	1028	32	0.231	0.231	1	0
	2850	34	0.147	0.151	1.027	0.081
W-1	1028	32	0.293	0.317	1.082	0.208
	2850	34	0.467	0.482	1.032	0.095
U-1	1028	32	0.555	0.591	1.065	0.166
	2850	34	0.883	0.910	1.031	0.091

To show the effect of ultrasonic oscillations on skin layer, the micrographs of surface of sample PS were taken (Figure 19). The mold temperature is 90°C. As shown in Figure 19, for the sample without ultrasonic oscillations, the weld line at the surface is not obvious (Figure 19 a). But for the samples induced ultrasonic oscillations, the weld line at the surface can be observed obviously (Figure 19 b, c), indicating that ultrasonic oscillations make against the fusion of melt at the skin. And compared with mode II, induced ultrasonic oscillations as mode I causes the weld line more obvious, indicating that induced ultrasonic oscillations as mode I is more disadvantageous to the fusion of the melt at the skin than that of mode II, which is consistent with the result that was observed by SEM.

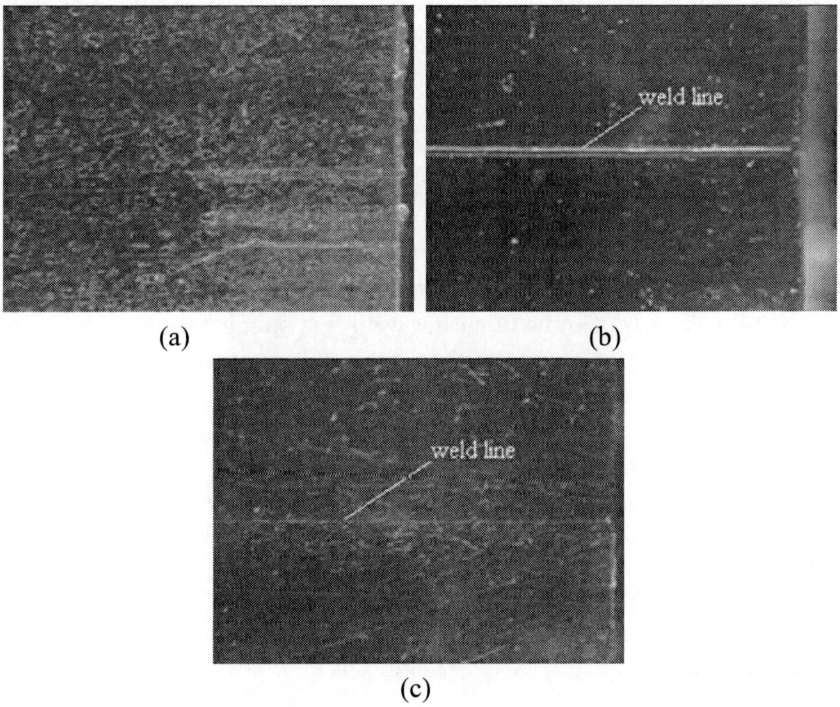

Figure 19. Effect of ultrasonic oscillations on surface morphology of PS with weld line; (a) without ultrasonic oscillations; (b) mode I; (c) mode II.

Figure 20 shows the relationship between the tensile strength of PS and the depth of cut at weld line region. Results show that for the specimens without ultrasonic oscillations the removing of the weld line skin is benefited to the increase of weld line strength. The weld line strength of the specimens with ultrasonic oscillations is higher than that without ultrasonic oscillations no matter how depth of cut is, indicating that ultrasonic oscillations can promote the molecular diffusion across weld line at the core, resulting in the increase of weld line strength. Compared with mode II, Induced ultrasonic oscillations as mode I is more benefit to the increase of weld line strength when the skin of weld line is removed, which is reverse to the testing results of specimens that the skin of weld line is not removed. It had been proved that ultrasonic oscillations can promote the molecular diffusion of the melts at the core and make against the fusion of the melts at the skin. For the specimens induced ultrasonic oscillations the remove of the skins means that the harm of ultrasonic oscillations to the weld line strength is cleared away. So for the specimens induced ultrasonic oscillations that the surface was removed, the more beneficial of the induced modes to promote the molecular diffusion, the more the increase of weld line strength. For induced ultrasonic oscillations as mode I the time that the melts at the core bear the ultrasonic oscillations is longer than that of mode II, resulting in the diffusion of melt more than that of mode II. That is the reason why induced ultrasonic oscillations as mode I is more beneficial to the increase of weld line strength than that of mode II when the skins of weld line were removed.

PS/HDPE Blends

As shown in Figure 21, compared with PS, melt temperature of PS/HDPE blends have great influence on weld line strength of PS/HDPE blends. In order to study the reason that the increase of melt temperature can greatly enhance the weld line strength of PS/HDPE blends, the tensile fractured surface of injection molded PS/HDPE blends parts with weld line were observed through SEM. As shown in Figure 22 a, the whole fracture surface is smooth when melt temperature is 195°C. But when melt temperature is 230°C the fracture surface is smooth at the skin and rough at the core, (Figure 22 b). And the fracture site of two specimens is different. At low melt temperature (195°C) the fracture path typically follows the weld line, resulting in the oxbow shape of the fracture surface. But at high melt temperature (230°C) the fracture path at the skin follows the weld line and deviates the weld line at the core. These can be explained as follows: It was known that the weld line strength would depend on whether the polymer chains have enough time to diffuse across the interface to form a strong bond and the bond area of the weld line. So the increase of melt temperature is benefited to the increase of interdiffusion of polymer at the weld line. Compared with straight line, oxbow of the weld line is also benefited to the increase of weld line strength. When the melt temperature is 230°C, the polymer chains at core form a strong bond so that the tensile fracture deviate from weld line. So the tensile strength gets improved. At the skin the fracture path follows the weld line due to the quick solidification of polymer melts that result in the poor bond. When the melt temperature is 195°C, the polymer chains don't have enough time to form a strong bond, resulting in the fracture path following the weld line.

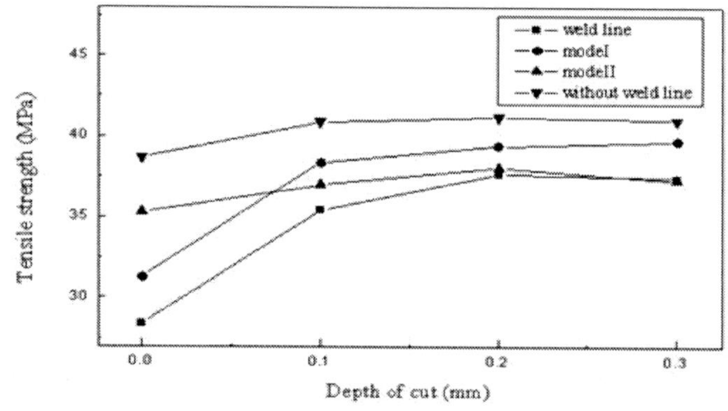

Figure 20. Relationship between tensile strength of injection molded PS parts and depth of cut under different process conditions

Compared with mode I, the induced ultrasonic oscillations as mode II can greatly enhance the weld line strength of PS/HDPE blends, which follows the same rule as that of PS.

In order to investigate the mechanism for ultrasonic improvement of weld line strength of PS/HDPE blend, the SEM micrographs of weld line for PS/HDPE blend were taken. As shown in Figure 23, the dispersed domains are spherical shape and the size of dispersed domains is about 1 μm in weld line, which is the same as that in the bulk, indicating that ultrasonic oscillations can't change the morphology of dispersed phase greatly in weld line region. So in the presence of ultrasonic oscillations the increase of weld line strength of PS/HDPE blend is not due to the change of dispersed phase morphology.

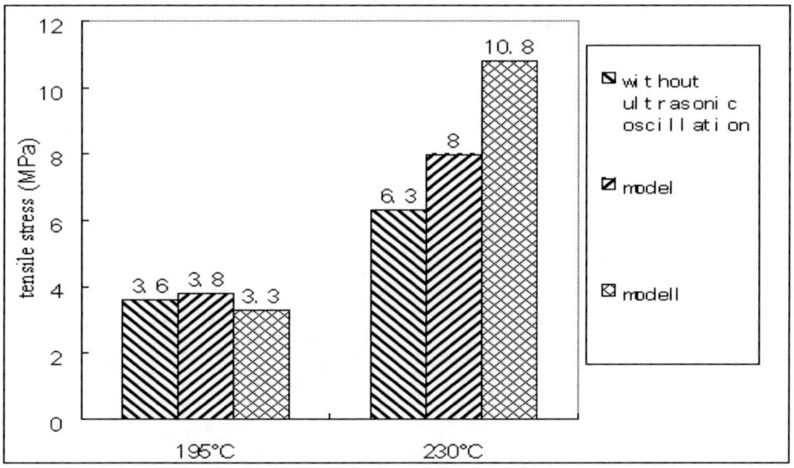

Figure 21. Tensile stress of PS/HDPE blends (90/10) with weld line under different process conditions

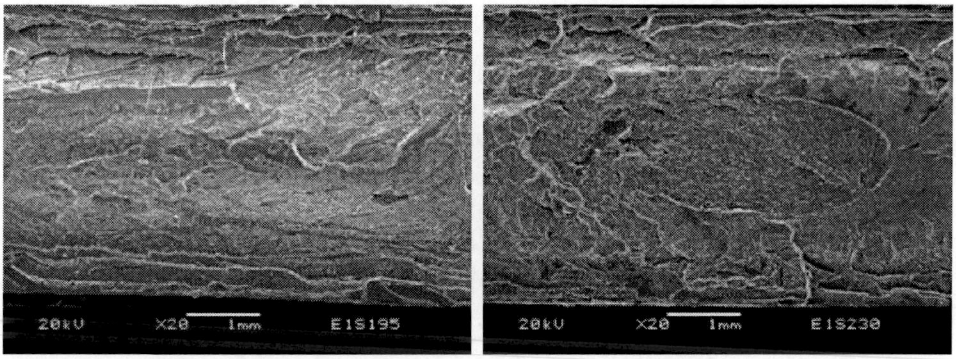

(a) (b)

Figure 22. SEM micrographs of fracture specimens for PS/HDPE blends at different melt temperatures; (a) 195°C; (b) 230°C

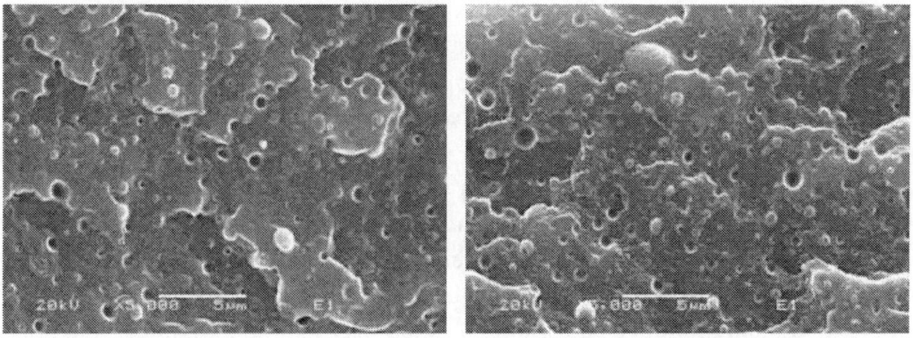

(a) (b)

Figure 23. SEM micrographs of blend PS/HDPE (90/10); (a) weld line; (b) bulk

Figure 24 shows the effect of ultrasonic oscillations and induced modes on the weld line surface of PS/HDPE blends. As shown in Figure 24, compared with the surface of the specimens without ultrasonic oscillations, in presence of ultrasonic oscillations the weld line on the surface was obvious, indicating that ultrasonic oscillations make against the fusion of melt at the skin, which is consistent with the result of PS. Induced ultrasonic oscillations as mode I cause the weld line more obvious than that of mode II, indicating that induced ultrasonic oscillations as mode I is more disadvantageous to the fusion of the melt at the skin than that of mode II.

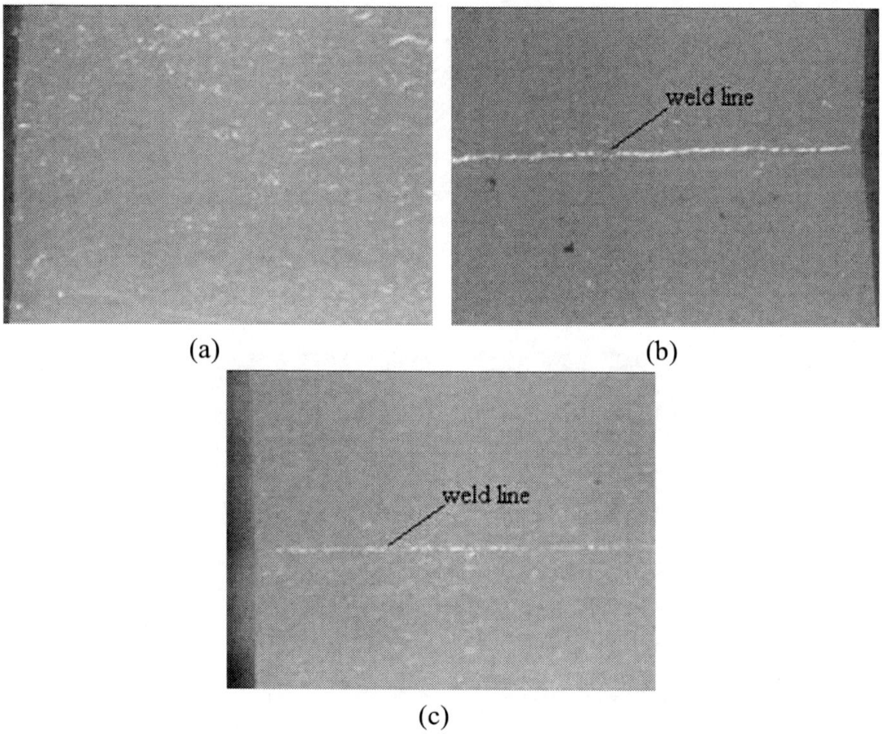

Figure 24. Effect of ultrasonic oscillations on surface morphology of PS/HDPE with weld line; (a) without ultrasonic oscillations; (b) mode I; (c) mode II.

The weld line skins of PS/HDPE blend were also removed different depths to study the mechanism for ultrasonic improvement of weld line strength. As shown in Figure 25, for the specimens without ultrasonic oscillations the removing of the skins is beneficial to the increase of weld line strength. The more the depth is cut, the stronger the weld line strength is. The presence of ultrasonic oscillations is beneficial to the improvement of weld line strength no matter how the depth of surface is cut, indicating that ultrasonic oscillations can promote the molecular diffusion across weld line at the core, resulting in the increase of weld line strength. Compared with mode I, Induced ultrasonic oscillations as mode II is more beneficial to the increase of weld line strength as the depths of cut is less than 0.4mm. But as the depth of cut is 0.8mm, induced ultrasonic oscillations as mode I are more beneficial to the increase of weld line strength than that of mode II, which follows the same rule and similarly explains as that of PS discussed above.

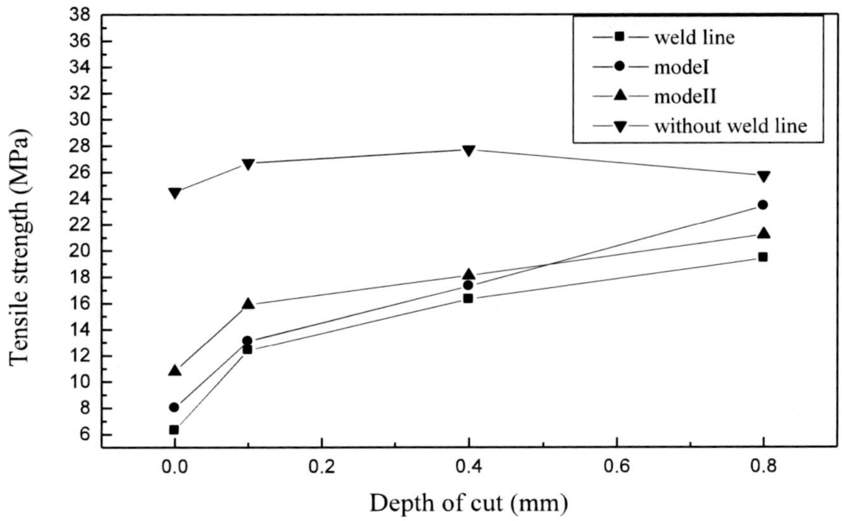

Figure 25. Relationship between tensile strength of injection molded PS/HDPE blends parts and depth of cut under different process conditions

So the mechanism for ultrasonic improvement of weld line strength of PS/HDPE blend can be concluded as follows: ultrasonic oscillations can improve the molecular diffusion across weld line at the core, which is beneficial to the increase of weld line strength, and make against the fusion of melt at the skin.

REFERENCES

[1] Fellahi, S; Meddad, A; Fisa, B; Favis. BD. Adv. *Polym. Technol.*, 1995, 14, 169.
[2] Bozarth, MJ; Hamill, JL. SPE ANTEC *Tech. Papers*, 1984, 30, 1091.
[3] Bangert, Kunstst. Ger. *Plast.*, 1985, 15, 325.
[4] Selden, R. *Polym. Eng. Sci.*, 1997, 37, 205.
[5] Mielewski, DF; Bauer, DR; Schmitz, PJ; Oene, HV. *Polym. Eng. Sci.*, 1998, 38, 2020.
[6] Tomari, K; Tonogal, S; Harada, T; Hamada, H; Lee, K; Morii, T; Mcakawa, Z. *Polym. Eng. Sci.*, 1990, 30, 931.
[7] Hobbs, SY. *Polym. Eng.,* 1974, 14, 621.
[8] Fisa, B; Rahmani, M. *Polym. Eng. Sci.*, 1991, 31, 1330.
[9] Murfitt, PS. *Fillers Conference*, London, UK; 1986.
[10] Savadori, A ; Pelliconi, R. Plast. *Rubber Process. Appl.*, 1983, 3, 215.
[11] Fisa, B; Rahmani, M. Polym. *Eng. Sci.*, 1997, 37, 1330.
[12] Thamm, RC. Rubber Chem. *Technol.*, 1977, 50, 24.
[13] Malguarnera, SC; Riggs, DC. *Polym. Plast. Tech. Eng.*, 1981, 17, 193.
[14] Mekhilef, N; Ait-Kadi, A; Ajji, A. *Polymer*, 1995, 36, 2033.
[15] Brahimi, B; Ait-Kadi, A; Ajji, A. *Polym. Eng. Sci.*, 1994, 34, 1202.
[16] Fisa, B ; Favis, BD; Bourgeois. S. *Polym. Eng. Sci.*, 1990, 27, 241.
[17] Mekhilef, N; Ait-Kadi, A; Ajji. A. *Adv. Polym. Technol.*, 1995, 14, 315.
[18] Guo, SY; Ait-Kadi, A. *J. Appl. Polym. Sci.*, 2002, 84, 1856.

[19] Son, Y; Ahn, KH; Char, K. *Polym. Eng. Sci.*, 2001, 41, 554.
[20] Jarus, D; Summers, JW; Hiltner, A; Baer, E. *Polymer*, 2000, 41, 3057.
[21] Kim, JK; Park, SH; Jeon, HK. *Polymer*, 2001, 42, 2209.
[22] Fellahi, S; Favis, BD; Fisa, B. *Polymer*, 1996, 37, 2615.
[23] Ramjumar, DHS; Bhattacharya, B; Vaidya, UR. *Eur. Polym. J.*, 1996, 32, 999.
[24] Ramjumar, DHS; Bhattacharya, B; Vaidya, UR. *Eur. Polym. J.*, 1997, 33, 729.
[25] Kim, SG; Suh, NP. *Polym. Eng. Sci.*, 1986, 26, 1200.
[26] Pecorini, J; Seo, KS. *Plast. Eng.*, 1996, 52, 31.
[27] Hamada, H; Tomari, K; Yamane, H; Senba, T; Hiragushi, M. *SPE ANTEC Tech.* Papers, 1997, 1, 1071.
[28] Flory, PJ. *"Principles of Polymer Chemistry"*, Cornell University Press, Ithaca, NY, 1978.
[29] Mosle, HG; Criens, RM; Dirk, H. *SPE ANTEC Tech.*, Papers, 1984, 30, 772.
[30] Tjader, T; Seppala, J; Jaaskelainen, P. *J. Materials Sci.*, 1998, 33, 923.
[31] Wenig, W; Stolzenberger. C. *J. Materials Sci.*, 1996, 31, 2487.
[32] Criens, RM; Mosle, HG. in *Failure of Plastics,* W; Brostow, RD. Corneliussen, eds., Hanser Verlag, Munich, 1986.
[33] Bucknall, CB. *Pure Appl. Chem.,* 1986, 58, 985.
[34] Kamel, MR; Moy. FH. *J. Appl. Polym. Sci.*, 1983, 28, 1787.
[35] Katti, SS; Schultz. JM. *Polym. Eng. Sci.*, 1980, 22, 197.
[36] Malguarnera, SC. *Polym. Plast. Tech. Eng.*, 1982, 18, 1.
[37] Malguarnera, SC; Manisali, AT; Riggs, DC. *Polym. Eng. Sci.*, 1981, 25, 1149.
[38] Singh, D; Mosle, HG; Kunz, M; Wenig, W. *J. Mater. Sci.*, 1990, 25, 4704.
[39] Wenig, W; Singh, D; Mosle. HG. Angew. *Makromol. Chem.*, 1990, 179, 35.
[40] Singh, D; Mosle. HG. *Makromol. Chem. Macrom. Symp.*, 1988, 20/21, 489.
[41] Bell, RG. Cook. CD. *Plast. Eng.*, 1979, 8, 18.
[42] Wendt, U. Kunstst. Ger. *Plast.*, 1988, 78, 123.
[43] Ersoy, GO. Nugay, N *Polymer*, 2004, 45, 1243.
[44] Lu, C; Guo, SY; Wen, L; Wang. JY. *Eur. Polym. J.*, 2004, 40, 2565.
[45] Guo, SY ; Ait-Kadi, A ; Bousmina, M. *Polymer*, 2004, 45, 2911.
[46] Bueche, F; Cashin, WM; Debye. P. *J. Chem. Phys.*, 1952, 20, 1956.
[47] Fox, TG; Allen. VR. *J. Chem. Phys.*, 1964, 41,344.
[48] Brochard, F; Jouffroy, J; Levinson, P. *Macromolecules*, 1983, 16 , 1638.
[49] Wu. S. *J. Polym. Sci. -Phys.*, 1989, 27, 723.
[50] Wu, S. *J. Polym. Sci. -phys.*, 1987, 25, 557.
[51] Tomari, K; Harada, T. *Polym. Eng. Sci.*, 1993, 33, 996.
[52] Piccarolo, S; Salu, M. Plast. *Rubber Process. Appl.*, 1988, 10, 11.
[53] Janickl, SL; Peter, RP. *SPE ANTEC Tech. Papers*, 1991, 37, 391.
[54] Gardner, G; Cross, C. *SPE ANTEC Tech. Paper*, 1992, 38, 2127.
[55] Malguarnera, SC ; Manisali, AI; Riggs, DC. *Polym. Eng. Sci.*, 1981, 21, 17.
[56] Titomanlio, G; Piccarolo, S; Rallis, A. *Polym. Eng. Sci.*, 1989, 29, 4.
[57] Kim, S; Suh, NP. *Polym. Eng. Sci.*, 1986, 26, 17.
[58] Ulcer, Y; Cakmak, M; Hsiung, CM. *J. Appl. Polym. Sci.*, 1995, 55.
[59] Fellahi, S; Fisa, B; Favis, BD. *J. Appl. Polym. Sci.*, 1995, 57, 1319.
[60] Grace, H. *Chem. Eng. Commun.*, 1982, 14, 225.
[61] Shieu, FS; Wang, BH; Wu, JY. *Plast. Rubber Compos. Process. Appl.*, 1997, 26, 230.

[62] Nichaeli, W; Galuschka, S. *SPE ANTEC Tech.*, 1984, 39, 534.
[63] Malloy, R; Gardner, G; Grossman, E. *SPE ANTEC Tech. Proc.*, 1993, 51, 521.
[64] Pour, BK; Sinner, TD. *ANTEC*, 1995, 495.
[65] Garder, JR; Cross, G. *Plast. Eng.*, 1993, 49, 29.
[66] Gardner, G; Malloy, R. *SPE ANTEC Tech. Papers*, 1994, 40, 626.
[67] Dooley, DJ. *ANTEC*, 1993, 414.
[68] Jasse, B; Koenig, JL. *J. Polym. Sci., Polym. Phys.*, 1979, 117, 799.

In: Injection Molding: Process, Design, and Applications
Editor: Phoebe H. Kauffer
ISBN: 978-1-61761-307-4

Chapter 5

MIM OF CO ALLOY FOR BIOMEDICAL APPLICATIONS

P. V. Muterlle[1*], M. Perina[2] and A. Molinari[1]
[1]University of Trento, Italy
[2]MIMEST spa, Pergine Valsugana, Trento, Italy

ABSTRACT

The cobalt alloys are used for surgical implant applications and their properties are influenced by the carbon content. They can be produced with different carbon contents. The increase in the carbon content results in an increase in hardness but also in a corresponding decrease in ductility. It is then important to find a proper combination of the carbon content and the processing route (sintering plus any post sintering treatments) to optimize mechanical properties. The strain hardening behavior of the material was underlined and justified by strain induced transformation of austenite into martensite. In addition, the role of carbides on both wear and corrosion resistance was unknown.

Keywords: Co29Cr6Mo alloys, mechanical properties, MIM, wear and corrosion.

1. INTRODUCTION

The mechanical properties of the Co-Cr-Mo-C castings depend on the carbon content which promotes the precipitation of carbides at the grain boundary and in the interdendritic regions [1, 2]. Carbides increase hardness, yield stress and wear resistance but reduce ductility since they are mainly localized in the grain boundary regions. On the other hand, a high carbon content increases the fluidity of the liquid alloy, improving castability. To modify

* Corresponding author: Dipartimento di Ingegneria dei Materiali e Tecnologia Industriali, Università di Trento, Mesiano 77, 38100, Trento, Italy. Present position - Faculdade de Tecnologia, Departamento de Engenharia Mecânica, Campus Universitário Darci Ribeiro, Universidade de Brasília, 70910-900, Brasília, Brazil. E-mail: pmuterlle@yahoo.com.br

the microstructure with the aim to increase ductility, a solution annealing treatment is carried out, which dissolves the carbide particles almost completely [3, 4]. The improved ductility is, however, accomplished by a decrease of strength. The constitution of the metallic matrix evolves with heat treatment, as well, but the major effect on properties is attributed to the change in the number of carbide particles and their distribution. This treatment can be followed by aging, to further modify the microstructure both precipitating carbides more uniformly and modifying the matrix constitution through the transformation of the austenite in martensite [5-7]. The effect of the microstructure on the mechanical properties has been studied by J. B. Vander Sarde et al. [8].

Biomedical prosthesis are mainly produced by either investment casting or hot forming, but recently metal injection molding (MIM) has been proposed as a cost-effective technological alternative [9, 10]. The Co-Cr-Mo-C alloys can be produced by MIM of prealloyed powders. The powders are mixed with a binder (around 50 vol. %) to produce a feedstock having the proper rheological properties to be injected at a moderate temperature and pressure in a rigid die and to keep the shape once the injection pressure has been removed [11]. After debinding, parts are highly porous and are sintered at high temperature (1300-1380°C) to obtain a full density. Powders with different carbon contents, between 0.05% and 0.35%, are commercially available.

MIM produces an as sintered microstructure largely different from that of cast alloys of similar composition. In particular, the as cast microstructure is quite coarse, with both dendritic and interdendritic carbides resulting from micro-segregation on solidification (coring) [12]. The as sintered microstructure, instead, contains quite large (several tenths of micrometers) equiaxed grains, with large eutectic cells and/or intergranular carbides, which amount depends on the carbon content. Even in the case of a similar chemical composition, the transformations during heat treatment could be influenced by the different distribution and composition of carbides.

Because of the negative effect of carbide, a carbon free alloy (0.05%C) is expected to possess a higher ductility than the higher carbon content (0.3-0.35% C), even if it needs for a higher sintering temperature to get full densification. Indeed, densification of the high carbon alloy is enhanced by the liquid phase resulting from two eutectic reactions [1]. These reactions do not occur in the carbon free alloy, then supersolidus sintering conditions must be imposed to get full density [13].

In the present work, alloys with a carbon content between 0.05%C and 0.35%C were studied. The microstructures were modified by heat treatment (solution annealing and aging) to study the phase transformations and their effect on mechanical properties. Tensile tests were carried out in all the conditions investigated, and the results were interpreted on the basis of the microstructural characteristics and on the deformation behaviour deduced by the analysis of the fracture surfaces. The effect of the microstructure on wear and corrosion resistance are also studied.

2. Materials and Experimental Procedures

2.1. Production of the Specimens

The specimens were produced by pre-alloyed Co-29Cr-6Mo (F75) powders with different carbon contents, produced by gas-atomization, (80% of particle sizes below 22μm). The powders were blended with a proprietary binder (polyolefin based), to produce the feedstock.

Test bars, according to ASTM E 8M-03 – Standard Flat Unmachined Tension Test Specimen for Powder Metallurgy (P/M) Products - were injection molded and debinded by a two steps process (80% of the total binder content was dissolved in water, while the remaining 20% was removed by thermal debinding in Ar/10%H_2 atmosphere). The brown parts were sintered in a graphitic vacuum furnace (at 10^{-2} bar) (manufactured by TAV Spa, Caravaggio, Italy) under the following conditions: 1 hour holding time, N_2 backfilling, free cooling at 1 bar of nitrogen flux (approximately 15 K/min). The sintered density is higher than 95% of the theoretical one. The materials investigated are listed in table 1.

2.2. Heat Treatments

The sintered materials were solution annealed at different temperatures (1200°C, 1220°C and 1250°C) with 4 hours of isothermal holding, in the same vacuum furnace used for sintering under argon backfilling and then gas quenched in 8 bar of nitrogen flux. After solution annealing, the samples were aged in a tubular furnace at 750°C in argon atmosphere for 3 and 20 hours and then air quenched.

2.3. Chemical Analysis

The carbon content was measured by LECO CS125.

2.4. Porosity and Density

Density was measured by Archimedes' method, according to standard ASTM B 328-96. Measurements were carried out with a precision balance (AdventurerSL - OHAUS) with a sensibility of 0.0001g.

Table 1. Materials and sintering conditions

Materials	Temperature of Sintering
F75 with 0.35%C	*Sintering at 1300°C*
F75 with 0.23%C	Sintering at 1300°C
F75 with 0.23%C	*Sintering at 1350°C*
F75 with 0.05%C	Sintering at 1380°C

2.5. Thermal Analysis

Thermal analyses were carried out by differential scanning calorimetry (DSC - Netzsch STA 409PC - Luxx) to study the transformation on heating of the sintered materials. The analyses were carried out in Ar with a heating rate of 0.33 K/sec and a cooling rate of 0.83 K/sec.

2.6. Metallography

Metallographic characterization was carried out by a LOM microscopy (Leica DC300) and by environmental scanning electron microscopy (ESEM, Philips XL 30) before and after electrolytic etching with 94 ml distilled water, 4.5 ml HNO_3 and 1.5 ml H_2O_2 solution for 3V and 4sec. The Staining method, a chemical color etching with 1 part (20% $KMnO_4$ + 80% water) + 1 part (8% NaOH + 92% water), was carried out after electrolytic etching to distinguish the different types of carbides ($Cr_{27}C_6$ brown, Cr_7C_3 pale yellow to light tan, M_6C red, green, yellow, blue) [2].

Microanalysis was carried out by energy dispersive X-ray spectroscopy (EDXS).

2.7. X-Ray Analyses

X-ray diffractometry (XRD) was used to investigate the cobalt matrix constitution (f.c.c./h.c.p. phases) and the type and amount of carbides. The diffraction patterns were collected using a Cu-Kα source and the experimental data were elaborated with the Rietveld method using the MAUD software ("Materials Analysis Using Diffraction") [14, 15].

2.8. Hardness and Microhardness

The microhardness was measured by the Vickers methods, according to ASTM384, with a MHT-4 machine (Anton Paar), on etched metallographic specimens, with a load of 0.5N (HV0.05).

The hardness was measured by the Vickers method, according to ASTM 18, with an Emco test machine, on etched metallographic specimens, with a load of 300N (HV30).

2.9. Tensile Tests

Tensile tests were carried out in an Instron 8516 SH 100 kN machine, at a strain rate of 0.1 s^{-1} and measuring strain with an axial extensometer with a gauge length of 12.5 mm. The morphology of the fracture surfaces was examined by ESEM.

2.10. Corrosion Tests

Two kinds of corrosion tests were carried out:

the measurement of open-circuit potential according to *ISO 16429 – Implants for surgery*
the potentiodynamic polarization.

The OCP measurements were carried out in a thermostatically controlled cell at 37°C. The electrolyte solutions employed consisted of 0.9% NaCl and as reference electrode a silver/silver chloride electrode (Ag/AgCl 3M) was used. The inert Argon gas was continuously flushed during 72 hours test.

The polarization test were carried out in the same environment and temperature, but no gas was flushed in this case. The electrochemical cell was a three electrode cell using platinum as counter and Ag/AgCl 3M as reference. Since the open circuit potential was very instable at the beginning of immersion, all the samples were immersed for 30 min before starting the potentiodynamic measurements. The potential range was from -0.25 mV vs OCP to 1 V with a scan rate of 0.1 mV/s. Two measurements were performed for each samples and representative curves will be reported.

For both tests the samples were first mechanically polished using SiC grinding paper and polished with 1 μm diamond solution to a mirror finish and then cleaned in alcohol and dried.

2.11. Wear Tests

The tests were performed in a block on disc configuration using an AMSLER A135 testing machine, see figure 1.

Co-29Cr-6Mo alloy and ultrahigh molecular weight polyethylene (UHMWPE) were used as block and disc materials, respectively. Tests were carried out with a sliding speed of 0.105m/s (50 rpm) and 300N applied load, corresponding to 24MPa nominal contact stress, this value is in accord to the range of conformity in total knees replacements and it varies from 10 to 45 MPa [16]. Water was used as lubricant. The total sliding distance was 3 km and the weight loss was measured after each test. The surface roughness of the samples was measured before and after wear tests using a Hommelwerke T8000 machine and the results analyzed with a Turbo Roughness V 2.86" program.

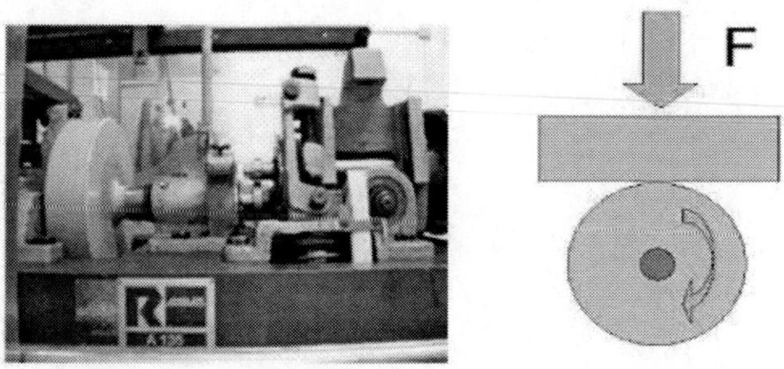

Figure 1. AMSLER A135 testing machine and the block on disc configuration.

3. Results and Discussion

As a preliminary work, the effect of the solution annealing temperature, in the range from 1200°C to 1250°C, on the microstructure, microhardness and hardness of a 0.35%C Co-Cr-Mo alloy produced by MIM of prealloyed powders was investigated. The high carbon material was chosen because it presents the higher quantity of carbides.

3.1. Study of Solution Annealing Temperature

Figure 2a shows the as sintered microstructure of the Co alloy. It contains a large amount of eutectic cells, as well as grain boundary precipitates. Three types of carbides were revealed by the color metallographic etching:

$M_{23}C_6$ is the main constituent of the eutectic cells (pearlite like);
M_6C and M_7C_3 are present in the eutectic cells and along grain boundaries.

The ESEM image (figure 2b) shows details of the morphology of these carbides ($M_{23}C_6$ in the centre, M_6C and M_7C_3 on the right).

In addition, the small and fragmented precipitates at the grain boundary are sigma phase.

The microstructure is quite coarse, the grain size is 115±12 μm, because of the high sintering temperature.

DSC analysis was carried out with the purpose to individuate the eutectic transformations leading to the formation of the liquid phase. Figure 3 shows the DSC curve. Two endothermic peaks are well evident, with the maximum around 1245 °C and 1265 °C, respectively. They correspond to two eutectic reactions involving the eutectic cells containing $M_{23}C_6$ (1245°C) and M_7C_3 carbide (1265°C). M_6C can both precipitate directly from the f.c.c. solid solution and form as a product of the transformation on heating of $M_{23}C_6$ [2, 4].

Three temperatures were selected for the solution annealing treatment: 1200°C, 1220°C e 1250°C. The first two temperatures are lower than the eutectic ones, the third is just in between.

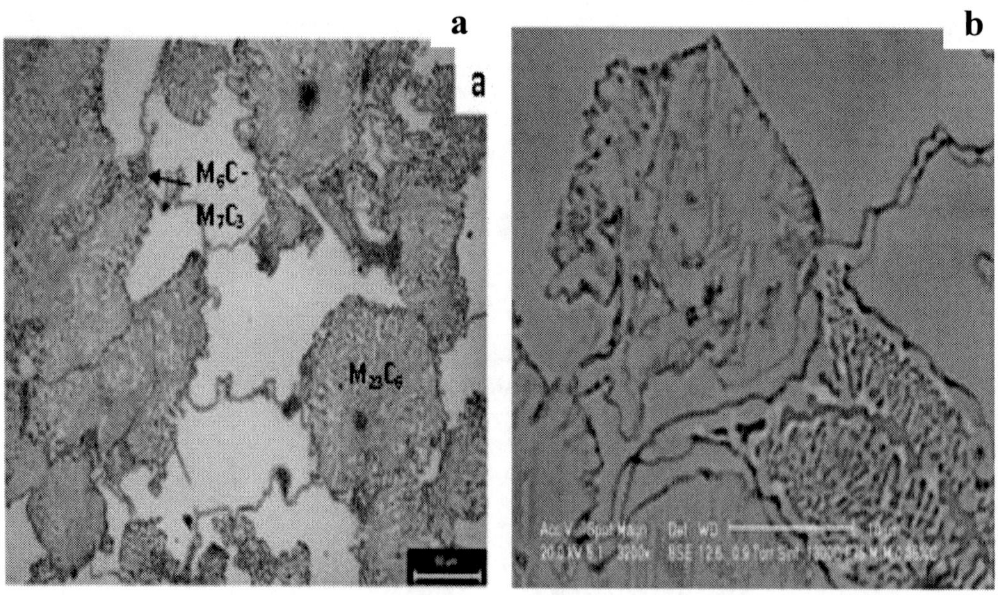

Figure 2. As sintered 0.35%C Co-Cr-Mo alloy: optical (a) and ESEM (b) micrographs [17]

The results of the microstructural characterization of the material, after solution annealing at the three temperatures are presented in the following.

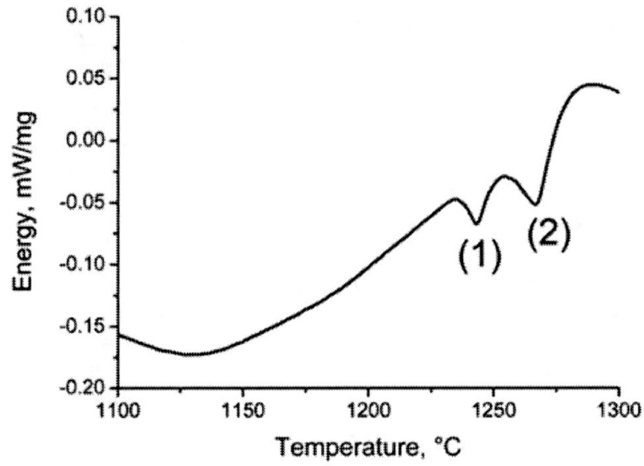

Figure 3. DSC curve of the 0.35%C as sintered alloy [17]

Solution annealing at 1200°C

Figure 4 shows the microstructure of the material solution annealed at 1200°C.

Solution annealing at 1200°C causes the partial solubilisation of the eutectic cells and the fragmentation of the grain boundary precipitates. The grain size of the metallic matrix does not change with heat treatment.

Solution annealing at 1220°C

Figure 5 shows the microstructure of the material solution annealed at 1220°C.

The treatment at 1220°C causes the complete solubilisation of the eutectic cell. $M_{23}C_6$ e M_7C_3 particles are still present at the grain boundary. Sigma phase is also still present, and the grain size of the metallic matrix does not change with the treatment. Kilner [1] found that the heat treatment at temperatures close to but below the melting point of the interdendritic constituent results in the appearance of $M_{23}C_6$ as the only interdendritic phase. Prolonged heat treatment (more than 24 hours) eventually dissolves this carbide resulting in a homogeneous alloy.

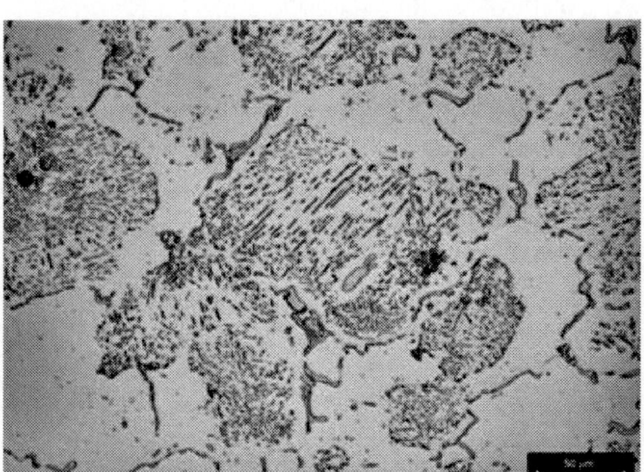

Figure 4. Microstructure of 1200°C solution annealed 0.35%C material [17]

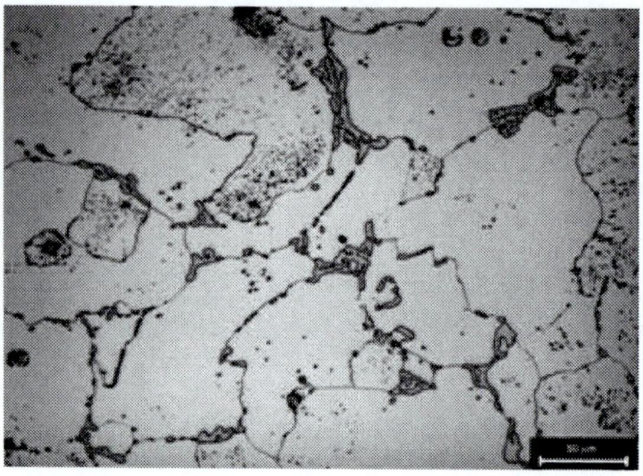

Figure 5. Microstructure of 1220°C solution annealed 0.35%C material

Solution annealing at 1250°C

Figure 6a shows the microstructure of the material solution annealed at 1250°C. The eutectic cells are completely solubilised, as well as most of the other carbides. Only a few $M_{23}C_6$, M_6C e M_7C_3 with a modified morphology are visible. In particular, SEM analysis (figure 6b) highlights globular particles of $M_{23}C_6/M_7C_3$ and eutectic M_6C with a very low dihedral angle. The grain size of the metallic matrix is slightly increased to 140±23 μm, because of the carbide solubilisation.

The heat treatment above the eutectic reaction was studied by Kilner [1]. He found a strikingly serrated appearance of the boundary between the interdendritic constituent and the matrix. This has been described as "starlike" phase. In addition to this phase, the grain boundary melting takes place, as in the present study, as demonstrated by the change of morphology of residual eutectic carbides.

Table 2 lists the matrix microhardness, and the hardness of materials investigated. Solution annealing causes an increase in microhardness (because of solution hardening by carbon and alloying elements), that does not show any significant influence of temperature. Hardness decreases after solution annealing, since it is influenced by carbides and by the grain size; this justifies the decrease with the solution annealing temperature, as well.

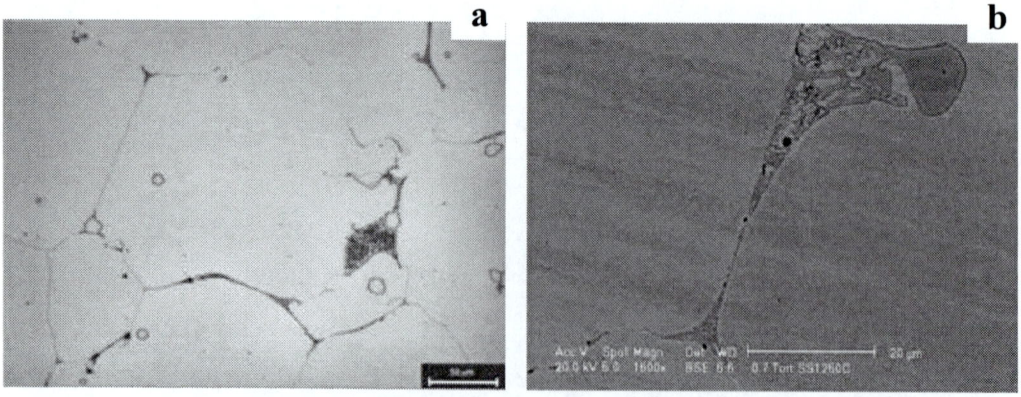

Figure 6. Optical (a) and SEM (b) microstructure after solution annealing 0.35%C material at 1250°C [17]

Table 2. Matrix Microhardness and Hardness

Material	HV0.05	HV30
0.35%C alloy – As Sintered	366 ± 13	347 ± 7
0.35%C alloy – SA at 1200°C	*421 ± 21*	*349 ± 5*
0.35%C alloy - SA at 1220°C	395 ± 12	313 ± 5
0.35%C alloy - SA at 1250°C	*400 ± 15*	*294 ± 5*

On the basis of the results of the microstructural and of microhardness and hardness tests, the solution annealing temperature for the Co-Cr-Mo materials was selected at 1220°C. The lower temperature (1200°C), does not significantly reduce the amount of carbides. The higher temperature (1250°C) results in an effective solubilisation of carbides but causes a decrease of hardness and grain growth. The material solution annealed at 1220°C is expected to have a much lower brittleness than that treated at 1200°C, and a greater strength than that treated at 1250°C.

3.2. Microstructures of Co-Cr-Mo Alloys

0.35%C Alloy

The density is 99% of the theoretical one. Figures 7 a and c show the microstructure of the as sintered and heat treated 0.35%C alloy sintered at 1300°C. The as sintered alloy, as described at the previous 3.1 section, contains both lamellar carbides ($Cr_{23}C_6$) within large eutectic cells and grain boundary carbides (Cr_7C_3 and M_6C), as confirmed by the colour metallographic etching, figure 7b, as well as fine grain boundary precipitates of sigma phase.

The $Cr_{23}C_6$ carbides are the product of the slow eutectic solidification of a liquid phase [3]. M_6C can both precipitate directly from the f.c.c. solid solution and form as a product of the transformation on heating of $M_{23}C_6$ [2, 18]. The formation of sigma phase was detected by Signorelli, in conjunction with, or subsequent to, $M_{23}C_6$ precipitation [2]. According to Shortsleeve [19], the precipitation of carbides promotes the formation of sigma phase, then a large quantity of carbides enhances the probability of the precipitation of sigma phase.

Most of the cellular carbides of the 0.35%C alloy are effectively dissolved by solution annealing, see figure 7b, and only a discontinuous network of grain boundary carbides ($Cr_{23}C_6$ and Cr_7C_3) is still present.

The ESEM images of the aged 0.35%C alloy at 750°C for 3 and 20 hours are shown in figures 8a and 8b, respectively. Aging promotes an intragranular star-shape (Widmanstätten type) precipitation and the formation of a fine constituent (very closely packed lamellae) which nucleates at the grain boundary and grows inside grains. The amount of these precipitates increases with the aging time.

The Widmanstätten precipitates are $M_{23}C_6$ particles [2, 6, 21, 22]. This demonstrates that solution annealing and aging are effective in promoting the formation of a homogeneous dispersion of carbides in the metallic matrix from the highly segregated microstructure of the as sintered materials. The fine lamellar constituent is described by K. Rajan [21, 22]. For short aging time, the f.c.c. phase transforms to a highly faulted h.c.p. phase with a martensitic mechanism involving interstitial diffusion. This hcp phase contains some carbides nucleated at the grain boundary (hcp_1). On increasing aging time, it evolves to hcp_2 which differs from hcp_1 for the lack of any crystallographic relationships with the parent phases and for the lower fault density. This evolution is coupled to an abundant precipitation of carbides.

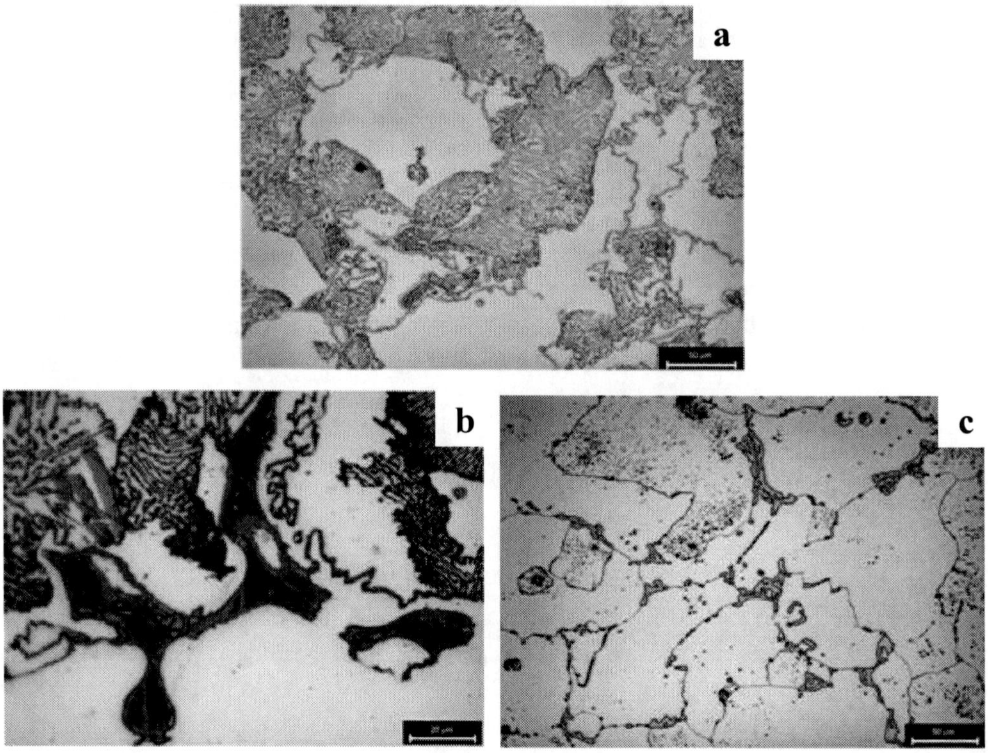

Figure 7. Optical micrographs of as sintered (a [20] and b) and solution annealed (c) 0,35%C material [20]

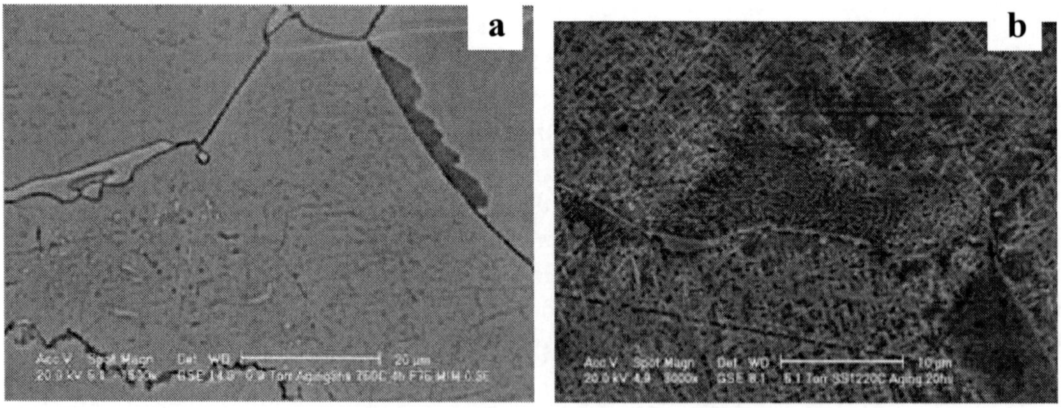

Figure 8. ESEM images of aged 3hours (a)[13] and 20hours (b) 0,35%C material [17]

0.23%C Alloy sintered at 1300°C

Density of this material is 7.55 g/cm^3, 91.1% of the theoretical value.

The microstructure of the as sintered and of the heat treated 0.23%C alloy sintered at 1300°C are reported in the figures 9a and 10a, respectively. This material contains only lamellar eutectic carbides of the $Cr_{23}C_6$ type, as confirmed by the colour metallographic etching.

The ESEM image, figure 9b, shows the grain boundary precipitates of sigma phase.

After solution annealing all the carbides and sigma phase are fully solubilised in the metallic matrix, figure 10a. The grain size increases from 75±16 μm to 90±13 μm after annealing.

On aging, the same transformations described previously occur, involving the intragranular Widmanstätten precipitation and the grain boundary precipitation of the hexagonal constituents. A detail of the microstructure after 20 hours of aging is shown in figure 10b.

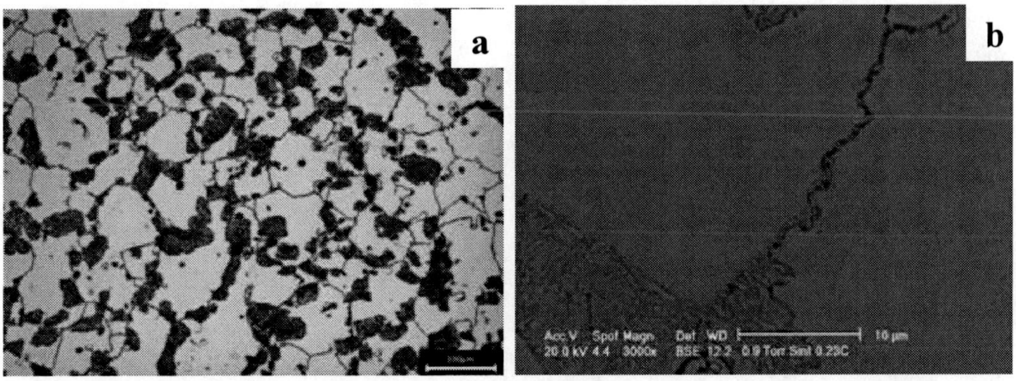

Figure 9. Optical and ESEM images of as sintered 0,23%C material sintered at 1300°C [[23].

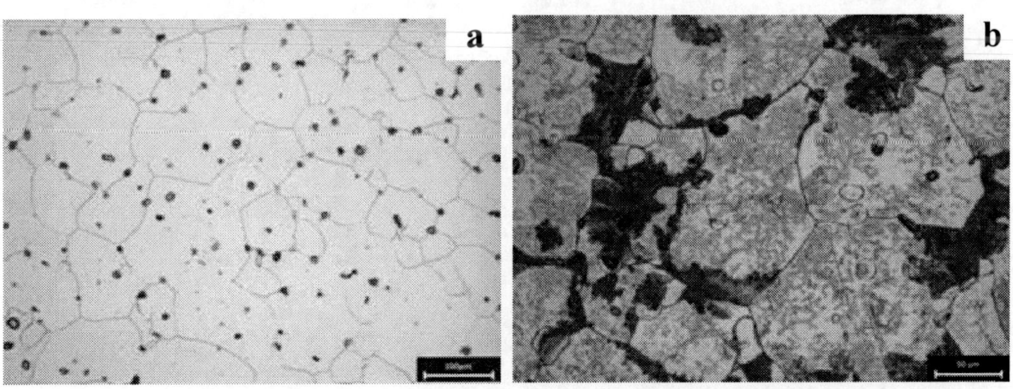

Figure 10. Optical micrographs of sol. annealed (a) and aged for 20hours (b) 0.23%C material sintered at 1300°C [23]

0.23%C Alloy sintered at 1350°C

The 0.23%C alloy sintered at 1350°C presents a higher density (98.7% of the theoretical one) than the same material sintered at 1300°C. The matrix grain size increases from 75±16 μm up to 193±25 μm.

Figure 11 shows the microstructures of the as sintered and of the solution annealed 0.23%C alloy (sintered at 1350°C). This sintered material contains again both lamellar $Cr_{23}C_6$ carbides and grain boundary M_6C carbides and sigma phase. After solution annealing, some small spheroidized $M_{23}C_6$ particles and a very low fraction of sigma phase are still present at the grain boundary, figure 12a. The spheroidised particles are the product of breaking-up and spheroidisation of the lamellar "pearlite-like" eutectic carbides [2].

Grain size remains almost unchanged after solution annealing, because of the grain boundary pinning exerted by the residual carbides.

After aging (figure 12b), both the intragranular Widmanstätten precipitates and the grain boundary precipitates of the hexagonal constituents are visible. The material sintered at 1300°C contains a higher fraction of hcp phases than the present material, likely because of the smaller grain size (larger grain boundary surface area which enhances nucleation) and the uncompleted solubilization of the as sintered carbides. It also shows a lower amount of Widmanstätten carbides, because of the concurrent carbide precipitation coupled to hcp phase formation. Grain size does not change with aging.

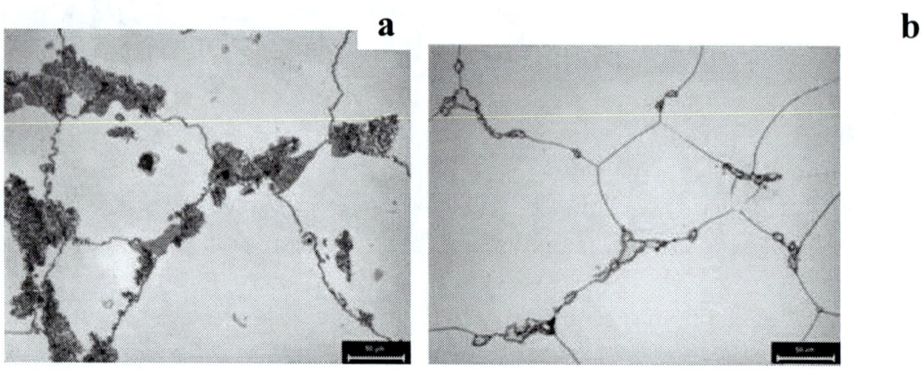

Figure 11. Optical micrographs of as sintered (a) and solution annealed (b) 0.23%C material sintered at 1350°C

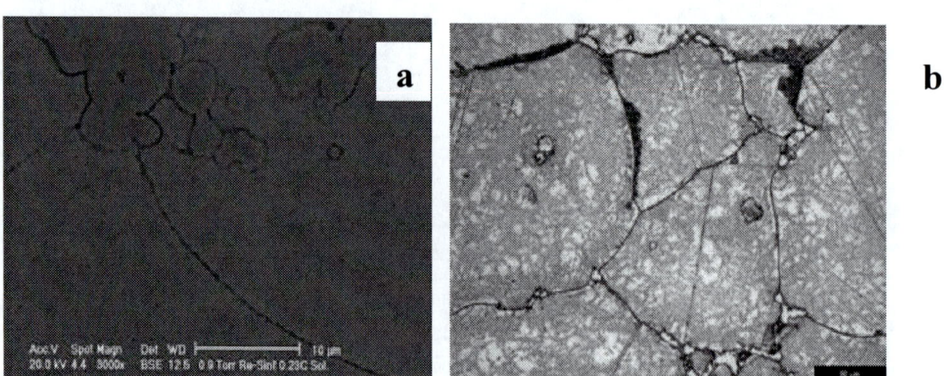

Figure 12. Optical and ESEM images of sol. annealed (a) and aged for 20hours (b) 0.23%C material sintered at 1350°C [23]

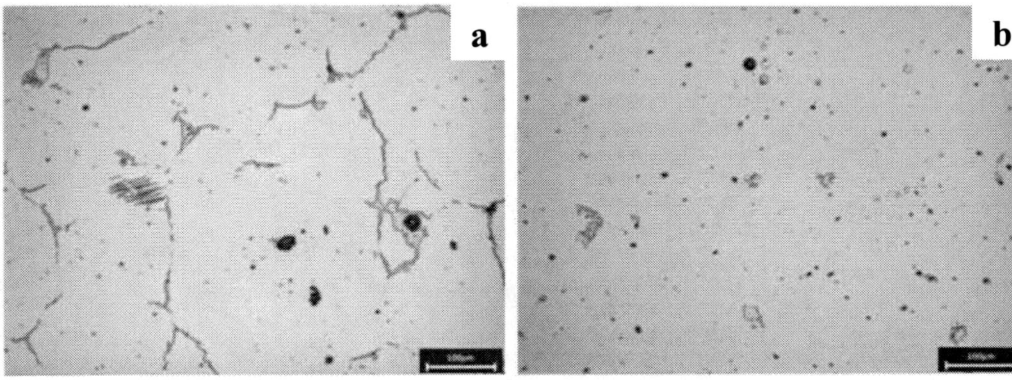

Figure 13. Optical micrographs of as sintered (a) and solution annealed (b) 0.05%C material

0.05%C Alloy

Figure 13 shows the microstructure of the as sintered and of the solution annealed 0.05%C alloy. This as sintered alloy contains a less quantity of carbides than the others, showing only discontinuous precipitates of $Cr_{23}C_6$ particles on the original particles boundaries. The sigma phase is absent, as expected, due to the low quantity of carbides. Solution annealing is very effective on this material; only a few $Cr_{23}C_6$ particles are visible in the microstructure after heat treatment.

On aging, the same transformations described previously occur, but these transformations are quite less pronounced. Figure 14, relevant to the 20 hours aged specimen, shows only a slight amount of grain boundary martensite.

3.3. Constitution of the Metallic Matrix

Figure 15 shows, as an example, the XRD pattern of 0.23%C alloy, both in the sintered at 1300°C and solution annealed conditions. It is possible to identify the h.c.p. and f.c.c. peaks in the patterns. Carbides are not detected by XRD because their mean peaks are in the same positions of those of the f.c.c. and h.c.p. phases.

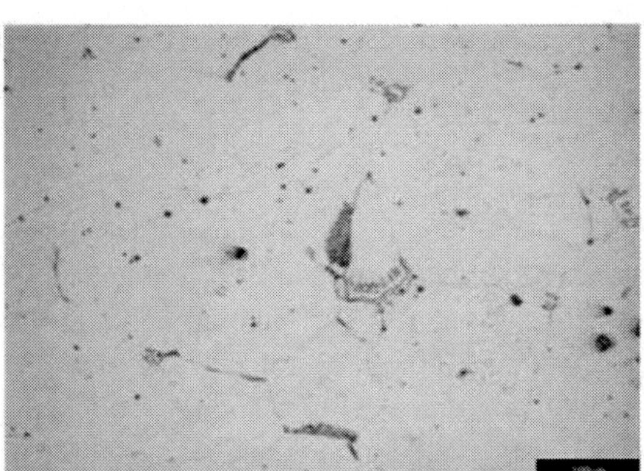

Figure 14. Optical micrograph of 0.05%C material aged for 20hours [13]

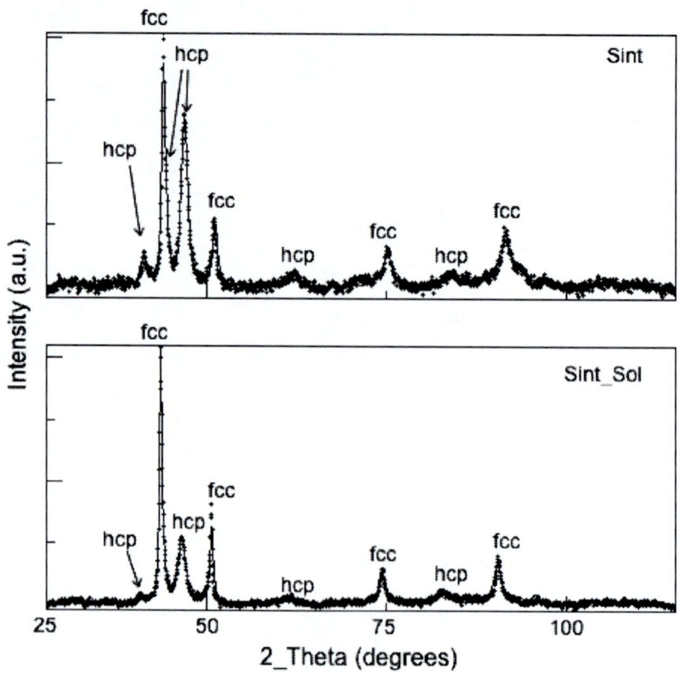

Figure 15. XRD patterns for 0.23%C alloy sintered at 1300°C and sintered + solution annealed [23]

Table 3. XRD analyses

Materials	Constitution of the metallic matrix	
	% fcc	*% hcp*
0.35%C alloy as sintered	-	*100*
0.35%C alloy + SA	*62±1.3*	*38±1.3*
0.23%C alloy - S1300°C	42±1.2	58±1.2
0.23%C alloy - S1300°C + SA	61±1.2	39±1.2
0.23%C alloy - S1350°C	*43±1.5*	*57±1.5*
0.23%C alloy - S1350°C + SA	*33±1.5*	*67±1.5*
0.05%C alloy as sintered	-	100
0.05%C alloy + SA	-	100

Table 3 reports the results of the quantitative XRD analyses on the as sintered and the solution annealed materials. The as atomized powder is fully f.c.c. in all cases, since the carbide precipitation is suppressed by the large solidification undercooling on atomization.

The 0.05%C alloy is fully h.c.p. in the as sintered condition since the low C content allows the α—>ε transformation to go to completion on slow cooling. The as sintered 0.35%C material is fully h.c.p. too, since carbon is almost completely bounded to carbides and the behavior on cooling is very similar to the previous material.

The 0.23%C as sintered alloys contains both f.c.c. and h.c.p. phases in the same relative proportions, despite the different microstructures. In the 1300°C materials, f.c.c. phase is stabilized by the residual interstitial carbon in solid solution and by the fine grain size. The main difference between the two materials is the presence of the Mo-rich M_6C carbides and a significantly larger grain sizes than those found after sintering at higher temperature. Whilst the carbon in solution tends to stabilize the f.c.c. phase, smaller grain size is favorable to the α—>ε transformation [12]. The two effects are then balanced, and the fraction of h.c.p. in the as sintered materials does not change with the sintering temperature

Solution annealing does not influence the matrix constitution in 0.05%C alloy, since there is no significant modification of the composition of the matrix. The extensive dissolution of carbides in 0.35%C is responsible for the stabilization of some f.c.c. phase in the final microstructure.

In the 0.23%C material sintered at 1300°C, solution annealing dissolves most of carbides which increases the concentration of interstitial carbon, therefore the f.c.c. fraction increases after heat treatment. The f.c.c. phase is then stabilized against the transformation in h.c.p. on quenching.

In 0.23%C material sintered at 1350°C, solution annealing dissolves preferentially the M_6C carbides, leading to an increase in the Mo content of the f.c.c. phase. Consequently, the transformation to h.c.p. upon quenching is enhanced, and the amount of f.c.c. phase decreases with heat treatment.

XRD analyses were not carried out on aged materials, because of the lower interest they present for mechanical properties as will be described below.

3.4. Hardness and Microhardness

Microhardness and hardness are reported in figures 16 to 19.

Both microhardness and hardness do not change with heat treatment in 0.05%C material, since microstructural characteristics are very similar because of the very low C content.

Solution annealing causes a slight increase in microhardness (because of solution hardening by carbon and alloying elements) in all the other materials. On aging, microhardness increases, in particular after 20 hours, because of the intragranular precipitation.

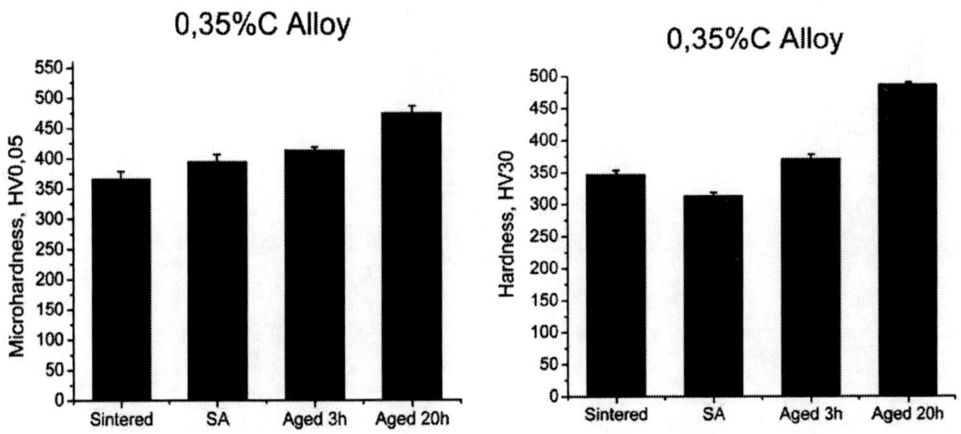

Figure 16. Microhardness and hardness of 0.35%C alloy sintered at 1300°C

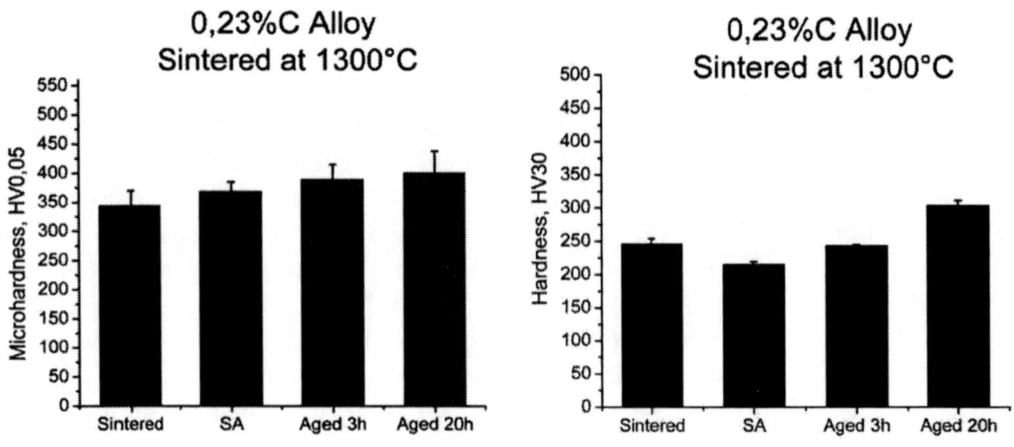

Figure 17. Microhardness and hardness of 0.23%C alloy sintered at 1300°C

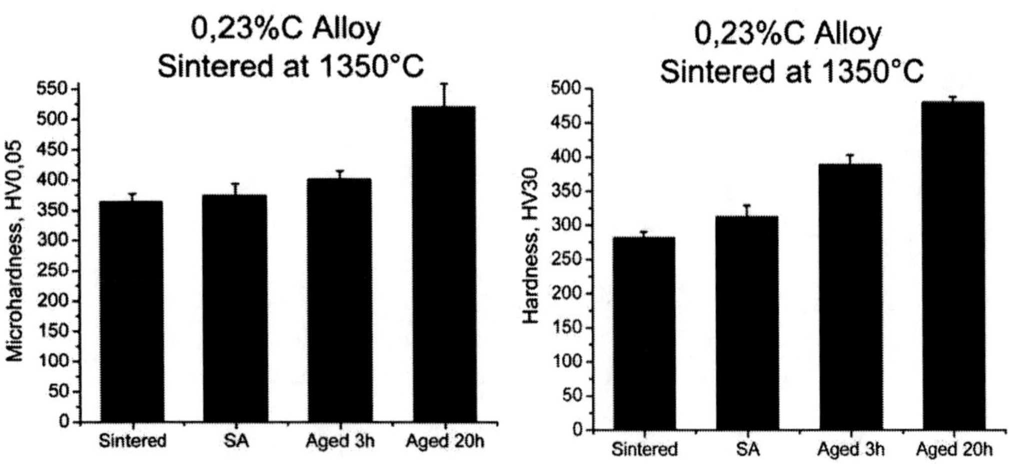

Figure 18. Microhardness and hardness of 0.23%C alloy sintered at 1350°C

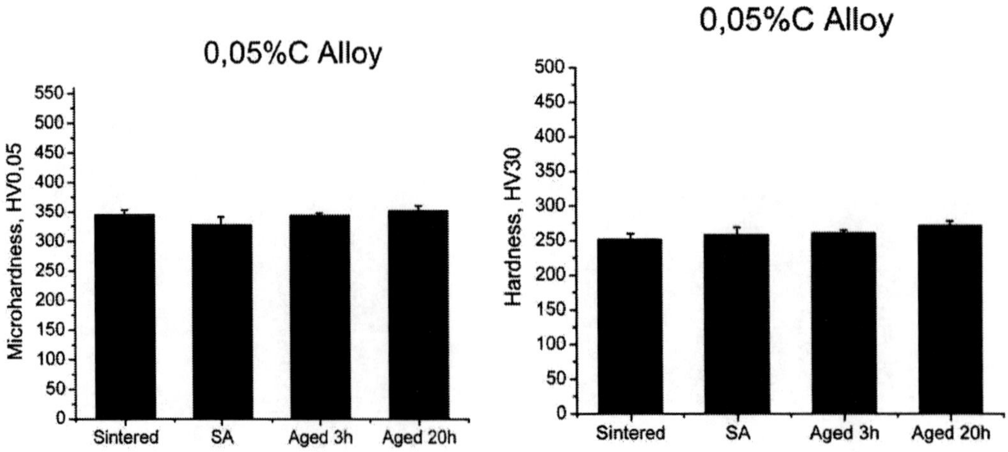

Figure 19. Microhardness and hardness of 0.05%C alloy sintered at 1380°C

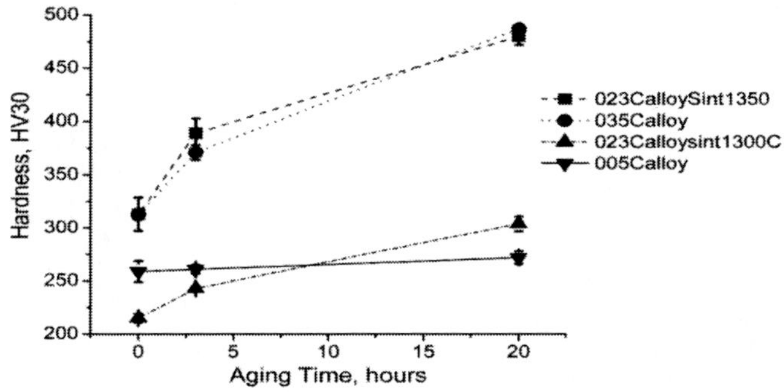

Figure 20. Hardness versus aging time profile

Hardness values of the 0.35%C (figure 16b) and 1300°C sintered 0.23%C (figure 17b) materials decrease with solution annealing because of grain growth and dissolution of carbides.

These two effects prevail on the solution hardening of the metallic matrix provided by the enrichment in alloy elements. Contrarily, solution hardening of the matrix causes a slight increase in hardness of the material 0.23%C sintered at 1350°C (figure 18b) since solution annealing does not eliminate all carbides and does not cause grain growth.

Figure 20 shows hardness versus aging time.

The increase in hardness during aging is due to the intragranular precipitation of the Widmanstätten carbides and of the grain boundary lamellar constituent. The trend is less pronounced for the 0.05%C material because of the less precipitation during aging. The lower density of the material 0.23%C sintered at 1300°C accounts for the lower hardness, with respect to that other materials.

3.5. Tensile Tests

The results of tensile tests are summarized in figures 21 to 24, and compared with the prescriptions of the ISO 5832-4 standard for orthopaedic implants.

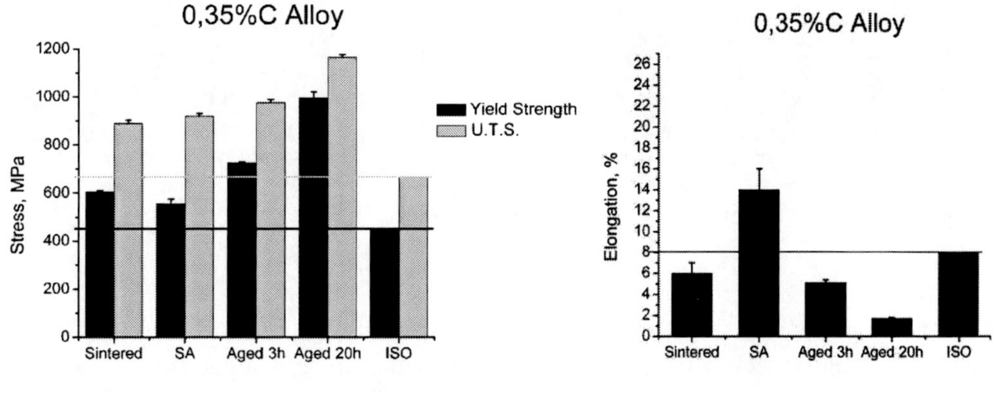

Figure 21. Tensile properties of 0.35%C alloy sintered at 1300°C

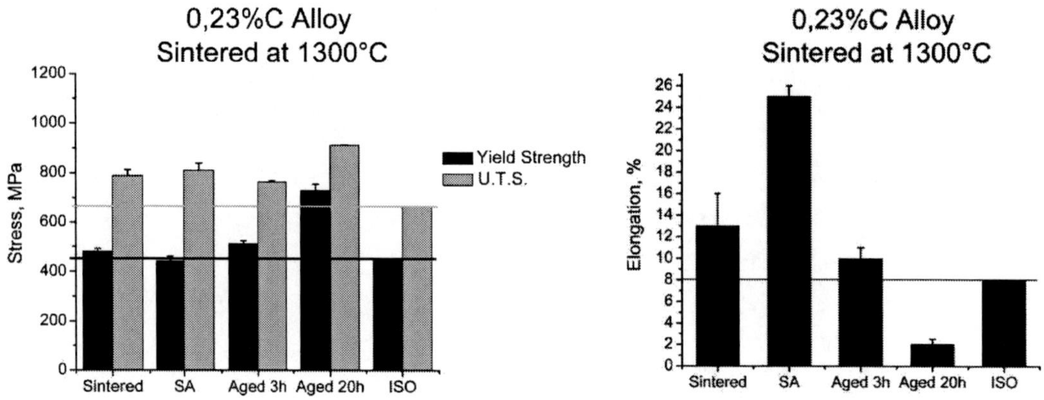

Figure 22. Tensile properties of 0.23%C alloy sintered at 1300°C

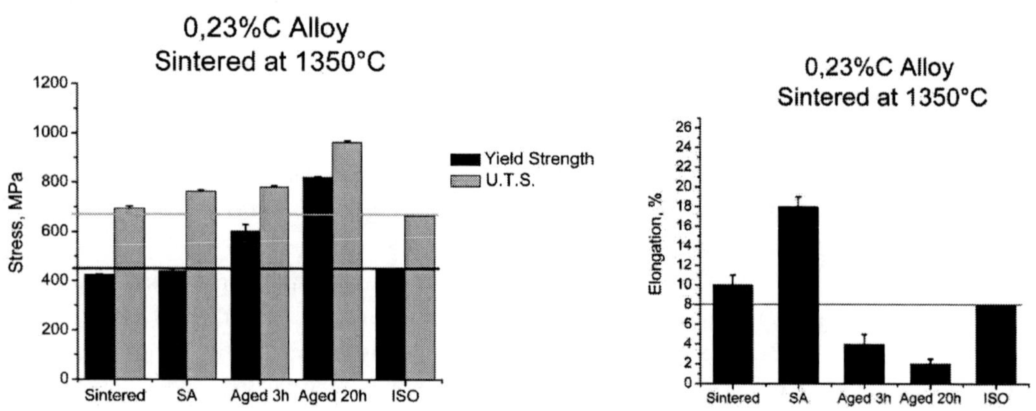

Figure 23. Tensile properties of 0.23%C alloy sintered at 1350°C

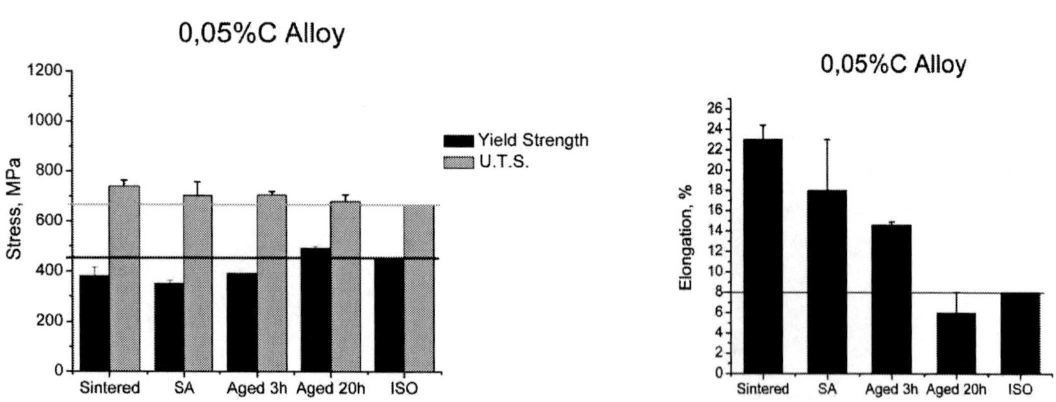

Figure 24. Tensile properties of 0.05%C alloy sintered at 1380°C

Solubilisation modifies the mechanical properties of the 0.23%C and the 0.35%C materials significantly; yield strength is decreased since the strengthening effect of carbides is stronger than that of the dissolved carbon and alloying elements in the matrix, and ductility and UTS are correspondingly increased because of the improved plasticity of the material. The 0.05%C material has both better strength and ductility in the as sintered condition, likely because the microstructure remains martensitic.

The 0.23%C material sintered at lower temperature presents both strength and ductility surprisingly better than those of that sintered at the higher one, despite the lower density. The higher yield stress may be correlated to the smaller grain size, according to the Hall-Petch relation. Even after solution annealing, the material sintered at lower temperature has better mechanical properties than the high temperature one. The different grain size and the residual grain boundary carbides are responsible for this difference.

After aging, strength still increases but ductility strongly decreases for all materials, because of the slip restrictions exerted by faults in h.c.p. martensite and by the Widmanstätten precipitates [6], but mainly of the grain boundary precipitates of the h.c.p_2. phase.

Figure 25 shows the effect of the carbon content on the yield strength and elongation of the as sintered materials. In principle, on increasing the carbon content the strength of the alloy increases, whilst ductility decreases, because of the effect of carbides on the deformation and fracture behavior [13, 23, 24], which promote an intergranular fracture [24].

With reference to the ISO standard, the solution annealed 0.35%C material matches all the specifications.

The material with 0.23%C sintered at 1300°C and solution annealed shows good mechanical properties too, but due to its high porosity cannot be considered for the specific applications. The same material, when sintered at 1350°C and solution annealed matches UTS and ε% of the ISO standard, whilst yield strength is slightly lower. The 0.05%C alloy has a yield strength much lower than that required by ISO specifications.

Figure 26 shows the summarized results of tensile tests of the 0.35%C material in a "yield strength – percent elongation" map.

So, solution annealing increases ductility, with a slight decrease of yield strength. On aging strength further increases, but the decrease in ductility is very pronounced because of the negative effect of the grain boundary hexagonal constituent [23].

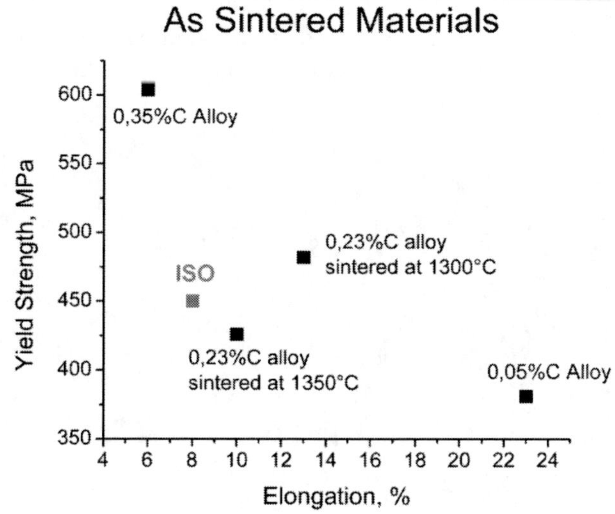

Figure 25. Graphic of yield strength vs. elongation for as sintered materials

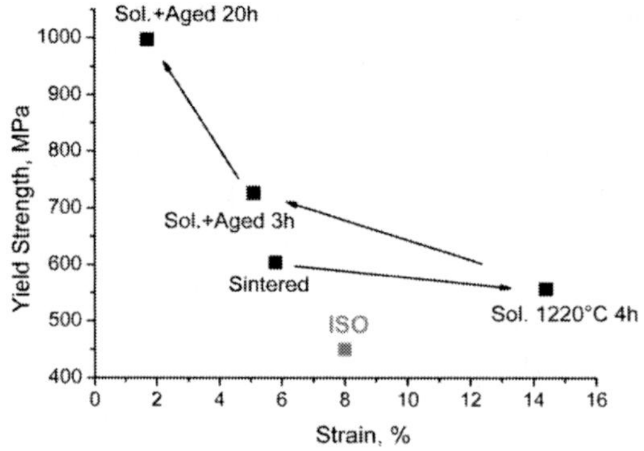

Figure 26. Yield strength and percent elongation at fracture for 0.35%C as sintered material, solution annealed and then aged. [17]

Fracture surfaces

In figure 27, the fracture surfaces for 0.35%C material are showed. In the as sintered material, figure 27a, cracks propagate along grain boundaries by cracking of grain boundary carbides and separation at the carbide/matrix interface. After solution annealing, figure 27b, the failure mode is still intergranular, but ductility of the alloy is increased due to the increased ability of the material to deform by localized shear in the regions near grain boundaries.

The different ductility for 0.23%C materials finds an explanation in the fracture behaviour. Figure 28 shows the fracture surfaces of the two as sintered materials.

Fracture is fully intergranular and associated with the presence of carbides and sigma phase which cause strain localization and favour the nucleation and propagation of cracks [4].

The fracture surface of the solution annealed 0.23%C material gives evidence of the improved ductility with respect to the as sintered ones (figure 29). Fracture involves the cobalt matrix, which shows striations similar to deformation bands in the regions between the zones of intense shear. Only in the material sintered at the higher temperature (figure 29b), a brittle morphology is visible, attributable to grain boundary carbides. The low amount of carbides, and their absence in the 1300°C sintered material, allows deformation to propagate through the metallic matrix by stacking fault formation and twining [4].

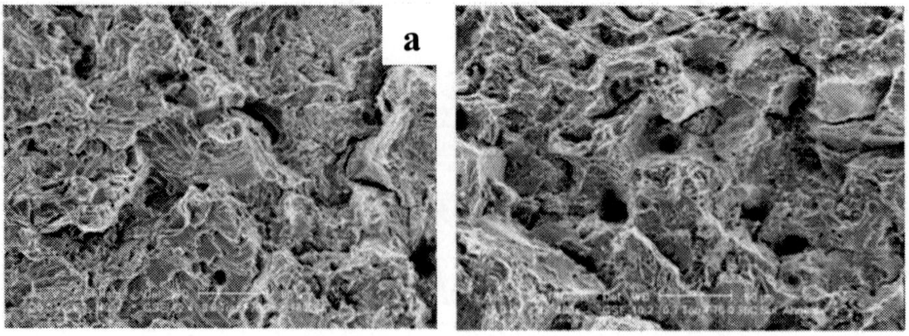

Figure 27. Fracture surface for 0.35%C material as sintered (a) and after solution annealing (b)

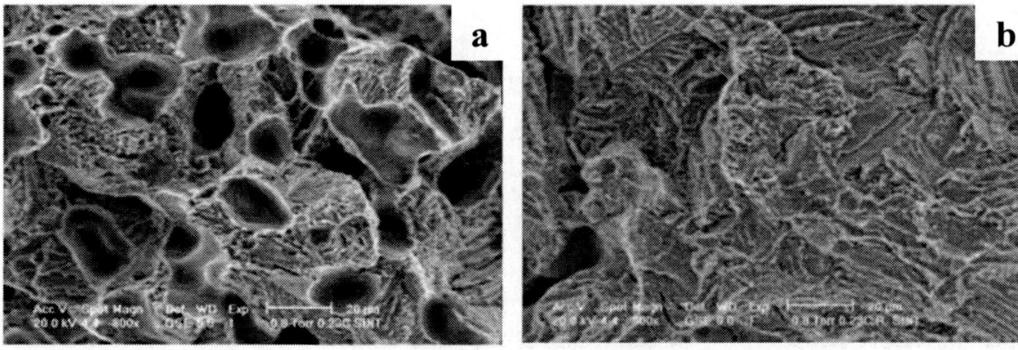

Figure 28. Fracture surface for as sintered 0.23%C material, sintered at 1300°C (a) and 1350°C (b)[23]

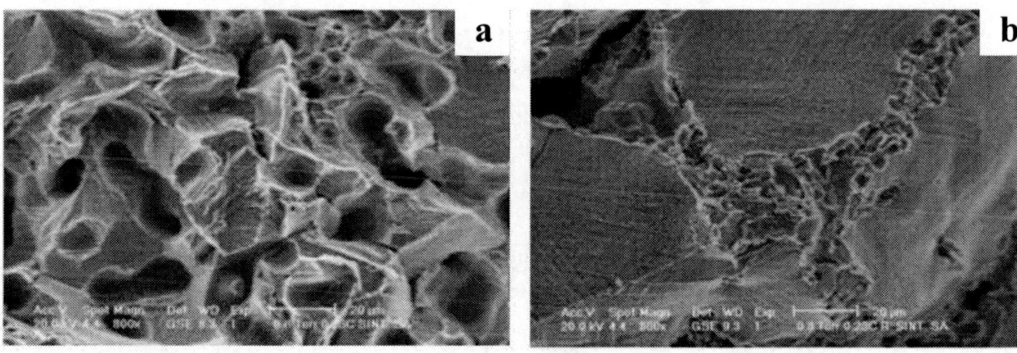

Figure 29. Fracture surface for the solution annealed 0.23%C material, sintered at 1300°C (a) and 1350°C (b) [23]

In aged specimens, the grain boundary precipitation of hcp_2 causes intergranular fracture, clearly visible on the fracture surfaces shown in figure 30.

The fracture surface of the 0.05%C as sintered material, figure 31a, is still intergranular with dimples, which accounts for the large ductility. The large decrease of ductility on aging is clearly demonstrated by the fracture surfaces of the as sintered and 20 hours aged materials, shown in figure 31.

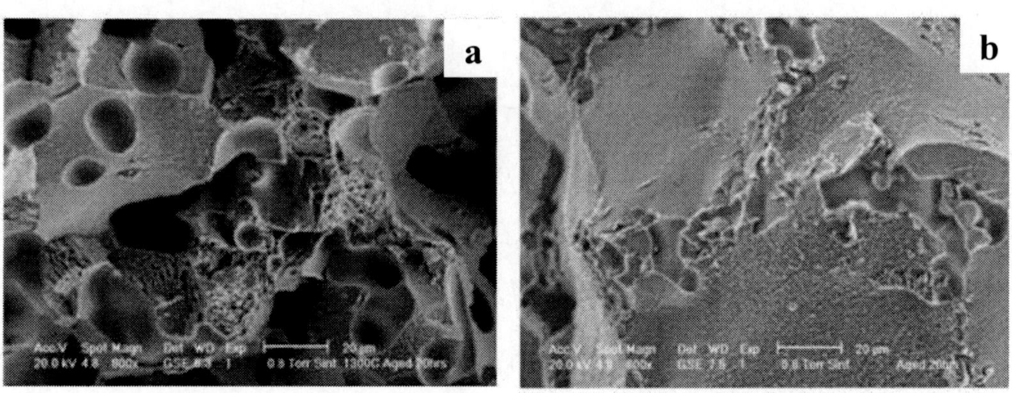

Figure 30. Fracture surface for the aged 0.23%C material, sintered at 1300°C (a) and 1350°C (b)[23]

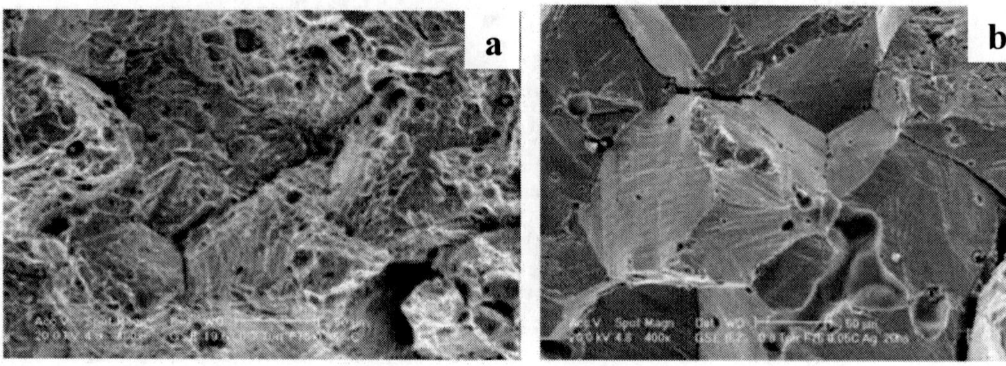

Figure 31. Fracture surface of 0.05%C material as sintered (a) and aged for 20hs (b) [13]

For all materials the fracture is intergranular and the best combination of mechanical properties is obtained by solubilization, which minimizes the amount of grain boundary phases, responsible of brittleness irrespective on their nature.

Strain hardenability

An interesting feature of these materials is the large difference between yield strength and UTS, which is representative of a high strain hardenability. Figure 32 shows the stress-strain curves for the materials that present the best mechanical properties of this study; the increase in the proof stress along the plastic field is pronounced, significant of a great resistance to plastic deformation [4]. This property is of great importance to resist overloading.

The microstructure of the solution annealed material is mainly austenitic (Table 3), but the amount of austenite close to the fracture surface after tensile tests is almost negligible, as shown by XRD analysis. Strain hardenability is therefore due to the strain induced transformation of austenite into martensite. The strain hardening coefficient cannot be calculated with the classical Ludwick-Hollomon model [25], since the log-log diagram of true stress versus true strain does not depict a linear correlation but it shows a continuous increase in the slope, as figure 33 demonstrates. The upward curvature is significant of the increasing strain hardenability with plastic strain [13].

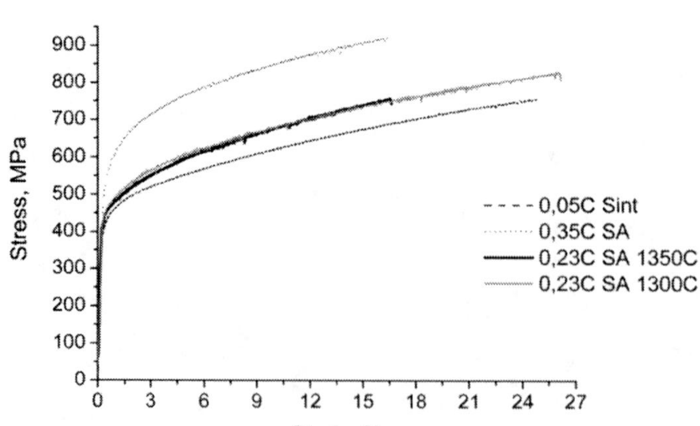

Figure 32. Tensile stress - strain curves for the materials

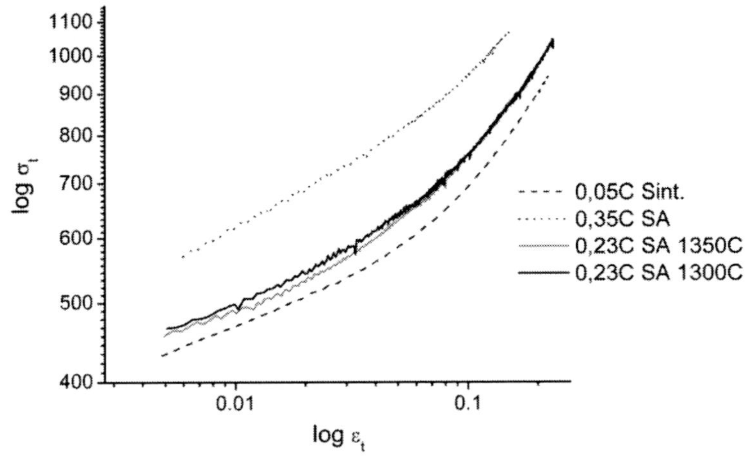

Figure 33. True stress – true strain plot of the as sintered 0.05%C material and solution annealed 0.35%C and 0.23%C materials

The cobalt alloy has a low stacking fault energy, which promotes stacking faults formation and twinning [4, 6, 21, 22]. Interaction of dislocations of limited mobility with other sessile dislocations, and with stacking faults and twins, can lead to very high strain hardening rates that produce intensive localized stresses which need to be relieved in order to allow the material to resist higher loads [4]. Moreover, the f.c.c. phase transforms into h.c.p. martensite on straining, providing an additional contribution to strain hardenability by the well known transformation induced plasticity - TRIP – phenomenon [13, 26].

Kim and Lin [27] proposed a three parameters model to determine the strain hardening coefficient for a metastable austenite in steel. The correlation between true stress and true strain, assumed to be quadratic, is:

$$\ln\sigma_t = a(\ln\varepsilon_t)^2 + b(\ln\varepsilon_t) + c \tag{1}$$

where a, b and c are constants to be determined by interpolation of experimental data. Experimental data show that a linear relationship can be assumed between strain hardening coefficient n and true strain, then the following equation is proposed by Kim and Lim [27]:

$$n = d\ln\sigma_t / d\ln\varepsilon_t = M\varepsilon_t + N \tag{2}$$

where M and N are temperature dependent constants. Upon integration of equation 2, the following flow equation is obtained:

$$\sigma_t = K\varepsilon_t^{\ N}\exp(M\varepsilon_t) \tag{3}$$

K is a constant depending on the material, whilst M describes how the flow stress increases with the plastic strain because of the strain induced transformation of austenite.

Figure 34 shows the results of fitting of equation (2) and Table 4 lists K, M and N parameters. The correlation parameter is included, as well, to evaluate the goodness of the model.

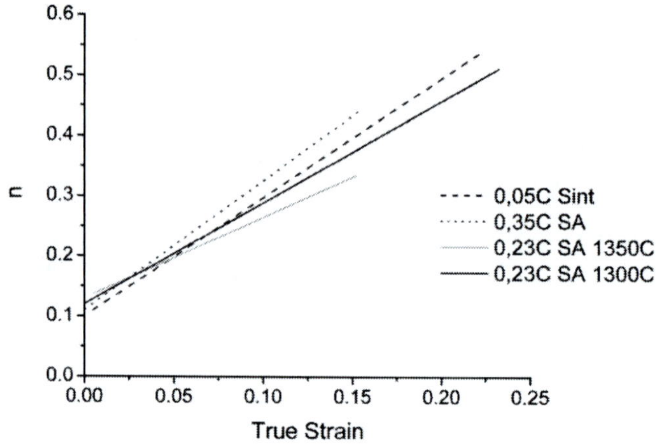

Figure 34. Strain hardening coefficient versus plastic strain for as sintered 0.05%C material and solution annealed 0.35%C and 0.23%C materials

Table 4. Parameters of the fitting of plastic field

Materials	K	M	N	R^2
0.35%C alloy – SA	*1120*	*1.35*	*0.13*	*0.9992*
0.23%C alloy – S1300°C + SA	839.43	1.69	0.12	0.9992
0.23%C alloy – S1350°C + SA	*775.29*	*2.16*	*0.11*	*0.9992*
0.05%C alloy – as sintered	706.50	1.98	0.10	0.9997

N, which represents the initial work hardening coefficient, is quite low in all the materials. The increase in strain hardening with strain is represented by M. It is higher in the 0.23%C material sintered at 1350°C and in the 0.05%C as sintered material, which contains the higher fraction of martensite having a great tendency to strain hardening, because of the large density of stacking faults which accompany its formation on heat treatment [8].

The high strain hardening can be the explanation for the good ductility of the solution annealing materials also in presence of the intergranular fracture, because the material has a great capacity to resist localized loads by deformation.

3.6. Wear Resistance

The influence on the wear resistance of MIM products is not yet established, so lubricated wear tests are carried out on all as sintered and solution annealing materials. On aged materials the tests are not made because of their less favorable mechanical properties.

As far as wear resistance is concerned, a study on a cast alloy concludes that carbides may cause abrasive wear of the counterface polymeric material, which could release debris at the metal-polyethylene interface [16]. The UHMWPE wear debris is mainly responsible for periprosthetic tissue reaction, osteolysis, and eventually, late aseptic loosening. In a metal-on-metal wear test, the effect of the constitution of the metallic matrix on wear was investigated, concluding that a fully h.c.p. martensitic material has a lower wear resistance than a fully austenitic, because of the more favourable friction coefficient [5].

In figure 35 the friction coefficient tracks for the 0.35%C materials are reported, as representative of the results obtained for the other materials, too. A gradual decrease of the friction coefficient towards a steady value occurs. The steady value is typical of a mixed lubricant condition, and is slightly but significantly lower for the solution annealed material.

As expected from the different hardness of the two counteracting materials, no mass loss of the Co alloys was measured, whilst some debris of UHMWPE are transferred to its surface, as figure 36 shows. This phenomenon was described by Walker [16] on a cast cobalt alloy. A slight but measurable mass loss of the polymer is then induced, comparable to literature data [28]. A higher quantity of transferred polymer in the solution annealed materials is found for all materials.

To quantify the wear phenomenon, the mass loss of UHMWPE and the increase in the surface roughness of the Co alloys, which is significant of the material transfer, are considered and reported in table 5. It has to be considered that the measure of the mass loss of the polymer, even with the use of a precision balance, is quite a hard task; therefore data must be considered as indicative.

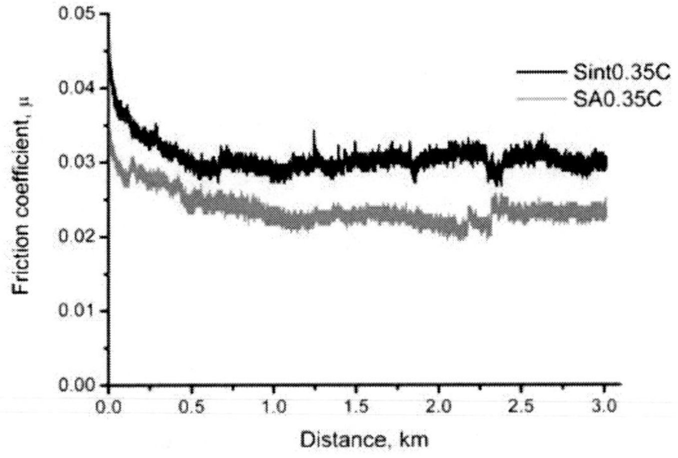

Figure 35. Friction coefficient vs. distance curve [20]

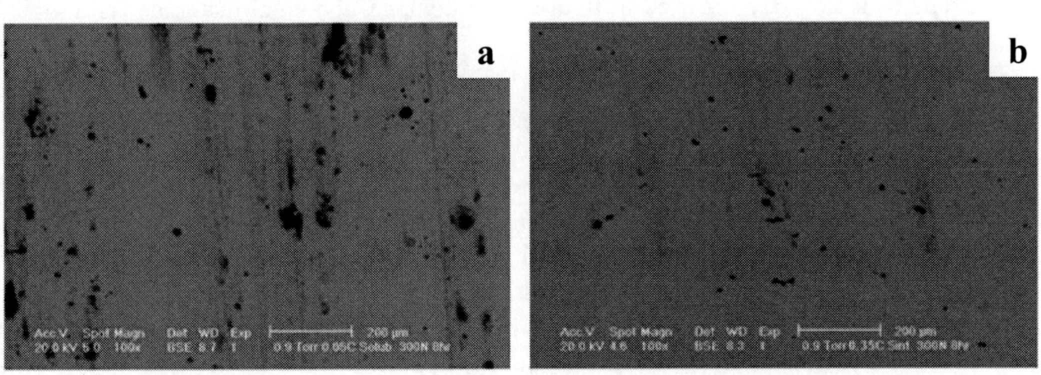

Figure 36. Wear traces for 0,05%C SA material (a) and 0,35%C sintered material (b)[20]

Table 5. Mass loss of UHMWPE versus surface roughness of the Co alloys

Materials	ΔRa cobalt alloy (μm)	Δm UHMWPE (mg)
0.35%C alloy – as Sintered	*0.028*	*0*
0.35%C alloy – SA	*0.031*	*-0.4*
0.23%C alloy - S1300°C	0.055	-1
0.23%C alloy - S1300°C + SA	0.087	-0.5
0.23%C alloy - S1350°C	*0.076*	*-0.9*
0.23%C alloy - S1350°C + SA	*0.100*	*-0.6*
0.05%C alloy – as sintered	0.061	-0.4
0.05%C alloy – SA	0.076	0

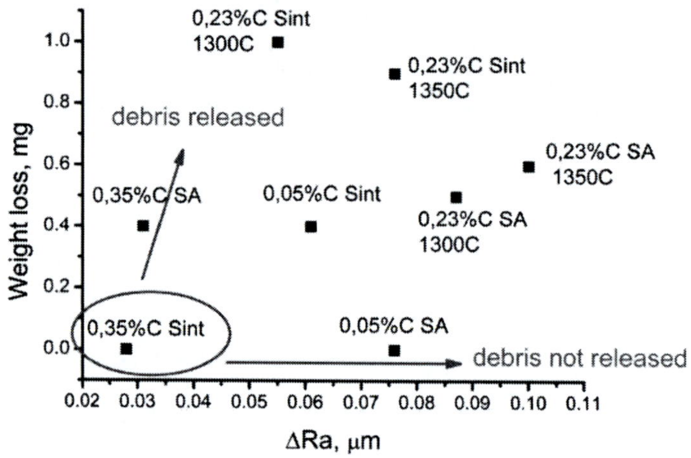

Figure 37. Weight loss and ΔRa of the investigated materials

The solution annealed materials cause a lower mass loss on the UHMWPE. They also evolve towards a higher surface roughness than the as sintered ones which, in combination with the lower steady value of the friction coefficient, could suggest that the transferred layer may give an additional contribution to lubrication. The 0.35%C material is the exception, because the polymer did not present a mass loss.

The data of table 5 are reported in figure 37 to compare the eight materials for a global evaluation.

The best wear behavior is displayed by the 0.35%C material, in the as sintered condition. It does not cause a measurable wear of the polymer and the surface roughness worsens very slightly. In presence of a lubricant, the abrasive effect of carbides is then almost completely avoided. All the other materials have a worse behavior. Among them, solution annealed 0.35%C and solution annealed 0.05%C represent two opposite situations. The former causes a measurable wear of the polymer, but roughness of the cobalt alloy does not worsen significantly; the latter does not cause a measurable mass loss of the polymer, but its surface roughness worsens significantly. This different behavior could be interpreted considering that debris of the polymer can either adhere on the metal surface (increasing roughness) or be removed by the lubricant (in this case the surface roughness does not change). Which of the two options might be preferred in application is a matter of discussion; in principle, the former could be preferred, which does not release debris from the tribological system and, at the same time, does not worsen the contact conditions, as the decrease of the steady friction coefficient indicates.

The role of carbides, which are suspected to abrade the counterface polymer, is minimized by the lubricant. However, the comparison between as sintered and solution annealed 0.35%C seems to indicate that the localized particles after heat treatment are more abrasive than the more homogeneously distributed carbides in the as sintered material.

So, wear tests show that the wear rate of the counterface UHMWPE (0,05÷0,4 mm/year) is comparable to that reported for a real prosthesis (0.1÷0.3 mm/year [28]), and is influenced by the Co alloy microstructure.

3.7. Corrosion Tests

Co-base alloys are known to possess a high resistance to corrosion in physical media, imparted by a passive oxide film that forms spontaneously on the alloy surface, and to their excellent mechanical properties. According to Georgette and Davidson [29], a more stable, uniform oxide layer would be expected with a more homogeneous matrix (annealed alloy) than with a highly dendritic as cast structure. Placko et al. [30] found a progressive dissolution of the matrix with preferential attack of the grain boundaries and regions adjacent to carbides due to sensitization. According to Montero-Ocampo and Rodriguez [31], the low C content ASTM-F75 as cast alloys resulted in lower release rates of corrosion products. Heat treatments may change the biocompatibility of the alloys through the modification of the electrochemical properties [32].

The influence on the corrosion resistance of MIM products is not yet established, so the corrosion tests are carried out for the as sintered and solution annealed materials.

Open circuit potential test

As shown in figures 38a, b, c and d, all the materials reach a similar OCP value just after immersion that was about -0.3 and -0.2 V, except the 0.23%C material sintered at 1300°C, but the behavior and the stabilization of the potential during time is quite different: the 0.35%C and 0.23%C sintered at 1300°C materials become much nobler during the first 10-15 hours of the immersion, showing an increase of about 0.3 and 0.5 V compared with the starting value and monitoring a change in the surface equilibrium due to the exposure environment. The 0.23%C material sintered at 1350°C, SA material sintered at 1300°C, and 0.5%C material on the contrary do not show big changes in the potential during immersion. After this initial transient two different behaviors are visible:

- a stabilization of the potential on a steady value, (0.05%C SA, 0.23%C Sint at 1350°C and SA, and 0.23%C SA sintered at 1300°C).
- a slight but continuous increase in the potential, towards more noble values (0.05%C Sint., 0.35%C Sint., 0.35%C SA and 0.23%C sintered at 1300°C).

The influence of the presence of carbides can be appreciated comparing the same material in the as sintered and solution annealed conditions, figure 38, even if they have a different matrix constitution. When the carbide content increases, the potential increases too.

In order to consider the effect of the matrix constitution on the OCP evolution, the 0.05%C SA, 0.023%C SA sintered at 1350°C and 0.35%C SA materials containing a decreasing amount of h.c.p. phase and almost no carbides are compared, figure 39. The difference in OCP values is quite evident and the more h.c.p. fraction (0.05%C SA > 0.23%C SA sint. at 1350°C > 0.35%C SA), the less noble is the electrochemical behavior.

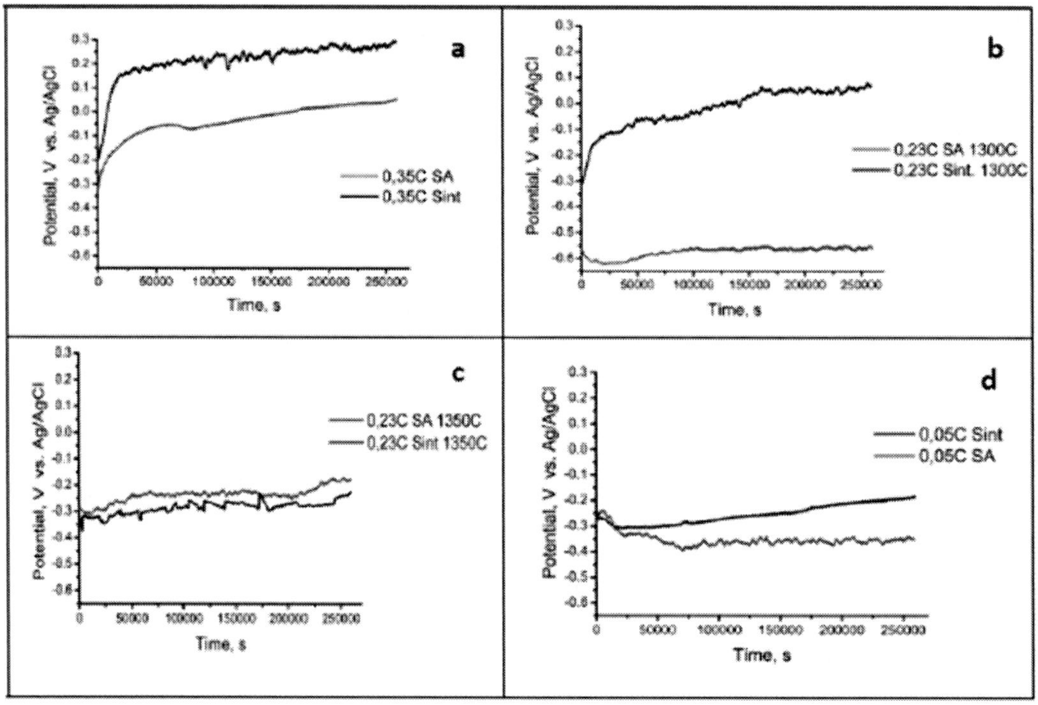

Figure 38. Open circuit curves for 0.35%C (a), 0.23%C (b and c) and 0.05%C (d) materials [20]

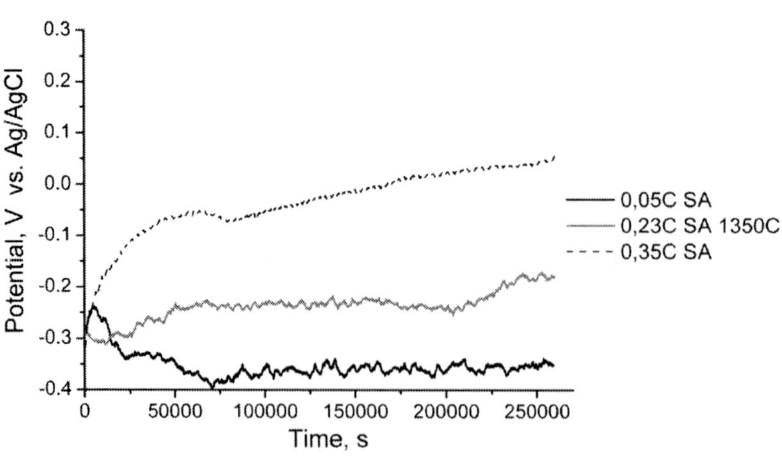

Figure 39. Effect of the matrix constitution on the OCP evolution

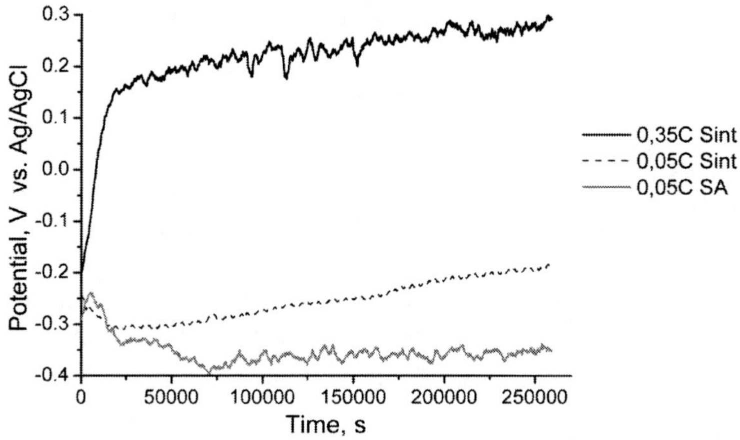

Figure 40. Effect of carbides on the OCP evolution

Considering samples with the same matrix phase, figure 40, the comparison between 0.05%C sintered and solution annealed materials and 0.35%C sintered confirms that the higher quantity of carbides tends to enhance nobility.

Potentiodynamic curves

In order to better evaluate the stability of the passivity and the electrochemical behavior of the samples potentodynamic curves were analyzed.

Some representative potentiodynamic curves are reported in figure 41.

All the materials show passive behavior in this environment and the main electrochemical parameters are summarized in table 6.

The sintered 0.35%C and 1350°C sintered and solution annealed 0.23%C materials present a nobler free corrosion potential than 0.05%C, as found in open circuit curves.

Regarding the passive corrosion density, the material with 0.23%C presents higher values than 0.35%C and 0.05%C, but the potential of breakdown is the same for all specimens.

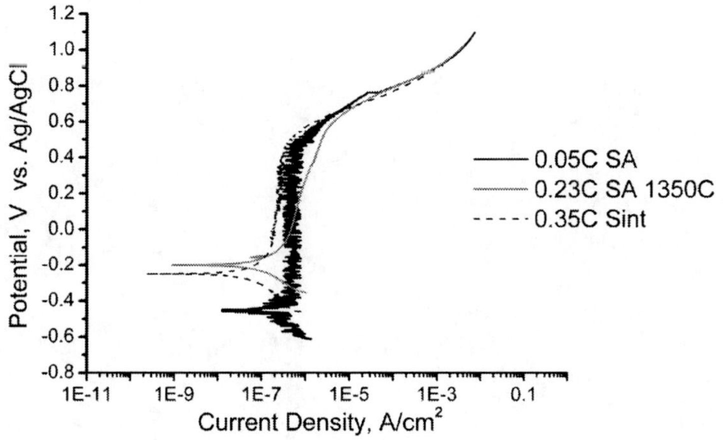

Figure 41. Potentiodynamic curves [20]

Table 6. Electrochemical parameters [20]

Materials	E_{corr} [V]	i_{pass} [A/cm^2]	E_b [V]
0.35%C alloy – as Sintered	*-0.24*	*3x10^{-7}*	*0.6*
0.23%C alloy - S1350°C + SA	-0.2	10^{-6}	0.6
0.05%C alloy – SA	*-0.48*	*5x10^{-7}*	*0.6*

To understand the morphology of the corrosion process, the specimens have been examined by optical microscopy just after breakdown and after the evolution of the corrosion process (figures 42 to 44). Based on the micrographs we can appraise:

Carbides do not oxidize;

The metal matrix oxidizes, the breakdown is due to the transpassivity and no pits are present;

The corrosive attack is localized at the carbides-matrix interface and leads to the complete removal of the carbide particles/eutectic cells.

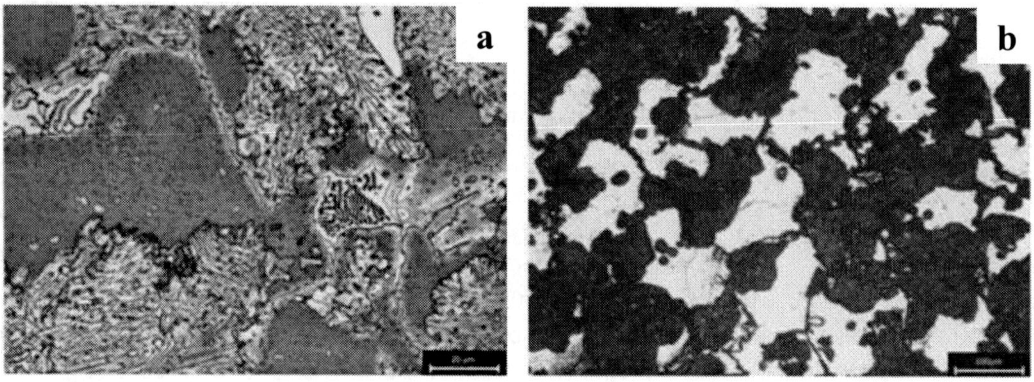

Figure 42. Corrosion micrographs of 0.35%C after breakdown (a) and the evolution of corrosion process (b) [20]

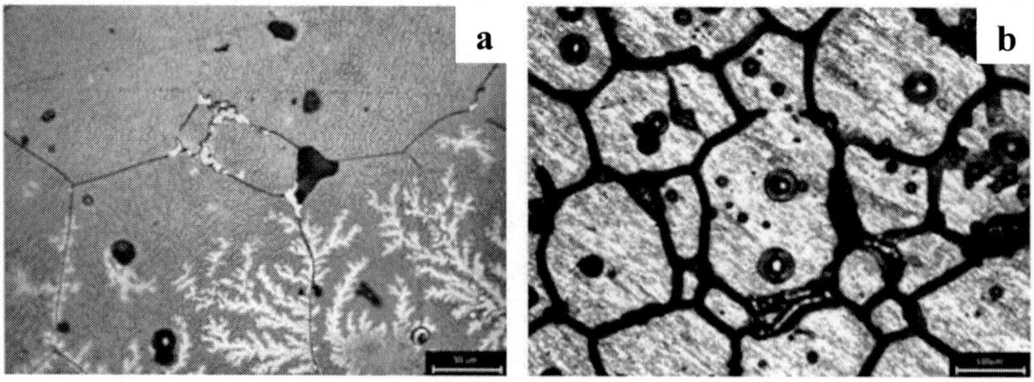

Figure 43. Corrosion micrographs of 0.23%C after breakdown (a) and the evolution of corrosion process (b) [20]

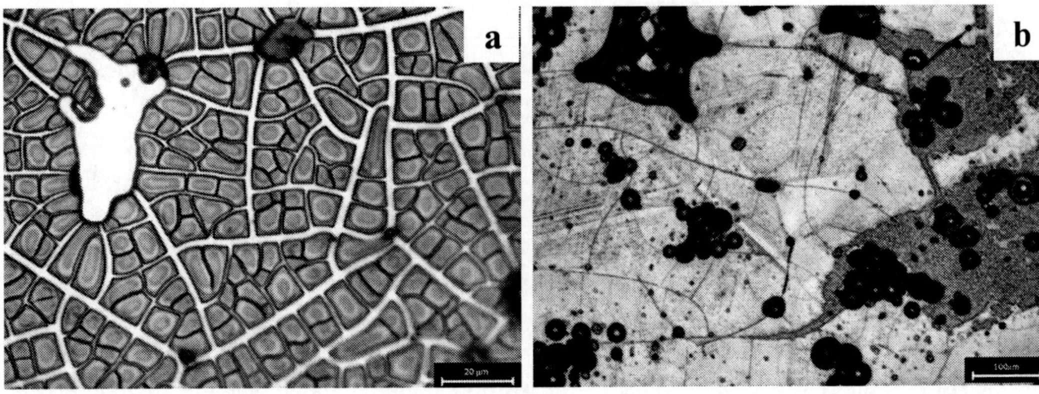

Figure 44. Corrosion micrographs of 0.05%C after breakdown (a) and the evolution of corrosion process (b) [20]

The composition of the layer is predominantly Cr_2O_3 oxide with some minor contribution of other oxides (Co and Mo oxides) [33]. According to Hodgson et al. [34], the passive behavior of Co-Cr-Mo is due to the formation of an oxide layer with high content of Cr (mainly as Cr III) and a smaller amount of $Cr(OH)_3$.

Moreover it may be appreciated that the corrosion of the 0.05%C occurs by localized attacks, while the 0.23%C samples show an intergranular corrosion morphology. The 0.35%C material shows a large corroded area corresponding to carbide/eutectic phase.

Therefore, although carbides enhance the electrochemical nobility, the presence of the carbide/eutectic phase leads to a different corrosion morphology when the traspassivity state is reached.

CONCLUSIONS

The aim of the research was the study of the influence of the microstructure of Co29Cr6Mo alloys produced by MIM on some specific properties related to the application in the biomedical field.

Near-full dense Co-Cr-Mo alloys with different carbon contents were obtained by MIM and sintering at temperature increasing from 1300°C up to 1380°C on decreasing the C content.

The mechanical properties are strongly influenced by the microstructural characteristics and in particular by carbides.

The content of carbides increases from 0.05%C to 0.23%C and to 0.35%C. After solution annealing the amount of carbides decreases significantly but grain size tends to increase.

Aging causes the re-precipitation of carbides within grains and of h.c.p.$_1$ and h.c.p.$_2$ at the grain boundaries.

Heat treatment changes the phase constitution of the matrix.

As far as mechanical properties are concerned:

Microhardness is affected by the matrix constitution, so after solution annealing, when C, Cr and Mo are in solid solution and after aging, when re-precipitation of the carbides takes place within the grain, microhardness increases for all materials.

Hardness is influenced by carbides, after solution annealing it decreases and after aging it increases. The hardness of the 0.05%C material does not change with heat treatments.

Tensile strength: The yield strength is strongly correlated with the quantity of carbon content in the alloy, the higher the carbon content, the higher σ_y. After solution annealing the strength of the materials slightly decreases, but on aging it increases due to the precipitates.

Ductility is inversely proportional to the carbon content, the lower the C content, the more ductile is the alloy. After solution annealing ductility increases for almost all materials and after aging it decreases again.

The best mechanical properties are obtained on the SA 0.35%C alloy, which match all the ISO requirements. The 0.23%C material presents good mechanical properties too, but due to its high porosity, it cannot be considered for orthopedics applications. When sintered at high temperature, it matches requirements although yield strength is very slightly lower than that requested.

Wear tests were carried out against a UHMWPE under lubricated conditions. The wear mechanism is characterized by the transfer of wear debris from the polymer to the metallic alloy, and is quantified by the mass loss of the former and the increase in the roughness of the latter. The material with the best wear properties is as sintered 0.35%C alloy, which does not cause a measurable mass loss of the polymer and displays only a very slight increase in the surface roughness.

Corrosion tests were carried out by measuring the Open Circuit Potential. All the materials have a fast transition to a noble behavior, with an effect of both the matrix constitution and the carbide amount. In particular, the higher the hcp fraction, the less noble is the electrochemical behavior, and a higher incidence of carbides tends to enhance nobility. On forcing corrosion by potentiodynamic tests, carbides tend to localize the attack responsible for transpassivation.

Wear and corrosion would suggest the use of the as sintered 0.35%C material, which is too brittle. The 0.35%C SA alloy presents the best mechanical properties while the corrosion behavior remains quite good, and the mass loss of the polymer is very low, and compatible with the application. This material could then represent the best solution for the production of the orthopedic prostheses by MIM of prealloyed powders.

REFERENCES

[1] Kilner, T; Pilliar, RM; Weatherly, GC; Allibert, C. "Phase Identification and Incipient Melting in a Cast Co-Cr Surgical Implant Alloy", *J. Biomed. Mater. Res.*, 1982, vol. 16, 63-79.

[2] Weeton, JW; Signorelli, RA. *Trans Am Soc Met*, 1955, 47, 815.

[3] Caudillo, M; Herrera-Trejo, M; Castro, MR; Ramirez, E; Gonzalez, CR; Juarez, J.I. *J. Biomed. Mater. Res*., 2002, 59(2), 378-385.
[4] Mancha, H; Gómez, M; Castro, M; Méndez, M; Méndez, J; Juárez, J. "Effect of Heat Treatment on the Mechanical Properties of an As-Cast ASTM F-75 Implant Alloy", *J. Mater. Synth Process*, 1996, vol. 4, no. 4, 217-226.
[5] Saldivar Garcia, AJ; Lopez, HF. *Journal of Biomedical Materials Research*, 2005, 74A, 269.
[6] Taylor, RNJ; Waterhouse, RB. *J. Mater. Sci*., 1983, 18, 3265-3280.
[7] Dobbs, HSl; Robertson, JLM. *J. Mater. Sci*., 1983, 18, 391-401.
[8] Vander Sande, JB; Coke, JR; Wulff, J. Met. *Trans*., 1976, 7A, 389-397.
[9] Tandon, R. in Cobalt-Base Alloys for Biomedical Applications, ASTM STP 1365, JA; Disegi, RL; Kennedy, R. Pilliar, eds., West Conshohocken, PA, ASTM, 1999, 3-10 .
[10] Johnson, L; Heaney, DF. *Metal Injection Molding of Co-28Cr-6Mo,* Advanced Powder Products, inc. www.advancedpowderproducts.com/publications.asp
[11] Billiet, RT; Billiet, TH. A practical guide to metal and ceramic injection moulding. *Elsevier*, Oxford, 2006.
[12] Betteridge, W. *Cobalt and Its Alloys*, Ellis Horwood Limited, UK, 1982.
[13] Muterlle, PV; D'Incau, M; Perina, M; Bardini, R; Molinari, A. *Advances in Powder Metallurgy & Particulate Materials*, 2008, 4, 183-190.
[14] Lutterotti, L; Matthies, S; Wenk, HR; Shultz, AS; Richardson, JW. *J. of Appl. Phys*., 1997, 81, 594-600.
[15] Lonardelli, I; Wenk, HR; Lutterotti, L; Goodwin, M. *Journal of Synchrotron Radiation*, 2005, 12, 354-360.
[16] Walker PS; Blunn, GW; Lilley, PA. *Journal of Biomedical Materials Research (Applied Biomaterials)*, 1996, vol. 33, 159-175.
[17] Muterlle, P; Lonardelli, I; Perina, M; Zendron, M; Bardini, R; Molinari, A. "Solution annealing and aging of a CoCrMo alloy produced by MIM", *International Journal of Powder Metallurgy*, accepted in 24 November 2009.
[18] Clemow, AJT; Daniell, BL. *J. Biomed. Mater. Res*., 1979, 13, 265.
[19] Shortsleeve, FJ; Nicholson, ME. *Transactions, American Society for Metals*, 1951, 43, 142.
[20] Vieira Muterlle, P; Zendron, M; Zanella, C; Perina, M; Bardini, R; Molinari, A. *Advances in Powder Metallurgy & Particulate Materials*, 2009, 4, 60-69.
[21] Rajan, K. Met. *Trans*., 1982, 13A, 1161-1166.
[22] Rajan, K. Met. *Trans*., 1984, 15A, 1335-1338.
[23] Muterlle, PV; Zendron, M; Perina, M; Bardini, R; Molinari, A. "Microstructure and tensile properties of metal injection molding Co-29Cr-6Mo-0,23C alloy", *Journal of Materials Science*, 2010, 45(4), 1091.
[24] Muterlle, PV; Perina, M; Mantovani, M; Molinari, A. *Metal Powder Report*, 2009, 64(2), 30.
[25] Dieter, GE. 'Mechanical Metallurgy', SI Metric ed. (adapted by D. Bacon) *Materials Science & Metallurgy*, Singapore, 1988.
[26] Salinas-Rodriguez, A; Rodriguez-Galicia, JL. *J. Biomed. Mater. Res*., 1996, 31, 409-419.
[27] Kim, YG; Lim, CY. Met. *Trans*., 1988, 19A, 1625.
[28] Dr. Mischler, S. Tribology and Implants. *Cours Biomatériaux*, novembre 2006.

[29] Georgette, FS; Davidson, JA. *Journal of Biomedical Materials Research*, 1986, 20, 1229.
[30] Placko, HE; Brown, SA; Payer, JH. *Journal of Biomedical Materials Research*, 1998, 39, 292.
[31] Montero-Ocampo, C; Rodriguez, AS. *Journal of Biomedical Materials Research*, 1995, 29, 441.
[32] Vidal, CV; Muñoz, AI. *Electrochimica Acta*, 2009, 54, 1798.
[33] Milosev, I; Strehblow, H. *Electrochimica Acta*, 2003, 48, 2767.
[34] Hodgso, AWE. et al., *Electrochimica Acta*, 2004, 49, 2167.

In: Injection Molding: Process, Design, and Applications ISBN: 978-1-61761-307-4
Editor: Phoebe H. Kauffer

Chapter 6

APPLICATION OF ULTRASONIC TECHNOLOGY IN INJECTION MOLDING PROCESS

Lei Xie*[*] *and Wangqing Wu
Institute of Polymer Materials and Plastics Engineering,
Clausthal University of Technology, Clausthal-Zellerfeld, Germany

ABSTRACT

In the past four decades, the ultrasonic technology has been introduced for polymers, composites and polymer-metal joint welding. Since the ultrasonic energy can vibrate the polymer melts without additional heating, this clean, reliable and efficient technology was fast spread and developed in a variety of polymer related areas, such as melting, mixing and welding. With the boom of sensor, actuator and transducer since 1980's, the ultrasonic is also involved in this technological development trend. The special features of ultrasonic in refection, diffraction and interference accomplished its applications in displacement metering and crack detection. This chapter demonstrates an overview on the state of the art in application of ultrasonic technology, focusing on polymer processing fields, particularly injection molding process. High ultrasonic frequency can produce heat and oscillation in materials from the result of high frequency stresses, which sparks the initiation of applying it for injection molding process to homogenize and increase the dispersion of the molten material, either in the liquid stage or in the solidifying stage. The special designed ultrasonic transducer provides a valid method to monitor the polymer melts filling during injection molding process. Therefore, the focus aspects of this chapter are placed on describing the main application principles of ultrasonic technology in equipment innovation, parts properties improvement and process on line monitoring of injection molding process.

Keywords: Injection Molding, Ultrasonic, Polymer, Polymer composites

[*] Corresponding author: Email: Lei.Xie@tu-clausthal.de

1. INTRODUCTION

Polymers and its composites materials' development in 20th century is the tremendous achievement of human being in material science and industry. Related technologies such as the ultrasonic technology which has been extensively introduced in polymers, composites and polymer-metal joint welding in the last four decades, have been greatly put forward by this material advancement. Great potentials of ultrasonic technology study and application are found because of the polymeric materials development. In the mean time polymeric materials are also benefit from the ultrasonic technology for its measurement and processing.

Ultrasonic is a high frequency mechanical wave defined as > 20 kHz. In general, there are two kinds of different applied ultrasonic technology in polymeric materials: high frequency and low energy ultrasound (Ultrasonic measurement) and low frequency and high energy ultrasound (Ultrasonic processing). Ultrasonic measurement typically utilizes ultrasound to penetrate a medium and measure the refection signature or supply focused energy in order to get the information like the inner defects, object dimensions and other physical chemistry properties etc. While the ultrasonic processing is mainly focus on the material processing with ultrasonic energy such as ultrasonic welding.

As one of the most common thermoplastic polymer processing process, injection molding process and quality control has been an active research area for many years. Both of ultrasonic measurement and ultrasonic processing have been applied to control the quality of injection molding process [1-5]. It was reported that ultrasonic sensors had been employed to investigate a wide range of process characteristics such as the melting behavior of the resin in the injection barrel, temperature and pressure variation in the nozzle, melt-front flow in the cavity, cavity pressure, solidification behavior in the cavity, polymer orientation, and some quality responses such as part dimension. As for the ultrasonic processing, the ultrasonic vibration had been applied to plastify polymer pellets, improve the polymer melt or even composites compounds flow properties [6-9]. Specifically, this chapter will demonstrate an overview on the state of the art in application of ultrasonic technology, focusing on polymer processing fields, particularly in injection molding process. Processing monitoring with ultrasonic measurement techniques and ultrasonic processing on polymer melt and composites compounding used in injection molding process will be discussed in following parts.

2. PROCESS MONITORING WITH ULTRASONIC MEASUREMENT TECHNOLOGY

In recent years, significant progresses have been made in the injection molding process. However, the emphasis has been on the machine design and new process development. Even though process monitoring and control are essential in improving the quality of molded parts and the efficiency of the injection molding process, the focus has been mainly on the machine itself and limited attention is paid to what happens inside the cavity. The behavior of the material inside the mold during the cycle was usually inferred from the characterization of the final part and mathematical models of the filling, packing, cooling, shrinkage, and warpage were developed based on these post observations [10,11]. In situ measurement provides an improved understanding of the process compared with computer simulations which are often based on simplified assumptions and is crucial for the model validation. Feedback from the cavity sensors to the control system allows direct supervision the process and assures consistent quality in the finished product [12,13]. The monitoring inside the cavity is mostly limited to conventional pressure and temperature sensors, both of which have several disadvantages. Both types of sensors require contact with the material and therefore holes have to be drilled into the mold. Pressure sensors can also be installed behind an ejector pin which puts a certain limit on the sensor's sensitivity. Furthermore, temperature sensors provide at best only surface temperatures while pressure sensors actually measure the force acting on the sensor's membrane, which includes the viscous forces, and not the bulk pressure in the material. Thus, the lack of effective sensors hinders the advancement of material process control and, as a result, demands for the development of such sensors for on-line measurement are high.

Ultrasonic sensors are very attractive for this purpose due to the ability of ultrasound to interrogate non-invasively, non-destructively, and rapidly the surface and internal regions of the material [14]. The basic idea of ultrasonic sensors is quite simple: they transmit acoustic waves and receive them after interaction of the ultrasonic wave and the investigated process. On its arrival at the receiver the ultrasound signal carries the information about the parameters to be measured (and unfortunately many other parameters too—which demands compensation). Ultrasound covers a frequency range from 20 kHz to about 1 GHz. For technical applications the range 20 kHz~10 MHz is the most important one.

2.1. Wave Parameters and Their Relation to Process Parameters

The propagation of a harmonic plane elastic wave can be described by the sonic alternating pressure

$$p(x,t) = p_0 e^{j\omega\left(t-\frac{x}{c}\right)} e^{-\alpha x} \tag{1}$$

where the frequency is $\omega = 2\pi f$, c is the speed of sound and α is the attenuation of sound. The parameters c and α determine the wave propagation and are specific for a substance[15]. This fact is used for analytical applications. Molecular interactions, phase transitions, molecular rearrangements and others are responsible for the c - and α - behavior. In an infinite solid medium the longitudinal speed of sound C_L and the transversal speed of sound C_{Tr} are given by equation (2) and (3) as the following[16]:

$$c_L^2 = c_L^2 \tag{2}$$

and

$$C_{Tr}^2 = \frac{G}{\rho} \tag{3}$$

where K is the bulk modulus, G the shear modulus and ρ the density. In a rod the longitudinal wave travels with the sound speed of C_{Lrod}, which is able to be calculated by equation (4):

$$C_{Lrod}^2 = \frac{E}{\rho} \tag{4}$$

where E is the Young's modulus[16]. The attenuation in a solid is determined by several absorption losses. In a polycrystalline solid reflection and scattering effects occur at the grain boundaries. In polymers relaxation processes can cause the absorption of sound. Many other effects are also responsible for the attenuation of sound in solids. Despite these several reasons for attenuation the order of magnitude of the absorption coefficients in solids is comparable and similar in different materials. This explains the early development and broad application of methods for NDT of solids on the basis of ultrasound. Only polymers represent an exception. Fluid systems show a more complex behavior in the ultrasonic range in comparison with other materials. In simple liquids the speed of sound is described by equation (5)

$$C_L^2 = \frac{1}{\rho\beta} \tag{5}$$

with β as the compressibility. β , ρ are substance-specific and integral parameters. In electrolytes for instance changes of c are very strongly influenced by slight variation of the ion concentration or the kind of ion. The attenuation of sound in fluid systems—such as suspensions, dispersions, emulsions, colloidal systems or aerosols—is caused by different mechanismss[18,19]:

$$\alpha = \alpha_0 + \alpha_{vis} + \alpha_{therm} + \alpha_{sc} + \alpha_{relax} \tag{6}$$

where α_0 is the 'classical' absorption of the emulsion or dispersion medium, α_{vis} describes the viscous losses, α_{therm} the thermal losses, α_{sc} the scattering losses and α_{relax} losses due to relaxation processes. The attenuation depends on different material parameters such as density ρ , viscosity η , thermal conductivity X , thermal capacity c_p , thermal expansion coefficient $\overline{a}$ of the different phases and the particle radius R or diameter D in the dispersion or suspension. They can be temperature and pressure dependent. The 'classical' absorption applies for simple homogeneous liquids. For α_0 the following equation is valid [20]:

$$\frac{\alpha_0}{f^2} = \frac{2\pi^2}{\rho c^3\left[\frac{4}{3}\eta_s + \eta_v\right]} \tag{7}$$

Where, is the shear viscosity and η_v is the bulk viscosity. The attenuation shows a square functional dependence on frequency. This is important for technological applications: at very high frequencies, for example higher than 10 MHz, the attenuation in many real liquid systems reaches a value where the sound is absorbed even at very small distances. The distance which can be passed by a sound wave is less than about 5 mm. On the other hand at lowfrequencies, for example 100 kHz, the attenuation is too small to be measured. The viscous losses are given by equation (8) as following:

$$\propto vis = \frac{\frac{\frac{\omega}{2c\left(k\frac{\rho'}{\rho}-1\right)^2}9}{4}\frac{\delta}{R\left(1+\frac{\delta}{R}\right)}}{\left[\frac{\rho'}{\rho}+\frac{1}{2}+\frac{9}{4\left(\frac{\delta}{R}\right)^2}+\frac{9}{4}\frac{\delta}{R\left(1+\frac{\delta}{R}\right)^2}\right]} \qquad (8)$$

with $\delta = \sqrt{2v/\omega}$, the concentration of the disperse phase k and the kinematic viscosity v. The main effect for the losses is caused by the friction between the particles in the suspension or emulsion (primed quantities) and the carrier medium such as water or oil. The scattering losses correspond to equation (9).

$$\propto_{sc} \approx k k_0^4 R^3 \left[\frac{1}{6}\left(\frac{\beta-\beta'}{\beta}\right)^2+\frac{1}{2}\frac{(1-\rho/\rho')^2}{(2+\rho/\rho')^2}\right] \qquad (9)$$

with the wave number $k_0 = \frac{2\pi}{\lambda}$. In dependence on the concentration of the dispersion particles the Rayleigh or Mie scattering is responsible for the measuring effect. A complex influence of particle radius and particle distribution on the attenuation can be observed. Scattering losses are also caused by gas bubbles in liquids, leading to the serious problem that bubble-loaded liquids are virtually opaque to ultrasound. Thermal losses are described by equation(10).

$$\alpha_{therm} = k\frac{3}{2R^2}cT\chi\rho\left[\frac{\overline{\alpha}}{\rho c_p}-\frac{\overline{\alpha}'}{\rho c_p'}\right]f. \qquad (10)$$

Where is the thermal conductivity, c_p is the specific heat and $f = f(R, \omega, \chi, \chi', c_p, c_p')$ is an analytical function of the different parameters. The main loss effect is caused by the difference in the thermal expansion coefficients. The thermal losses can strongly influence the attenuation behaviour of the system under investigation. It is difficult to know or to evaluate the values of the thermal parameters of the different components. Therefore thermal losses were often neglected in the past. In gaseous media the speed of sound and the ultrasonic absorption are described by the classical thermodynamic relationships. For the speed of sound c for instance can be calculated by equation(11) as following [21]:

$$c^2 = \left(\frac{dP}{dP}\right)_{ad} \quad (11)$$

is valid. *c* and *α* react very sensitively in dependence on changes of state of the gaseous medium. The acoustic impedance *Z* is a third material-specific acoustic parameter:

$$Z = \rho c \quad (12)$$

It determines the behaviour at interfaces between different materials. For normal incidence of a compressional wave the reflection coefficient *R* and transmission coefficient *T* for the sonic pressure of the acoustic wave are given by

$$R = \frac{Z_1 - Z_2}{Z_1 + Z_2} \quad (13)$$

$$T = 1 - R \quad (14)$$

with Z_1 and Z_2 the acoustic impedances of the material the wave is travelling in and the material the wave is reflected by, respectively. Similar to **c**, Z, is influenced by the substance composition. The frequency f of the harmonic ultrasonic wave is changed by the Doppler effect when the wave is reflected by a reflector that is moving towards the source of the wave with speed v Then the frequency shift Δf between the incident and reflected waves is

$$\Delta f \approx 2f_0 \frac{v}{c} \quad (15)$$

for $v \ll c$ with the frequency of the incident wave f_0. This effect is widely used in flow measurement when particles in a liquid are used as moved reflectors.

2.2. Measurement of Wave Parameters

The previous section has shown that c , α , Z and Δf are the parameters of the acoustic wave, that are related to process parameters. The determination of the speed of sound and absorption can be realized in different ways. In principle, all methods are based on the following two equations, which can be derived from equation (1):

$$\alpha = \frac{1}{\Delta x} ln\left(\frac{P_0}{P}\right) = \frac{1}{\Delta x} ln\left(\frac{U_0}{U}\right) \quad (16)$$

$$c = \lambda f = \frac{\Delta x}{\Delta t} \quad (17)$$

U and U_0 are the electrical voltages, which are proportional to the alternating sonic pressures p and p_0. Δx is the path length and Δt is the transit time. Consequently the absorption measurement is reduced to an amplitude measurement. However, the attenuation measurement becomes more complicated due to diffraction in the measurement equipment, which cannot be avoided. Nevertheless, in principle this is a very interesting and important method to determine concentrations, particle diameters, particle size distributions etc according to equation (9). The speed of sound is easily and reliably measured by determination of the transit time of an ultrasound pulse along a known path length. Because time measurements can be made at high resolution the ultrasonic parameter speed of sound is often known or determined very precisely. Alternatively, can be determined using interferometric principles with the wave length λ. The acoustic impedance can be determined from the reflection coefficient R using equation (13):

$$Z_2 = Z_1 \frac{1+R}{1-R} \tag{18}$$

with $\frac{R = A_r}{A_i}$ determined by the amplitudes of the reflected and incident wave. Consequently, Z is determined by amplitude measurements and is under the same accuracy restrictions as the attenuation coefficient. Summarizing, it can be concluded that in most cases the ultrasonic parameters are experimentally determined by time measurement (c , f , Δf), amplitude measurement (α , Z, R, T) or more rarely by interferometric principles (λ , φ). The determination of the spectral amplitude distribution can play a role if scattering effects are considered.

2.3. Ultrasonic Sensor Classification

The detection of each of the parameters of the ultrasonic wave requires specific transducers and a dedicated electronic system for signal acquisition and parameter extraction. Whereas in the past many ultrasonic systems could be realized only as scientific instruments, due to modern microelectronics they are now feasible as compact devices at low cost.

Sound generation and detection in process applications is usually performed using piezoceramic transducers. They can be used in temperature ranges up to 300 °C and have the required long-term stability and reliability. Industrial applications of piezoelectric polymers are reviewed in [22]; however, there is no significant usage due to the limited temperature range (maximum 80 °C). Composite transducer materials are under intense development; however, currently specified operating temperatures do not exceed 100 °C.

An overview of possible transducer configurations that create a sensor is shown in Figure 1. Two categories can be distinguished: First, the ultrasonic transducers actively transmit acoustic waves and receive them later. This is the case in (1), (2) and (4). The phase, frequency and amplitude parameters of the wave are changed in (1) during interaction with reflecting particles or bubbles (1a) in a liquid, a reflector (1b) in a liquid or gas or the liquid substance under investigation (1c). The Doppler frequency shift allows particle velocity measurement (1a), the phase information is used for distance measurements (1b) and the amplitude information (1c) is related to the acoustic impedance and thus density of the substance. The impedance allows concentration measurements in substances with attenuation too high for sound transmission. The phase, amplitude and amplitude spectrum parameters are changed in (2) during the interaction of the substance and wave. Here the wave carries information about the substance between the transmitter and receiver. This allows concentration measurements via speed of sound measurement, particle distribution measurements via attenuation measurement and flow measurements via phase shift in upstream and downstream directions. The resonance frequency of a resonator in (4) is changed by deposition of a material onto the sensor surface or by interaction of a sensitive layer on the sensor with surrounding species in a gas or a liquid. This principle is the basis for the so-called acoustic microsensors. They play an increasing role in the sensor scenery. The principle is simple and allows extremely high resolutions. Acoustic microsensors have attained increasing importance in recent years due to the progress in chemical layer synthetics and deposition methods. The second category operates with a transducer that receives passively the sound generated by the process under investigation (3), providing general information about the process condition. The detected signals must be related to a particular process by dedicated algorithms.

These configuration possibilities in Figure 1 lead to the classification of ultrasonic sensing techniques by the form of interaction between the observed process parameter and the ultrasonic wave, which is given in Table 1. The table highlights one fundamental problem of ultrasonic measurements: while there are only three parameters to be measured, there is a large matrix of origins (applications) causing changes of these three parameters. This fact calls for intelligent sensors, that extract and evaluate the information carried by the ultrasonic signal efficiently, reliably and with high accuracy and resolution by dedicated hardware. Software algorithms based on models for the ultrasonic propagation and the interaction between the ultrasonic wave and physical or chemical variables of interest are employed for this aim. Furthermore, ultrasonic measurements are only meaningful when state parameters such as temperature and pressure are measured simultaneously with the ultrasonic parameters at high accuracy for compensation.

After this general consideration of the function and configuration of ultrasonic sensors, their attractive features can be summarized as:

non-invasive measurement,
in-line measurement,
rapid response, usually a fraction of a second,
low power consumption,
excellent long-term stability and
High resolution and accuracy.

The most challenging issues facing ultrasonic sensors are that

the exact knowledge of the acoustic properties of the substances is necessary for most ultrasonic measurements,

substances under investigation must be acoustically transparent for transmission and some reflection techniques,
ultrasonic measurements are highly disturbed when gas bubbles in liquids are present,
ultrasonic signals tend to be complicated and need relatively complex signal processing,
only integral information along the entire sound path is delivered and
attenuation of sound increases with frequency.

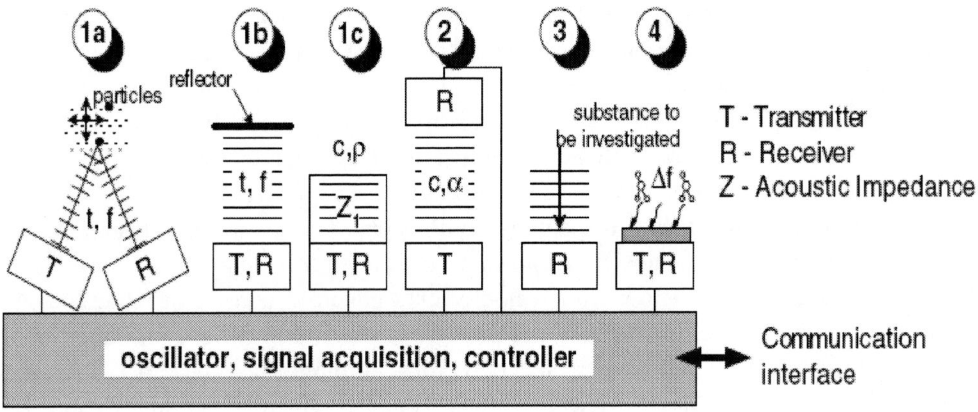

Figure 1. Transducer configurations of an ultrasonic sensor system. (1) Configuration for reflection techniques. (2) Configuration for transmission techniques. (3) Configuration for emission techniques. (4) Configuration for resonance techniques. [22]

Table 1. Classification of ultrasonic sensors by the technique by which the sensor interacts with the process [22]

	Technique	Parameters	Application
1	Ultrasonic reflection techniques	Phase Frequency Amplitude	• Distance, level, position, speed of sound • Object structure and presence of objects • Density, viscosity, concentration • Motion, velocity (Doppler effect)
2	Ultrasonic transmission techniques	Phase Amplitude	• Concentrations in multi-component systems • Particle size distribution in suspension, emulsions • Volume and mass flow, velocity • Density, viscosity • Temperature
3	Ultrasonic emission techniques	Amplitude Frequency	• Process condition monitoring
4	Ultrasonic resonance techniques	Frequency Amplitude	• Mass • Viscosity, viscoelasticity, density • Specific chemical or biological species • Multi-component analysis

2.4. Applications of Ultrasonic Sensors in Process Monitoring

The injection molding process can be considered as consisting of three stages: (1) the filling stage during which the material is injected into the cavity; (2) the packing/holding stage where additional material is forced into the cavity under high pressure until the gate is frozen to compensate for the shrinkage due to continuous cooling; and (3) the solidification stage during which the part is cooled further until it is sufficiently solidified. The part is then ejected and the cycle restarts. Some of the information to be monitored is specific to a single stage while others are important throughout the whole cycle. These requirements are described in more detail below.

During the filling stage, the flow front advancement and the flow front velocity are critical information. The flow front advancement indicates whether the cavity is properly filled. The flow front position can be used to control the plunger speed so as to allow a smooth transition from the filling to packing/holding phase to avoid part flashing and mold damage due to high impact. Since the viscosity of the material is dependent on the local flow rate, the measurement of the flow front velocity may ensure a consistent material behavior throughout the filling phase. In addition, the flow front velocity may have an important effect on the quality of the weld lines where the two fronts impinge on each other. Pressure and temperature sensors can be used to detect the flow front arrival. However, all pressure sensors have a threshold below which no reading is obtained. Therefore, the measured flow front arrival time is usually slightly longer than the real time. The advantage of ultrasonic technique in on-line monitoring of the arrival and advancement of the molten polymer flow front was demonstrated by Cao [23]. The research group of P.D. Coates in University of Bradford developed an ultrasonic measurement system for micro injection molding process monitoring based on piezoelectric sol-gel sensors, which is conjunct with the micro injection molding machine of Battenfeld Microsystem 50 and able to measure the flow front position, cooling dynamics, shrinkage, morphology changes during solidification and cycle repeatability. The measurement of polymer melts filling position and cooling dynamics of this ultrasonic system was validated by the other flow visualization unit on the same machine [24]. A. Sato et al. used a well designed ultrasonic injection molding device to investigate the effect of ultrasonic oscillation on micro structures' filling and replication during micro injection molding process [25].

During the packing/holding stage, the gate freezing time is considered as particularly important. At this point, no more material can be added into the cavity and further application of the holding pressure results in a waste of energy.

Upon further cooling during the solidification phase, as the material begins to shrink from the wall, an air gap is formed between the wall and the part giving rise to a significant change in heat transfer behavior. Effects of the gap formation on injection molding of plastics have been mentioned by many researchers [26,27], but little efforts were devoted to its monitoring. The formation of the gap has several consequences. An increase in thermal contact resistance between the part and the cavity wall reduces the heat transfer efficiency resulting in longer cycle time. Depending on the time the gap is formed and whether a gap is formed on one side or both sides of the part, this might result in a non-uniform temperature distribution in the thickness direction which induces residual stresses and warpage of the part. Ultrasonic technique was previously utilized to monitor the gap development during injection molding; however, the time when detachment occurred was not clearly identified [28]. Nishiwaki et al.

demonstrated the principle of identifying both detachments from external mold surface and internal mold insert [29].

Moreover, the material properties such as density and elastic contants change rapidly during solidification as the material passes from the liquid/rubbery state to a solid/glassy state. These properties of the solidifying material, if available, can be used as input parameters to the process control system to achieve the required final parts' quality. With semi-crystalline material such as HDPE, the density and ultrasonic velocity undergo a sharp change at the temperature of crystallization. By identifying such distinct characteristics, Nishiwaki, et al. have observed the two solid/liquid interfaces inside the polymer, using ultrasonic techniques [29].

A pulse/echo ultrasonic technique was used for on-line monitoring of the injection molding process by S.-S. Wen et al. [5] . In their study, 5 MHz longitudinal ultrasonic transducers (UT's) are used. The acquision system includes an oscilloscope, two pulsers and receivers with frequency range from 0.5~20MHz. A digitizer card is used to digitize the signals. The maximum sampling rate is 100 MHz for single channel (one UT) and 50 MHz for dual channels setup (2 UTs). The digitizer card has 8MB on-board memory and 8 bits A/D resolution. Data acquisition and signal processing were carried out using LabView software. The maximum acquision rate achievable for the overall system is 1000 acquisitions per second. A 150-ton co-injection molding machine was used in the experiments. The material employed in this study is an injection grade high-density polyethylene (HDPE). The mold is in the form of an open top box with non-uniform cavity thickness. An unfilled part and a finished part are shown in Figure 2. UT's were attached to external mold wall at different location as shown in Figure 3.

The results can be summarized in Figure 4. The Γ curve on the top was monitored from a thin and flat section of the molded box, while the Γ curve at the bottom was from a thick and curved section. Point 1 in Figure 4, where Γ's start to reduce indicate the arrival of melt front at the monitoring points. This occurs since the interface condition changes before and after melt arrival. Point 2 in Figure 4, where Γ's start to undergo a sharp change, indicate the end of cavity filling. The sharp drop results from the improvement in interface condition due to compression from high pressure. This drop identifies the complete filling of the cavity which was monitored with 100% success rate in this study. Point 3, where Γ's start to bounce gradually back to their maximum, indicate a pressure drop from either local part detachment, plunger retraction, or gate freezing. If the gradual increase does not occur simultaneously at different positions, it represents local part detachment due to the isolation from the gate caused by the complete solidification of the surrounding of the monitoring point. If the gradual increase is simultaneous at many locations, it may represent gate freezing. Plunger retraction will cause a sudden and linear increase corresponding to the sudden pressure change. Therefore section 4 in Figure 4 represents the change of Γ's due to the plunger retraction.

The solidification behavior of the material was also observed. The development of the two solid-liquid interfaces was detected. The thickness of each solid and liquid layer was estimated. In addition, the solid/liquid interface echo is indicated the locations within the molded part where temperature were around the crystallization temperature (130°C for HDPE) with high accuracy (±5°C). This temperature is a good reference for the derivation of the temperature profile inside the molded part or the validation of CAE software.

Table 2. Comparison between ultrasonic and pressure sensors for on-line monitoring of the injection molding process [5]

Ultrasonic Sensor	Pressure Sensor
Non-intrusive	Intrusive
Array configuration easily implemented	Array configuration difficult to implement
Capable of monitoring flow front arrival	Capable of monitoring flow front arrival
Capable of monitoring completion of filling	Capable of monitoring completion of filling
Capable of monitoring solidification at the same position	No such capability
Capable of monitoring part detachment on both sides	Capable of monitoring part detachment on one side only
Capable of monitoring material properties	No such capability

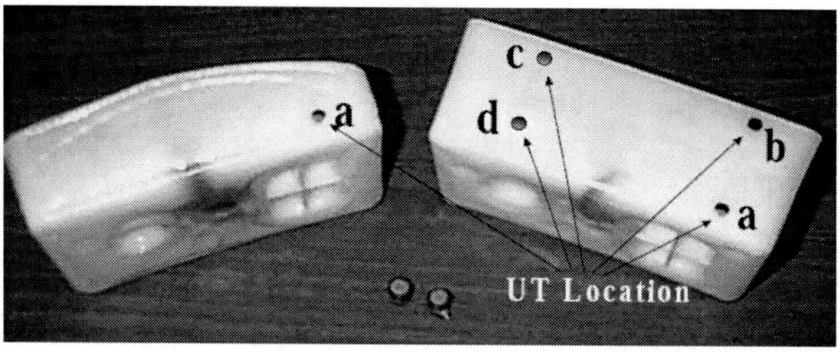

Figure 2. Molded boxes from incomplete and complete cavity filling with monitoring location as ‘a’, ‘b’, ‘c’ and‘d’ [5]

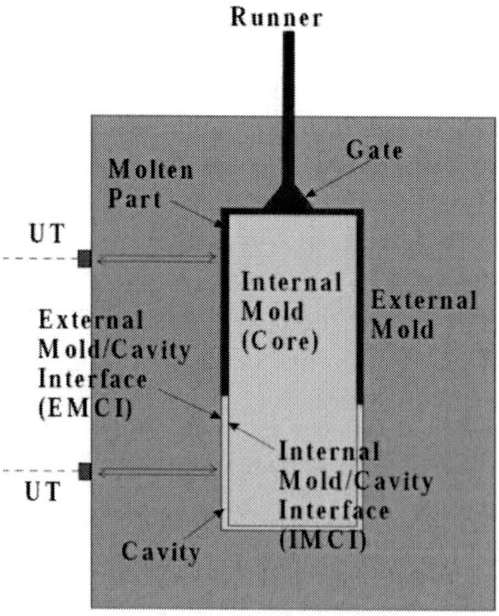

Figure 3. Installation of UT’s on the outside surface of the external mold [5]

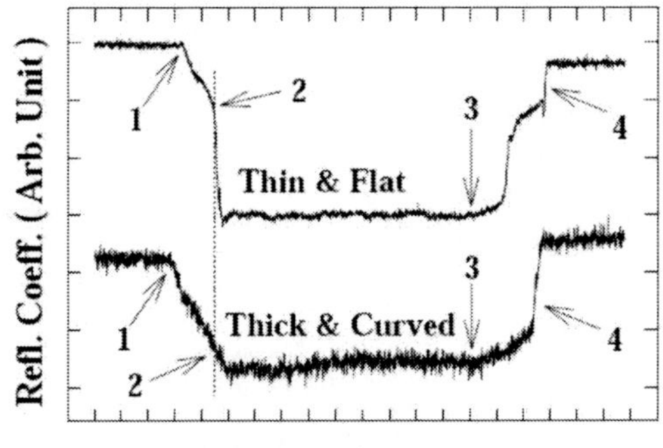

Figure 4. Summary of on-line monitoring of injection molding using reflection coefficients: 1) Flow front arrival, 2) End of filling, 3) Part detachment or gate freezing, 4) Plunger retraction [5]

From the above experimental results, a comparison between conventional ultrasonic and pressure sensors for on-line monitoring capability of the injection molding process is then summarized by S.-S. Wen et al [5] in Table.

3. Process Improvement with Ultrasonic Processing Technology

3.1. Ultrasonic Plastification

The ultrasonic plastification technology was developed especially for micro injection molding (MIM) process which is one of the most suitable processes for replicating microstructures with medium, even large production scales. Melt flow properties play an important part in MIM because of the microstructures cavity size down to micron. Two challenges are confronted by MIM: melt filling in micro cavity and the material wastage caused by the huge difference between the big sprues and micro parts.

To open new dimensions for the minimum weight (0.01 g), IKV, Aachen with its partners have developed a new micro injection molding machine [6]. Different plasticizing concepts are analysed. A test unit was build to prove the ability to operate as a plasticizing device and to optimize the necessary components and process parameters as shown in Figure 5. It was shown that plastification by ultrasonics is a very promising idea for very small amounts of plastics. In ultrasonic plastification the heat is generated at the contact surfaces of the ultrasonic horn and the polymer pellets and the friction among the pellets because of the high frequency vibration. In MIM a defined amount of material must be plasticized quickly and homogeneously from a semi-finished product or pellets. After plastification, the melt can be injected by a special plunger or the ultrasonic sonotrode itselt. The plastification results regarding homogenization and morphology of the molten mass have been evaluated by microscopy as shown in Figure 6. The material has crystallized very regularly with a homogenous structure.

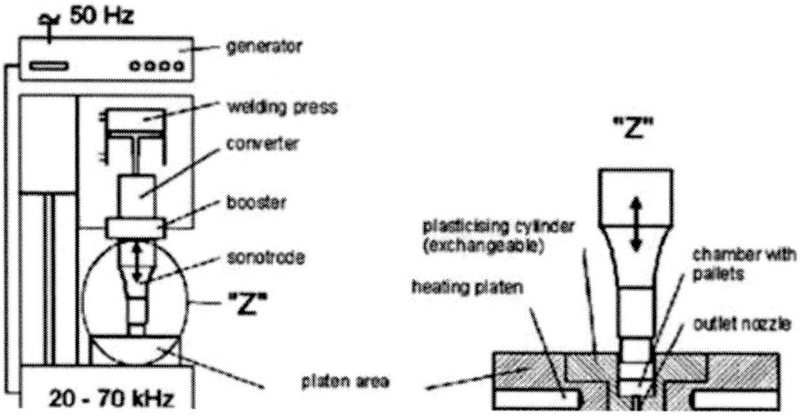

Figure 5.ultrasonic plastification system [6]

Figure 6. Morphology of molten mass plasticized by ultrasonics (POM) [6]

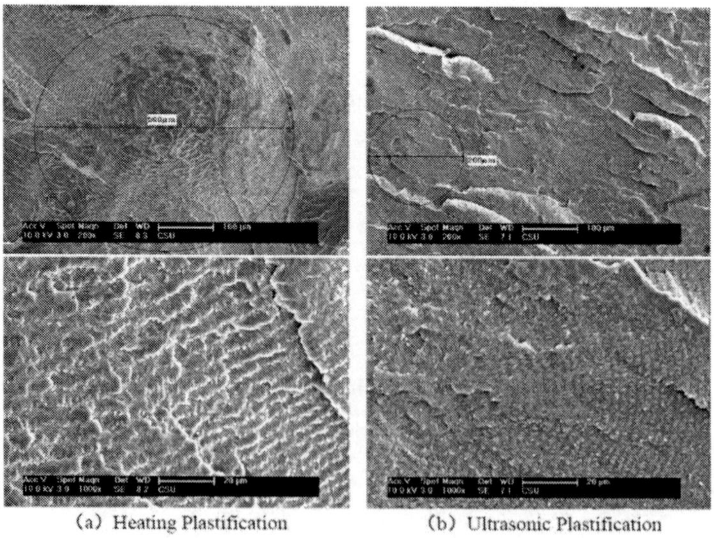

Figure 7. SEM images of plastic sample's cross sections under various ultrasonic plastification conditions[7]

B. Jiang et al [7]. studied the ultrasonic plastification process both by theoretically analysis and experiment. Ultrasonic plastification testing equipment was developed and has been successfully utilized to investigate the process. The microcosmic mechanism analysis reveals that the introduction of ultrasonic vibration field can obviously decrease the polymer melt's dynamic apparent viscosity, improve flow properties, accelerate polymer's molecule untie tangle and tropism, improve the melting rate, decrease melting temperature and the energy consumption; The key factors of the quick polymer plastification can be attributed to the high temperature stream and high pressure shock wave produced by ultrasonic cavitation field Two methods of polymer plastification, heating and ultrasonic plastification, are employed and different sizes of spherocrystal are obtaind. By mean of ultrasonic plastification, the diameter of the spherocrystal is about 200~300 μm smaller than that by heating plastification, and the microstructure is also finer and more homogenous. With ultrasonic vibration, polymer pallets are melting because of the acute friction among themselves induced by the horn. The melting pool which first formed at the horn's end extends gradually around the horn by the double action of ultrasonic vibration and plastification pressure. The barrel temperature has also an obvious effect on the melting efficiency. Barrel heating system with the same temperature ramp as the temperature rising caused by ultrasonic vibration was recommended. Figure 7 shows the plastic parts' broken cross section with different ultrasonic plastification conditions.

3.2. Composites Preparation with Ultrasonic Assistant

Filled polymer composites have been widely processed in normal injection molding process as well as in micro injection molding process. The preparation of these polymer composites is normally implemented by extrusion or kneading processes. In order to achieve polymer composites with well-dispersed and purified fillers in polymer matrix, many methods have been tried, like optimizing compounding process parameters (melting temperature, mixing time and rotation speed of screws etc.), adding stabilizer for fillers into the composites or modifying the surface of the filler. Ultrasonic technology is also applied in this field to improve the quality of composites used in injection molding process.

Cellulosic purification in natural fibers

Ultrasonic treatment is one major way to remove non-cellulosic components from wood fibers. It usually performed associated with the other two kinds of chemical treatments, since it has been found that ultrasound has the ability to separate lignin and hemicelluloses from wood fiber even without any chemicals being used. Even with added chemicals, ultrasound can assist in reducing the amount of chemical usage and shorten the reaction time. Thus ultrasound treatment is a relatively environmentally friendly process, and it has great potential in commercial usage.

Ultrasound-assisted extraction is well established in the processing of plant raw materials, particularly for extraction of low molecular substances and depolymerizing macromolecules. Recently, ultrasound has been reported to improve pectin technology from apple pressings [30] and pharmaceutically active compounds from Salvia officinalis [31], and increase of the yield of xylans from corn hulls [32] and corn cobs [32] without significant changes in their structural and molecular properties [34, 35].

Cavitation is induced in an aqueous suspension when the liquid is treated with ultrasound [36]. Then water decomposes into free radicals as shown below:

$$H_2O \rightarrow \bullet OH + \bullet H$$

Seino et al. [37] used electron spin resonance (ESR) methods to trap and characterize unstable radicals which were generated by ultrasound at 45 kHz and 100W in a lignin/Dimethylsulfoxide solution. They concluded that some specific linkages in lignin are homogeneously cleaved by the ultrasonic irradiation.

Tan et al. [38], using a very high power ultrasound of 1200W at 19 kHz, concluded that under the ultrasonic irradiation there is a formation of carbonyl groups (C=O) on the lignin surface. Pranovich et al. [39] studied the sonochemistry of lignin compounds under 20 kHz frequency, and reported that the hydroxyl radical attacked the aromatic ring and then formed various products from lignin.

Sun and coworkers [34. 35] have applied ultrasound in the extraction of non-cellulose component (such as lignin and hemicellulose) from wheat straw to reach higher extraction yields. In this research, after an ultrasound assisted alkali treatment, nearly 50 % of lignin and 70 % of hemicellulose were extracted from wheat straw. Laine and Goring [40] using the pulp fiber as the matrix studied the influence on the physical and chemical properties after ultrasound irradiation. They reported the oxidation of carbohydrate hydroxyls on the fiber surface increased the carbonyl group content of the fiber, and the porosity of fiber wall increased after ultrasound treatment. The treatment of cellulosic fibers with ultrasound has also been used in various areas such as pulping, debarking, defibration, bleaching, stock preparation and grafting, beating, impregnation and penetration.

Table 3. The yield of lignin fraction (% dry matter) from wheat straw [34]

Lignin fraction	Ultrasonic time (min)							
	0	5	10	15	20	25	30	35
Total solubilized lignins	7.5	7.5	7.5	7.6	7.6	7.8	8.0	8.4
Acid-insoluble lignins	4.9	5.0	5.1	5.1	5.3	5.2	5.8	6.1
Acid-soluble lignins	1.0	0.9	1.0	1.1	0.9	1.2	1.1	1.2
Lignin associated in isolated	1.6	1.4	1.4	1.4	1.4	1.4	1.1	1.1
Hemicelluloses								

Table 4. The molecular weight of extracted lignin [34]

	Ultrasonic time (min)							
	0	5	10	15	20	25	30	35
$\overline{M}_w$	2890	3010	3130	3470	3570	3300	3070	2760
$\overline{M}_n$	1490	1540	1550	1620	1620	1340	1150	1040
$\overline{M}_w / \overline{M}_n$	1.94	1.95	2.02	2.14	2.20	2.46	2.66	2.65

Although the high-frequency ultrasonic operation is quieter than low-frequency ultrasound, and high-frequency (650 kHz) ultrasound is effective on oxidation reactions, but so far, only the low frequency (20 kHz) ultrasound has been applied to wood fibers. The main reason is that the lower frequency ultrasound produces more violent cavitation, and this cavitation helps to split fiber into fibrils and destroy the links between cellulose and the non-cellulosic components. Further, compared to high-frequency ultrasound, low-frequency ultrasound has a better ability to penetrate. This makes low-frequency ultrasound break the fiber from the center and produce homogeneous fibrils.

Sun and coworkers [34, 35] compared the classic and ultrasound-assisted extraction process of wheat straw. In their conclusion, they indicated that the ultrasound-assisted procedures are superior to the traditional procedures. For the extraction of lignin, with 0.5M KOH solution and 35 min sonication time, the ultrasonic irradiation promoted the extraction of lignin from wheat straw, and the yield and purity increased with sonication time (Table 3). The lignin fraction obtained by the ultrasound-assisted alkali extraction shows a higher molecular weight (Table 4). They also indicated that under ultrasonic alkali treatment there is no significant change in lignin composition and its structure. For the hemicellulose, under 0.5M NaOH in 60% methanol at 60 oC for 2.5h, with a sonication time from 5-35 min resulted in an increasing yield from 2.9 to 9.2% (Table 5). And just as in the case of lignin and hemicellulose don't have significant change after ultrasound-assisted extraction.

Table 5. The yield of hemicellulose from wheat straw [34]

	Ultrasonic time (min)							
	0	5	10	15	20	25	30	35
Hemicelluloses	12.3	13.4	13.8	14.6	15.6	15.6	15.6	15.8
Lignin	10.5	11.6	11.8	12.8	13.3	13.3	13.4	13.5
Residue	75	72.8	72.2	70.4	68.9	68.9	68.8	68.5

Filler dispersion in polymer matrix

Polymer nanocomposites have opened a new horizon for a promising class of hybrid material by incorporation of particulate fillers into polymer matrices to improve or modify properties of neat polymers. Over the past several decades, a number of studies of the effects of ultrasound on polymers have been performed and reported. It was observed that long-chain polymer molecules can be ruptured by high intensity ultrasound during melt extrusion [41]. The breakage of the C-C bond by ultrasound leads to the formation of long-chain radical [42]. In polymer-filler systems, polymeric radicals may terminate on the clay surface or combine with the surface modifying agent of organoclay forming a chemical bond.

Several studies about the application of ultrasound in nanocomposite preparation were published within the last 5 years. These studies were devoted to preparation and characterization of nickel-polystyrene nanocomposites [43], conductive polyaniline-nanosilica particle composites [44] under static ultrasonic treatment conditions, and silica agglomerate breakdown in a continuous ultrasonic extruder [45]. They all found that ultrasound increases the dispersion of filler. The most recent attempts are to develop polymer/clay nanocomposites via melt intercalation in an intensive batch mixer [46-49] and a stationary cup [50] equipped with power ultrasound. However, a prolong ultrasonic treatment (10~20 min) was used in these studies, apparently leading to a substantial degradation of polymer matrix. A continuous method to achieve rapid intercalation at short residence times (7~21 s) and partial exfoliation of PP/clay nanocomposite without chemical modification of the matrix was developed [51].

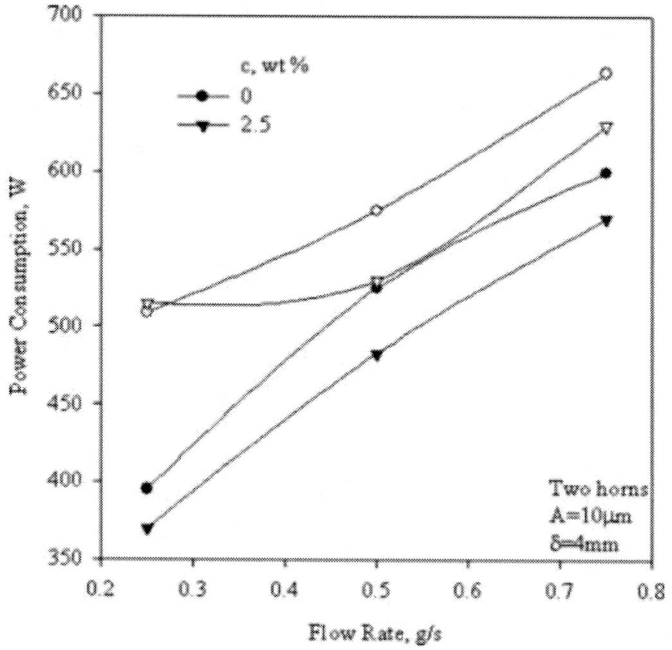

Figure 8. Ultrasonic power consumption as a function of flow rate for neat PP and PP/clay nanocomposites (solid symbols) and neat HDPE and HDPE/clay nanocomposite (open symbols) [52]

Sergey Lapshin et al. [52] studied the effect of residence time of the polymer in the ultrasonic treatment zone on controlling the structure and properties of polyolefin/clay nanocomposites. High-density polyethylene and isotactic polypropylene were compared. The recorded ultrasonic power consumption is the total power consumption, a part of which is dissipated as heat; while the rest is utilized to disperse clay filler and promote polymer intercalation into clay inter gallery spacing. However, it is not possible to determine the exact portion of power in these two cases. The only thing that can be recorded is the initial power consumption of the system when the horn is running without loading and this loss was subtracted from the recorded values of power consumption to give the values used in Figure 8. An increase of flow rate leads to an increase in power consumption which is an indication that more energy is being transmitted into the system at higher flow rates.

4. Improvement of Properties of Injection Molded Parts by Ultrasonic Assistant

As a mechanical wave, the ultrasonic oscillation can pass through many materials and vibrate them to change the materials' physical, mechanical even chemical properties. Therefore, the ultrasonic technology is also attempted to to homogenize and increase the dispersion of the molten material, either in the liquid stage or in the solidifying stage, either at a macroscopic or microscopic level in injection molding process.

Lemelson et al. described an apparatus and method for controlling the internal structure of plastics in a mold by application of ultrasonic energy to the solidifying material to affect beneficial control of the crystalline structure formed thereof upon solidification [53]. Atsushi Sato et al. developed an ultrasonic injection molding (UIM) system, which applies ultrasonic waves to injection molding, as a precision injection molding technology. Consequently, replication during the packing and holding stages is facilitated by the UIM. Moreover, evaluation of the residual optical strain of concave lens revealed that the strain was much lower in UIM than in conventional molding. The decreased strain was attributed to the local heat generation by the ultrasonic waves [54]. Chang Lu et al. investigated the effect of melt temperature, ultrasonic oscillations, and induced ultrasonic oscillation modes on weld line strength of polystyrene (PS) and polystyrene/polyethylene (PS/HDPE) (90/10) blend, the results show that for PS/HDPE blends, the presence of ultrasonic oscillations can improve the weld line strength when the melt temperature is 230°C, but when the melt temperature is 195°C, the induced ultrasonic oscillations hardly enhance the weld line strength [55]. L Xie et al. designed and constructed an ultrasonic device which is able to be integrated in the injection mold to transfer the ultrasonic oscillation energy into the polymer melts during their filling into mold cavities. With this device, they successfully improved mechanical properties of wood/pp composites parts formed by injection molding process [56]; furthermore some meaningful results in reinforcing the strength of the weld line defect in micro injection molding process were also obtained [57].

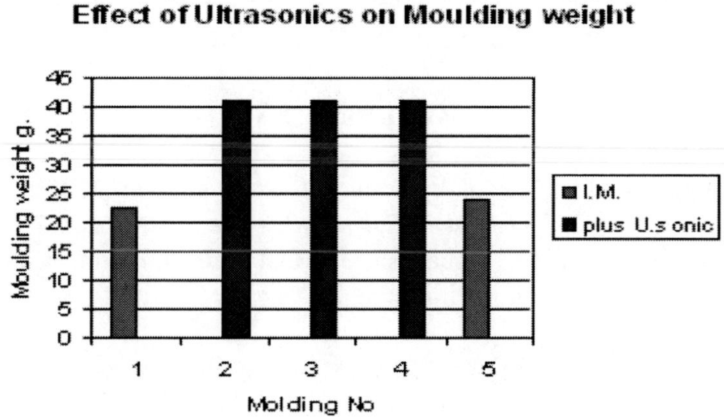

Figure 9. Effect of ultrasonic on injection molding parts weight

In addition, a previous European Funded CRAFT project was looking at producing highly filled parts and during the development work, it was found that the flow of unfilled polymer into a mould cavity could be increased by using ultrasonic energy. Using the same processing conditions, the moulding weight injected into an under-filled cavity was increased by 50% through the application of ultrasonic vibration to the polymer melt, shown in Figure 9. Another EU project named Ultra-Melt-Innovative energy saving development for the injection molding sector was proposed by researchers from Unite Kingdom, Germany, Italy and Spain. The use of ultrasonic vibration was investigated by applying it just prior to and during injection procedure. The ultrasonic is stopped once hold pressure is applied. Injection time speed and pressure remained unchanged. Mouldings were produced with and without the application of ultrasonic vibration.

5. Future Trends

The current position of ultrasonic technology in the field of ultrasonic measurement and ultrasonic processing on polymer melt and composites compounding in injection molding can be compared with that of ultrasound techniques when they were first introduced for medical applications. When the technology reached a required level, it generated an 'explosion' in medical diagnostics through its ability to visualize internal parts of a patient's body, analyze blood streams and other features. Now it is difficult to imagine a modern hospital without a set of ultrasonic instruments. Similarly, there will be a broad penetration of ultrasonic technology into injection molding process in various fields of monitoring and processing. Novel ultrasonic spectroscopy has the ability to perform a wide range of analyses, which other methods cannot do, as well as to increase the effectiveness and reduce the cost of a number of analytical tasks currently performed by traditional techniques. Product quality can also have a great improvement with novel ultrasonic processing technology, which cannot realize by normal injection molding. It could be prospected that a totally creative novel injection molding process based on ultrasonic technology is possible to be constructed which will comprises of ultrasonic plasticization unit, ultrasonic injection unit, ultrasonic demolding unit and process monitoring and quality controlling unit based on ultrasonic.

References

[1] He, B; Zhang, X; Zhang, Q. et al. Real time ultrasonic monitoring of the injection-molding process. *Journal of Applied Polymer Science*, 2008, vol. 107, 94-101.

[2] Ono, Y; Jen, CK; Cheng, CC; Kobayashi, M. Real-time monitoring of injection molding for microfluidic devices using ultrasound. *Polymer Engieering & Science*, 2005, vol. 45(4), 606-612.

[3] Micheaeli, W; Starke, C. Ultrasonic investigations of the thernmoplastic injection molding process. *Polymer Testing*, 2005, vol 24(2) 205-209.

[4] Coates, PD; Earnes, SE; Sibley, MG. et al. In-process vibrational spectroscopy and ultrasound measurement in polymer melt extrusion. *Polymer*, 2003, vol.44(9), 5937-5949.

[5] Wen, SSL; Jen, CK; Nguyen, KT. Advances in on-line monitoring of the injection molding process using ultrasonic techniques. *Int's Polymer Processing*, 1999, vol. XIV, 175-183.

[6] Michaeli, W; Spennemann, A; Gaertner, R. New plastification concepts for micro injection molding. *Microsystem technologies*, 2008, 8, 55-57.

[7] Jiang, B; Hu, J; Wu, W. et al. Research on the polymer ultrasonic plastification. *Advanced materials research*, 2010, vol. 87-88, 542-549.
[8] Lu, C; Yu, X; Guo, S. Ultrasonic improvement of weld line strength of injection molded polystyrene and polystyrene/polyethylene Blend Parts. *Polymer Engineering and Science*, vol. 45(12), 1666-1672.
[9] Sato, A; Ito, H; Koyama, K. Study of Application of ultrasonic wave to injection molding. *Polymer engineering and science*, vol. 49(4), 768-773.
[10] Kennedy, P. *Flow analysis of injection molds*. Hanser Publishers, Munich, Vienna, New York. 1995.
[11] Rafizadeh, M; Patterson, WI; Kamal, MR. *Int. Polym. Process*, 1996, 11, 352.
[12] Thomas, CL. Ph.D. *dissertation*, Department of Mechanical Engineering, Drexel University, Philadelphia, 1993.
[13] Johannaber, F. *Injection Molding Machines – a User's Guide*, 3rd Edition, Hanser Publishers, Munich, Vienna New York, 1994.
[14] Lynnworth, LC. *Ultrasonic Measurement for process control, Theory, Techniques, Applications,* Academic Press Inc., San Diego.
[15] Kino G S. *Acoustic Waves: Devices, Imaging, and Analog Signal Processing*, Englewood Cliffs, 1987, NJ: Prentice-Hall
[16] Krautkraemer J and Krautkraemer H. *Ultrasonic Testing of Materials,* 1990, Berlin: Springer
[17] Shutilov V A. *Fundamental Physics of Ultrasound,* 1988, New York: Gordon and Breach
[18] Povey M J W. *Ultrasonic Techniques for Fluids Characterization,* 1997, San Diego: Academic
[19] Hauptmann P and Rothe B. Zur Ultraschallabsorption in kolloiddispersen Polymersystemen, *Acta Polymerica ,1981,*32:215–20
[20] Schaaffs W. *Molekularakustik, 1967,* Berlin: Springer
[21] Kuttruff H. *Physik und Technik des Ultraschalls, 1988,* Stuttgart: Hirzel
[22] Hauptmann, P; Hoppe, N; Puettmer, A. Application of ultrasonic sensors in the process industry. *Meas. Sci. Technol.*, 2002, 13, R73-R83.
[23] Cao, B. *Master thesis*, Depart of Electrical Engineering, McGill University, Montreal, 1996.
[24] Brown, EC; Whiteside, BR; Spares, R; Coates, PD. *ANTEC*, 2009, Chicago, USA.
[25] Sato, A; Sakaguchi, H; Ito, H; Koyama, K. *ANTEC*, 2009, Chicago, USA.
[26] Nishiwaki, N; Cui, A; Konno, M; Hori, S. Proc. *Seikei-kakou*, 1993, 5, 870.
[27] Yu, CJ; Sunderland, JE; Poli, C. *Polym. Eng. Sci*, 1990, 30, 1599.
[28] Wang, H; Cao, B; Jen, CK. et al. *Polym. Eng.Sci*, 1997, 37, 363.
[29] Nischiwaki, N; Hori, S; Shimazaki, K. et al. Proc. The 1st Pacific Symp. *On Flow Visualization and Image Proc.*, 1, 225.
[30] Panchev, IN; Kirtchev, NA; Kratchanov, CG. On the production of low esterified pectins by acid maceration of pectic raw material with ultrasound treatment, *Food Hydrocolloids*, 1994, 8, 9-17.

[31] Salisova, M; Toma, S; Mason, T. Compararison of convensional and ultrasonically assisted extractions of pharmaceutically active compounds from Salvia officinalis, *J. Ultrason Sonochem*, 1997, 4, 131-134.
[32] Ebringerova, A; Hromadkova, Z. The effect of ultrasound on the structure and properties of the water-soluble corn hull heteroxylan, *Ultrason Sonochem*, 1997, 4, 305-309.
[33] Hromadkova, Z; Kovacikova, J; Ebringerova, A. Study of the classical and ultrasound-assisted extraction of the corn cob xylan, *Ind Crops Prod.*, 1999, 9, 101-109.
[34] Sun, R; Sun, XF; Xu, XP. Effect of ultrasound on the physicochemical properties of organosolv lignins from wheat straw, *J. Appl. Polym. Sci.*, 2002, 84(13), 2512-2522.
[35] Sun, R; Tomkinson, J. Comparative study of lignin isolated by alkali and ultrasound-assisted extraction from wheat straw, *Ultrasonic Sonochemistry*, 2002, vol. 9, 85-93.
[36] Petrier, C; Lamy, MF; Francony, A; Benahcene, A; David, B; Renaudin, V; Gondrexon, N. Sonochemical degradation of phenol in dilute aqueous solutions: comparison of the reaction rates at 20 and 487 kHz, *J. Phys. Chem.*, 1994, 98(41), 10514-10520.
[37] Seino, T; Yoshioka, A; Fujiwara, M; Chen, KL; Erata, T; Tabata, M; Takai, M. ESR Studies of radicals generated by ultrasonic irradiation of lignin solution. An Application of the Spin Trapping Method, *Wood Sci. Technol.*, 2001, 35(1-2), 97-106.
[38] Tan, GM; Yasuda, S; Terashima, N. The Effect of ultrasonic irradiation on delignification reactions. ii. Behavior of lignin under ultrasonic irradiation, *Mokuzai Gakkaishi*, 1985, 31(5), 388-396.
[39] Pranovich, AV; Reunanen, M; Holmbom, B. Sonochemistry of Lignin Model Compounds in Water. *Advances in Lignocellulosics Chemistry for Ecologically Friendly Pulping and Bleaching Technologies, European Workshop on Lignocellulosics and Pulp*, 5th, University of Aveiro, Aveiro, Port., 1998, Aug. 30-Sept. 2, 421-424.
[40] Laine, JE; Goring, DAI. Influence of ultrasonic irradiation on the properties of cellulosic fibers, Cell. *Chem. Technol.*, 1997, 11(5), 561-567.
[41] Isayev, AI; Wong, CM; Zeng, X. Adv. *Polym. Technol.*, 1990, 10, 31.
[42] Basedow, AM. *Adv. Polym. Sci.*, 1977, 22, 83.
[43] Kumar, R; Koltypin, Y; Palchik, O; Gedanken, A. *J. Appl. Polym. Sci.*, 2002, 86, 160.
[44] Xia, H; Wang, Q. *J. Appl. Polym. Sci.*, 2003, 87, 1811.
[45] Isayev, AI; Hong, CK; Kim, KJ. Rubber. *Chem. Technol.*, 2003, 76, 923.
[46] Ryu, J; Lee, P; Kim, H; Lee, J. *SPE ANTEC*, 2001, 2, 2135.
[47] Ryu, J; Lee, P; Kim, H; Lee, J. *SPE ANTEC*, 2002, 2, 2240.
[48] Ryu, J; Park, S; Kim, H; Lee, J. Materi. *Sci. Eng. C.*, 2002, 24, 285.
[49] Lee, EC; Mielewski, DF; Baird, RJ. *Polym.Eng. Sci.*, 2004, 44, 1773.
[50] Lee, EC; Mielewski, DF; Baird, RJ. *Polym. Eng. Sci.*, 2004, 44, 1773.
[51] Lapshin, S; Isayev, A. *SPE ANTEC*, 2005, 2, 1911.
[52] Lapshin, S; Swain, SK; Isayev. AI. Ultrasound Aided Extrusion Process for Preparation of Polyolefin-Clay Nanocomposites. *Journal of Vinyl and Additive Technology*, 2008, vol. 13(1), 40-45.
[53] Lemelson, J. U.*S. Patent*, 1981, 4, 288, 398.
[54] Atsushi Sato, Hiroshi Ito,Kiyohito Koyama. Study of Application of Ultrasonic Wave to Injection Molding. *Polym Eng Sci.*, 2009, DOI 10.1002/pen

[55] Chang, Lu; Xiaofeng, Yu; Shaoyun, Guo. Ultrasonic improvement of weld line strength of injection-molded polystyrene and polystyrene polyethylene blend parts. *Polym Eng Sci.*, 2005, DOI 10.1002/pen.20456.

[56] Xie, L; Grueneberg, T; Steuernagel, L; Ziegmann, G; Militz, H. Improvement of mechanical properties of injection molded wood/polypropylene composites parts with ultrasonic oscillation assistant, *Pacific Rim International Conference on Advanced Materials and Processing*, 1-5, August, 2010, Cairns Australia.

[57] Xie, L; Ziegmann, G; Jiang, BY. Reinforcement of micro injection molded weld line strength with ultrasonic oscillation, Microsystem Technologies, DOI: 10.1007/s00542-009-0928-9.

In: Injection Molding: Process, Design, and Applications ISBN: 978-1-61761-307-4
Editor: Phoebe H. Kauffer

Chapter 7

AN INTEGRATED QUANTITATIVE FRAMEWORK FOR SUPPORTING PRODUCT DESIGN IN THE MOLD SECTOR

Irene Ferreira*[*1]*,
***José A. Cabral*[2] *and Pedro Saraiva*[3]**
[1]Polytechnic Institute of Leiria, Portugal Campus 2,
Morro do Lena - Alto do Vieiro, 2411-901 Leiria, Apartado 4163
[2]Engineering Faculty of University of Porto, Portugal
[3]Chemical Engineering Department, University of Coimbra, Portugal

ABSTRACT

The injection mold is a high precision tool responsible for the production of most plastic parts used everywhere. Its design is considered critically important for the quality of the product and efficient processing, as well as determinant for the economics of the entire injection molding process. However, typically, no formal engineering analysis is carried out during the mold design stage. In fact, traditionally, designers rely on their skills and intuition, following a set of general guidelines. This does not ensure that the final mold design is acceptable or the best option. At the same time, mold makers are now highly pressured to shorten both leading times and cost, as well as to accomplish higher levels of mold performance. For these reasons, it is imperative to adopt new methods and tools that allow for faster and higher integrated mold design. To that end, a new global approach, based on the integration of well-known quantitative techniques, such as Design for Six Sigma (DFSS), Structural Equation Modeling (SEM), Axiomatic Design (AD) and Multidisciplinary Design Optimization (MDO) is presented. Although some of these methods have been largely explored, individually or in combination with other methodologies, a quantitative integration of all aspects of design, in such a way that the whole process becomes logical and comprehensible, has not yet been considered. To that end, the DFSS methodology, through its IDOV roadmap, was adopted. It is based on the ICOV Yang and El-Haik proposal, establishing four stages for the design process:

* Corresponding author: Email: irene.ferreira@ipleiria.pt

Identify, which aims to define customers' requirements/expectations; *Design*, where the creation of a product concept, and its system-level design, is performed; *Optimization*, in which all the detailed design, through product optimization, is handled; and finally, *Validation*, where all product design decisions are validated, in order to verify if the new designed entity indeed meets customer and other requirements. As a result, this approach tackles the design of an injection mold in a global and quantitative approach, starting with a full understanding of customer requirements and converting them into optimal mold solutions. In order to validate it, an integrated platform was developed, where all different analysis modules were inserted and optimized through an overseeing code system. The results attained highlight the great potential of the proposed framework to achieve mold design improvements, with consequent reduction of rework and time savings for the entire mold design process.

INTRODUCTION

The injection mould is a high precision tool, responsible for the production of most plastic parts used everywhere. Its design is considered critically important to product quality and efficient processing, as well as determinant for the economics of the entire injection moulding process. At present, the rapid change in business environments requires faster mould design and manufacturing, in order to reduce the time-to-market of plastic parts, along with higher quality, greater efficiency and lower costs. In this sense, and in order to achieve high levels of product quality, in less time, both academia and industry have been looking for new methods to address mould design projects. A great amount of scientific research work has been done in this field, especially involving the design of particular mould components (e.g. feeding system (Lam, Britton, & Liu, 2004; Lam & Jin, 2001; K.S. Lee & Lin, 2006; Pandelidis & Zou, 1990; Shen, Yu, Li, & Li, 2004) and cooling system (Lam, Zhai, Tai, & Fok, 2004; Qiao, 2006)). However, due to the high complexity and mould component interactions, this approach may be considered to be insufficient (Ferreira, Weck, Saraiva, & Cabral, 2010). In fact, it will only be possible to explore the design space adequately by the exploitation of coupled relations amongst mould components. In this sense, it is essential to develop integrative approaches for mould design in order to optimize a mould, seen as a global system, through its functional systems integration. For that reason, a conceptual framework based on Design for Six Sigma (DFSS) (Creveling, J.L.Slutsky, & Jr., 2003; Yang & El-Haik, 2003); European Customer Satisfaction Index (ECSI) (Fornell & Bookstein, 1982; Tenenhaus, 2003); Axiomatic Design (AD) (Suh, 1990); and Multidisciplinary Design Optimization (MDO) (AIAA, 1991), was developed. This framework, which tackles the design of an injection mould in a global and quantitative approach, aims to guide and systematize its design process (Ferreira, Cabral, & Saraiva, 2009; Ferreira, et al., 2010). Though some of the previous methods have been largely explored, individually or in combination with other methodologies, a quantitative integration of all aspects of design, in such a way that the whole process becomes logical and comprehensible, has not yet been considered. To that end, the design process must be broken down, first into phases, and then into distinct steps, each with its own supporting methods (Krishnan & Ulrich, 2001). This objective was achieved by the DFSS methodology, through an IDOV roadmap. The IDOV roadmap, which is based on the ICOV phases (Identify, Conceive, Optimize and Verify), according to the proposal of Yang and El-Haik (Yang & El-Haik, 2003), establishes four stages for the design process: *Identify*, which aims to define customers' requirements/ expectations; *Design*, where the creation of a product concept, and its system-level design, is performed; *Optimization*, in which all the detailed design, through product optimization, is handled; and finally, *Validation*, where all product design decisions are validated, in order to verify if the new designed entity indeed meets customer and other requirements.

To support the *Identify* stage, ECSI was adopted as a reliable and independent way of assessing Customer Satisfaction (CS) and retention (Tomarken & Waller, 2005). Hence, an ECSI model, specific to the injection mould industry, was designed and validated (I. Ferreira, J. A. Cabral, & P. Saraiva, 2008b; Ferreira, Cabral, & Saraiva, 2007). The estimation of this model was performed through a component-based approach, based upon Partial Least Squares (PLS). The results obtained indicate that the theoretical model has very good data explanation capability (Ferreira, et al., 2008b). Based on such a model, it was possible to build one single objective function regarding customers' satisfaction levels, which was defined as a weighted function of specific customer attributes (I. Ferreira, J. Cabral, & P. Saraiva, 2008a).

The next stage, ***Design*** stage, was supported by AD. Its main objective is to generate physical solutions, characterized by Design Parameters (DPs), by mapping the Functional Requirements (FRs) of products onto the corresponding DPs, in order to get an uncoupled design. One first attempt to apply AD to injection mould design was previously carried out, which pointed toward improvement opportunities (Ferreira, Cabral, & Saraiva, 2006). In fact, typically, due mostly to technological and time reasons, mould design solutions are highly coupled (not accomplishing the AD first axiom, which states that independence of FRs should be assured). In this sense, even the design decisions made at this conceptual design stage should promote the independence of FRs, although, if some remaining coupled relations subsist, they are not considered to be prohibitive.

Afterwards, the detailed design of moulds is undertaken in the *Optimization* stage. This stage was supported by MDO, which is considered an appropriate methodology to design complex systems through an exploitation of coupling phenomena (Sobieszczanski-Sobieski & Haftka, 1997; Weck, 2004). To that end, a framework based on MDO, aimed at optimizing the mould design as a system, was built through the integration of four main mould subsystems: Structural, Feeding, Ejection and Heat-Exchange modules.

Finally, in the last ***Validation*** stage, the new designed entity, generated by our MDO framework, was tested, in order to evaluate if it allows one to reach higher levels of CS. This task was achieved by comparing the behaviour of the design solutions generated by the developed framework with the behaviour predicted by numerical simulation codes, and to predict the impact of each solution performance on CS. Based on data gathered for a plastic part, it was possible to confirm the high potential of the proposed framework to achieve mould design improvements.

BACKGROUND

Currently, product development is assumed to be at the new frontier for achieving competitive advantage in today's rapidly changing business environments (Brown & Eisenhardt, 1995; Nebiyeloul-Kifle, 2005). This fact is especially true because decisions made during early stages of design have the greatest impact over total cost and quality of the system (Smaling, 2005; Yang & El-Haik, 2003). Typically, the design process follows a recursive process, characterized by trial and error, where decisions are mainly supported based on intuition, empiricism and the so-called handbook method. Due to these poor practices, the paradigm of product development is expensive, unpredictable and prone to failures, where the loss caused by selecting wrong design solutions affects the whole process and is harder to recover in later stages (Kim and Cochran 2000). Only with a more systematic, scientific and rational approach to the design process will it be possible to mitigate these limitations. For that reason, it is imperative to adopt new methods and tools for product development, allowing for faster and higher integrated product design, in order to design optimal products prior to their launch (Baake, Stratil, & Haussmann, 1999; Balakrishnan & Jacob, 1996).

In this sense, several product development approaches were studied, in order to develop a global and strongly quantitative methodology that can work together with the intuitive non-quantitative and creative side of the design process. As a result, a new approach for product development support was proposed (Ferreira, et al., 2008a; Ferreira, et al., 2007, 2009). In order to validate it, the developed methodology was applied to the design of metallic moulds for plastic injection. This case study is particularly important, regarding its significant contribution to Portuguese GDP, as well as to the volume of exports, since Portugal is one of the world's largest producers of advanced tools for injection (Cefamol, 2009). At the same time, some works previously carried out in the injection moulding field pointed out the need for new approaches to support the mould design process (Ferreira, 2002; Ferreira, Cabral, & Saraiva, 2001, 2003). In fact, the design and manufacture of injection moulds is a costly process, dominated by empiricism, where generally no formal engineering analyses are performed. The conventional method for making moulds involves a great amount of errors that are translated as wastes of time and money, resulting in very expensive products and long manufacturing periods. Currently these practices are clearly inefficient, and that justifies the effort to develop new approaches to support the design of moulds, in order to reach faster mould design and manufacturing, along with higher quality, greater efficiency and lower costs.

A large amount of scientific research has been done on mould design and its related fields over the last years, mostly based upon Knowledge-Based (KB) methods. This approach is justified by the extensive empirical knowledge about mould component functions. Examples of work in this area are IKB-MOULD (Chan, Yan, Xiang, & Cheok, 2003), IKMOULD (Mok, Chin, & Ho, 2001), ESMOULD (Chin & Wong, 1996), amongst others (K. S. Lee, Li, Fuh, Zhang, & Nee, 1997; R.-S. Lee, Chen, & Lee, 1997; Lou, Jiang, & Ruan, 2004). According to Chan et al. (Chan, et al., 2003), one emergent area of research in the injection moulding field attempts to automatically generate the design of mould tool components (Lam, Britton, et al., 2004; Lam & Jin, 2001; Lam, Zhai, et al., 2004; K.S. Lee & Lin, 2006; Pandelidis & Zou, 1990; Qiao, 2006; Shen, et al., 2004). However, this approach has been considered to be feasible only for the automatic generation of particular parts of mould design (Chan, et al., 2003; Low, 2003). In fact, due to the high complexity and mold component interactions, only with a global and integrative approach will it be possible to exploit the synergies of interacting phenomena and to adequately explore the design space in order to reach optimal mould designs (note that this solution is quite different from the solution gathered by the integration of partially optimal mould components). Furthermore, it is also essential to link the level of CS with the search for optimal solutions. These two aspects were the main focus of the developed approach, which aims to constitute an alternative to the traditional procedure to design injection moulds.

INTEGRATED QUANTITATIVE FRAMEWORK

The development of a fully integrated optimization framework, regarding the design of injection mould tools, is the main focus of this chapter. This framework, which is mainly based on a Design for Six Sigma (DFSS) approach, encompasses a four stage roadmap, designated as IDOV.

The First Stage: Identify

Customer satisfaction (CS) and retention are considered as key issues for organizations in today's competitive market place, turning them into a vital concern to achieve customer loyalty (Cassel, Hackl, & Westlund, 2000; Grigoroudis, Nikolopoulou, & Zopounidis, 2008). In this sense, it is necessary to develop reliable and independent ways of assessing CS, allowing the comparison between companies within the same sector and/or that operate in the same country, or at a macroeconomic level. The ECSI model is a framework, adapted from the Swedish Customer Satisfaction Barometer and the American Customer Satisfaction Index, which aims to harmonize the Customer Satisfaction Indexes (CSI) in Europe (M. J. Vilares, Almeida, & Coelho, 2005). As a Structural Equation Model (SEM), ECSI is based upon sets of linear equations used to specify CS in terms of their presumed cause-and-effect variables (Dunn, Everitt, & Pickles, 1993; Hair, Tatham, Anderson, & Black, 1998). These hypothesized relationships are then translated into mathematical models, which are tested against empirical data. Generally these models are established for variables, latent variables or constructs, which cannot be measured directly (CS is a typical example of this kind of variable). Thus, these variables must be inferred through observing or measuring specific features that operationally define them, the so-called manifest variables or indicators. Consequently, the ECSI model is composed of two models: the Structural Model, or Inner Model, and the Measurement model, or Outer Model. The first one specifies the relations between the constructs, while the Measurement model includes the potential interrelationships between constructs and their indicators. Technically, the Structural Model can be described through Eq. 1:

$$\eta = \beta\eta + \gamma\xi + \upsilon \qquad \text{Eq. 1}$$

where η is the vector of endogenous latent variables, and ξ is a vector of exogenous latent variables. The exogenous variables are exclusively influenced by factors lying outside the model, while variables that are hypothesized to be influenced from inside the model are endogenous variables. The coefficients of the structural model, β and γ, give the direct impact on a latent variable when there is a unit change in an antecedent latent variable. If the antecedent variable is an exogenous variable, the direct impact is represented by γ, while β represents the direct impact over endogenous variables derived by a unit variation of another endogenous variable. The vector of specification residuals for the endogenous latent variables η is represent by υ. On the side of the Measurement Model, there are three possible types of relations between latent variables and its indicators. When the observed variables are assumed to be the reflex of the latent variables, the model is reflective (Eq. 2 and Eq. 3).

$$x = \Lambda_x \xi + \delta \qquad \text{Eq. 2}$$

$$y = \Lambda_y \eta + \varepsilon \qquad \text{Eq. 3}$$

here x and y are the exogenous and endogenous manifest vectors, respectively, Λ_x and Λ_y are the correspondent weight matrices (loadings), and, finally, δ and ε are measurement error vectors. The formative model is used when the observed variables are assumed to cause or form the latent variables (Eq. 4 and Eq. 5):

$$\eta_i = \sum_{l=1}^{H} \lambda_{\eta_{il}} y_l + \delta_{\eta_i} \qquad \text{Eq. 4}$$

$$\xi = \sum_{l=1}^{G} \lambda_{\xi_l} x_l + \delta_{\xi} \qquad \text{Eq. 5}$$

Where $\lambda_{\eta_{il}}$ and λ_{ξ_l} are coefficients of the formative model, H and G are the number of manifest variables associated with variables η and ξ, respectively. Finally, δ_{η_i} and δ_{ξ} are the specification errors.The mixed model combines both models.

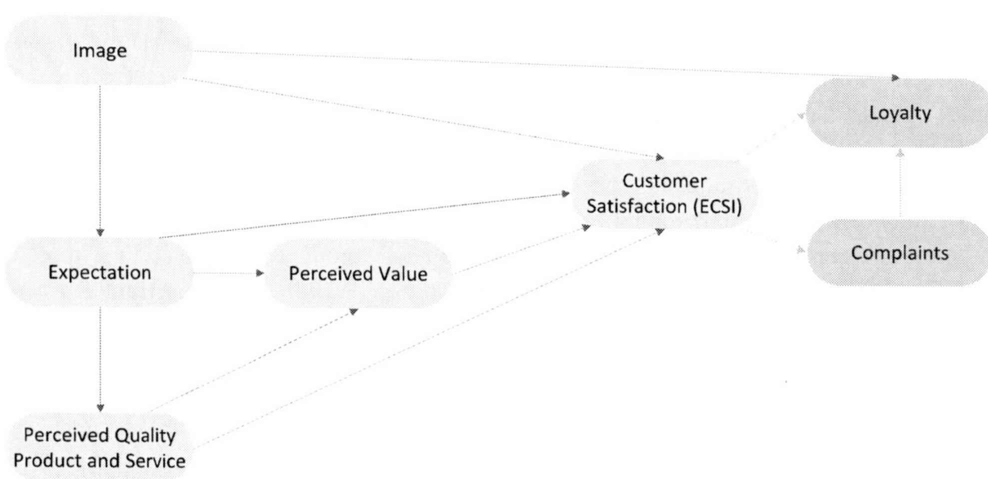

Figure 1. ECSI model adopted by Portugal

The ECSI model adopted by Portugal encompasses seven latent variables (Figure 1), where CS is linked to four drivers (Image, Expectations, Perceived Quality and Perceived Value), and two main consequences (Loyalty and Complaints) (www.ipq.pt/ecsi). As antecedents of CS, one has *Image*, which embraces the global idea that customers have of the product or company; *Expectations*, which includes the information that customers have acquired in the past regarding products and services offered by the company; *Perceived Quality*, which integrates product quality and service quality and corresponds to the evaluation of recent consumption experiences of products and associated services, respectively; and *Perceived Value*, which is the perceived level of product quality taking into account the price paid for it. The consequences of CS are *Complaints*, which evaluate the frequency and management of complaints, and *Loyalty*, which measures a long term customer's commitment and his/her re-purchasing intention.

In order to build an ECSI model, able to assess the particularities of the injection mould sector, it was important to identify the concepts to be measured, regarding *Perceived Quality* of Product and Service, as well as to elicit a comprehensive set of questions, that are potentially relevant in measuring these concepts. In this sense, a qualitative phase was previously carried out, through the conduction of semi-structured interviews. The information gathered from these interviews, where an illustrative sample of customers of Portuguese mould makers was inquired, allowed us to identify six factors that may contribute towards *Perceived Quality* of Product and Service (Figure 2). These factors are Quality of mould design, Quality of the mould construction, Cooperation, Resources, Response Capacity and Contracts (Ferreira, et al., 2007).

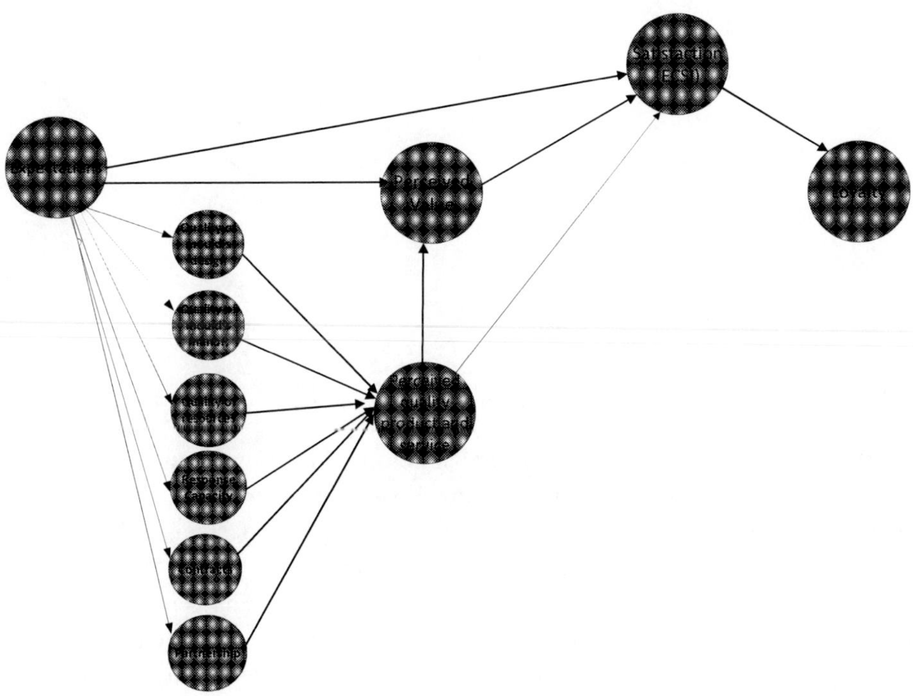

Figure 2. Hypothesized ECSI model for the Portuguese mould's maker sector

Afterwards, generic questions covering multiple indicators (at least two) were developed for each one of the latent variables, in order to construct a standardised questionnaire. Data collection was performed by sending a questionnaire directly to Portuguese injection companies (CAE 25240) during the spring and summer of 2007. A total of 108 surveys were replied, out of the total number of 489 mould companies which operated in Portugal at that time (Ferreira, et al., 2008b).

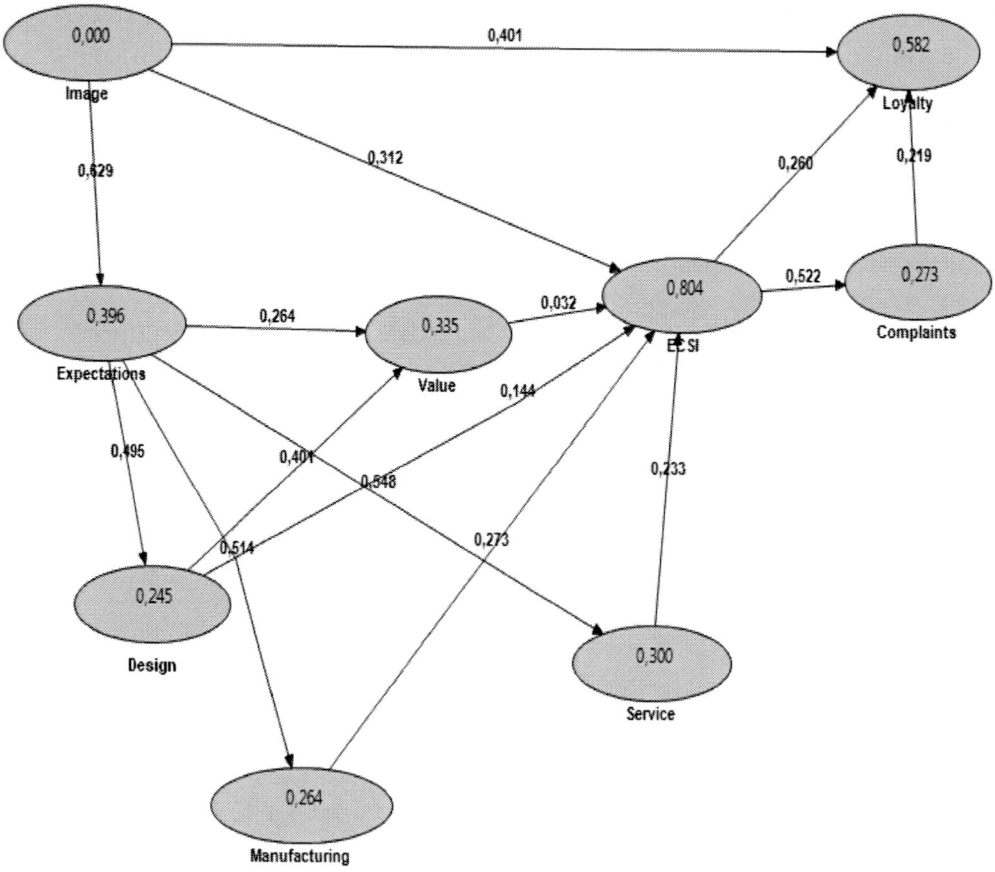

Figure 3. Estimated model for the Portuguese mould makers industry

The estimation of direct, indirect, and total structural effects amongst latent variables in a general structural equation model can be accomplished using two main types of methods: *Covariance-based* and *Component-based* methods (Fornell & Bookstein, 1982; Hsu, Chen, & Hsieh, 2006; Loughlin & Coenders, 2002; O'Loughlin & Coenders, 2004; Tenenhaus, 2003; M. Vilares & Coelho, 2005). Covariance-based techniques estimate path coefficients and loadings by minimizing the difference between observed (obtained by the data gathered) and predicted variance-covariance (defined by the hypothesized model) matrices. Some different estimation procedures can be adopted, where the most widely used is Maximum Likelihood (ML). The component-based approach, according to a PLS technique, estimates parameters similarly to principal components, through multiple regression. In fact, PLS algorithms iteratively generate estimates of latent scores based on inner relations (between latent variables) and outer relations (relations between a latent variable and its associated indicators), until the two relations converge. Due to the PLS advantages over covariance-based methods (Ball, Vilares, & Coelho, 2003; M. J. Vilares, et al., 2005), namely the fact that PLS does not rely on strict assumptions about the data (specifically regarding normality assumptions) and can be used for formative models estimation (Cassel, et al., 2000), this technique was preferred to estimate our CSI models. In this sense, a first model estimation was performed (Ferreira, et al., 2008a; Ferreira, et al., 2008b), which indicates very good capacity of CS explanation (based on the obtained determination coefficient (R^2) for CS) (Figure 3).

The value of R^2 is very satisfactory, especially considering the complexity of the model. Considering the total effect (direct plus indirect) on the CS Index, it is possible to verify that the most important variable over CS is Image (0.53). Value has no significant impact on CS (0.032), something that is not a very typical situation in other ECSI studies. This evidence may indicate that the mould's value is not a main preoccupation for customers. Furthermore, it is also possible to define ECSI structural equations as follows:

$$ECSI = \beta\_50 + 0.35 Expectations + 0.53 Image + 0.16 Design + 0.27\ Manufacturing + 0.23\ Service + 0.032\ Value + (\_5$$

Eq. 6

Regarding each particular contribution for mould design quality, the relative outer weight values were also determined for each indicator of this variable (Table 1).

Table 1. Relative outer weights for the indicators of quality of design

Indicators of Quality of mould design	Relative weights
The capacity of the mould's design in meeting product requirements	0.20
The mould's design capacity according to specific injection process	0.19
The use of adequate constructive solutions	0.23
The companies' accessibility in discussing the mould's design	0.18
The overall quality of mould's design	0.19

For each one of the previous indicators, a team of seven mould designers defined the respective Customer Attributes (CAs). These CAs are typically required by injection mould customers, when they order the mould (Figure 4).

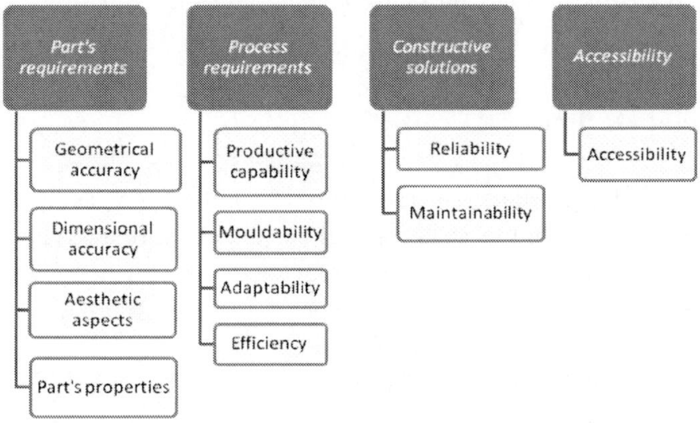

Figure 4. Typical CAs regarding injection mould design

To determine the relative priority of each CA, the Analytical Hierarchical Process (AHP) was adopted. This technique is widely used for addressing multi-criteria decision-making problems, since it assures the consistency and stability of the subsequent decisions (Lu, Madu, Kuei, & Winokur, 1994). In order to get a meaningful group preference, and assuming that each decision-maker is of equal importance, the Aggregating Individual Judgment (AIJ) approach was used (Raharjo, Xie, Goh, & Brombacher, 2007). Hence, each attribute was ranked according to its relative importance to customers in order to build a weighted objective function (Table 2).

Table 2. Relative priority of each CA

	Customer's attributes (CAs)	Ranking
Part requirements	Geometrical accuracy	0.436
	Dimensional accuracy	0.234
	Aesthetic aspects	0.198
	Properties	0.132
Process requirements	Productive capability	0.422
	Mouldability	0.289
	Adaptability	0.235
	Efficiency	0.054
Constructive solutions	Maintainability	0.568
	Reliability of solutions	0.432
Accessibility	Accessibility	1.000

Based upon these values, it is possible to mathematically express CS as a function of CAs (Eq. 7):

$$\begin{aligned} CS &= 0.2Part's + 0.19\,\mathrm{Pr}\,ocess + 0.23Solutions + 0.18Accessibility \\ &= 0.2(0.436Geometrical + 0.234Dimensional + 0.198Aesthetic + 0.132\,\mathrm{Pr}\,operties) \\ &+ 0.19(0.422Capability + 0.289Mouldability + 0.235Adaptability + 0.054Efficiency) \\ &+ 0.23(0.568Ma\,\mathrm{int}\,ainability + 0.432\,\mathrm{Re}\,liability) + 0.18Accessibility \end{aligned} \quad \text{Eq. 7}$$

The Second Stage: Design

Afterwards, in the Design stage, an initial mould solution is generated. To that end, the creation of a product concept, and its system-level design, supported by the AD methodology, will be undertaken. In this sense, the CAs, previously established, were translated into specific requirements, the FRs, which correspond to the minimum set of functional requirements states in the functional domain (Table 3).

According to the AD theory (Suh, 1990), the creation of product solutions must be accomplished by mapping FRs with the proper DPs, in order to respect the Independence Axiom (i.e. maintain the independence of the FR). In fact, the designer must choose a correct set of DP, that are able to satisfy the FR and assure FR independence (i.e. uncoupled design).

However, some authors believe that there are cases where functional independence may not be feasible (e.g. (Crawley, et al., 2004)). In fact, mostly due to technological, cost and time reasons, uncoupled design solutions are not common in injection moulds. In this sense, though seeking for the independence of FRs, even if some remaining coupled relations subsist, they are not considered prohibitive at this stage (they will be dealt with in the optimization stage). Hence, following the zig-zagging approach established by AD, the FRs-DPs mapping was performed for the upper levels of mould design (Figure 5 and Figure 6).

Table 3. Mapping between CAs and FRs

	Functional Requirements (FRs)
Geometrical accuracy	Deflection
Dimensional accuracy	Tolerance
Aesthetic aspects	Visual marks
Properties	Specific property
Productive capability	Cycle time
Mouldability	Pressure range
Adaptability	Mould's size
Efficiency	Volume of scrap
Maintainability	Mean Time to Repair (MTTR)
Reliability of solutions	Mean Time Between Failure (MTBF)
Accessibility	Information content

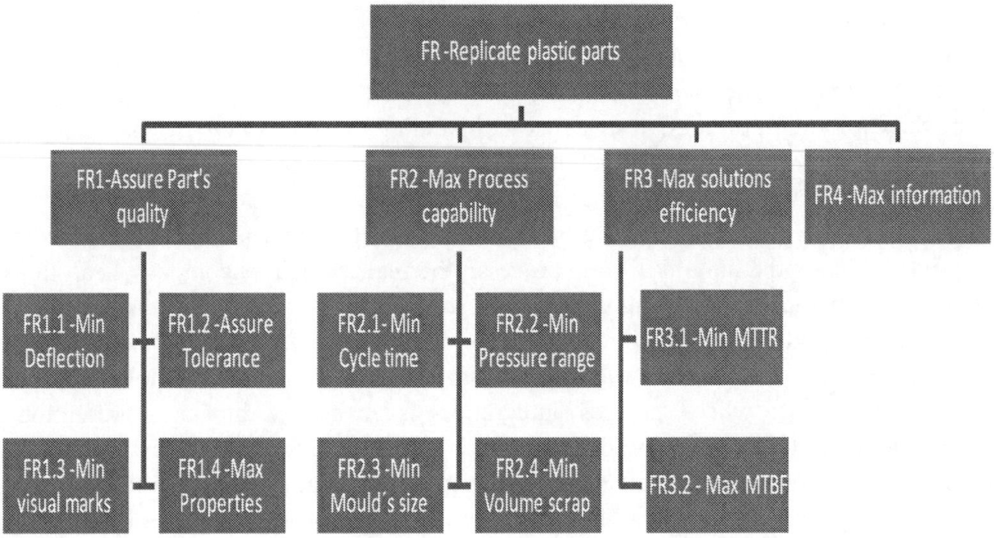

Figure 5 - FRs defined for top design levels

For each level of decomposition, the respective design matrixes were developed using X and 0 to express the relationships between FRs and their associated DPs, where X indicates a mapping relationship and 0 a lack of mapping relationships (Figure 7). Based on these figures, it is possible to verify that injection mould design is a highly coupled solution, as expected.

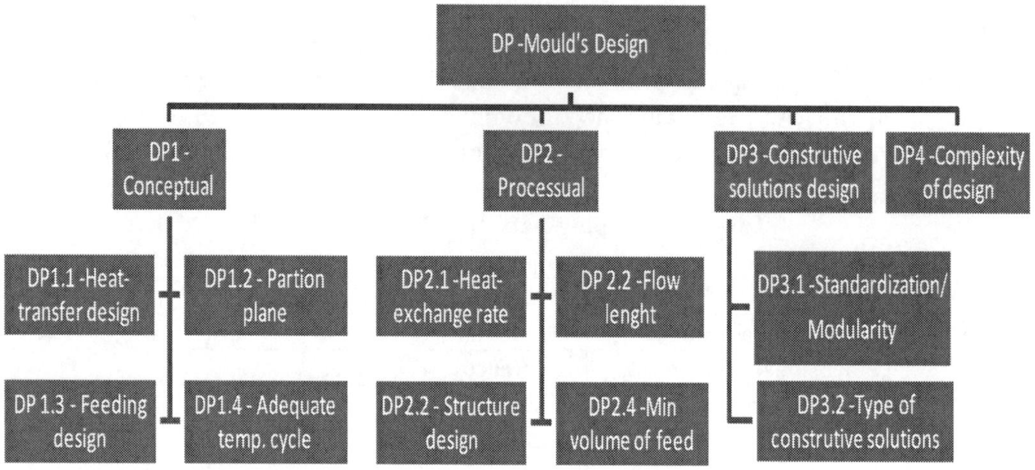

Figure 6 - DPs defined for top design levels

The Third Stage: Optimization

Since MDO is considered to be a powerful approach, that exploits the synergies of the interdisciplinary couplings through a systematic and mathematically-based approach (AIAA, 1991), this methodology was adopted to support the optimization stage. The goal of MDO is to find the optimal design of complex systems, achievable by a systematic exploration of the alternatives generated conceptually, which then lead to the optimal state. To that end, MDO adopts quantitative mathematical models, where some formal optimization algorithms facilitate the exploration of large design spaces, including those that may be characterized by discrete variables or discontinuous functions (Korte, Weston, & Zang, 1997).

A framework based on MDO, that aims to optimize mould design as a system, was developed. It involves the following multidisciplinary phenomena: rheological, which seek to model and evaluate the mould filling process; thermal, which encompass heat transfer phenomena; mechanical, concerning the mould's physical movements; and structural, aiming to minimize the mould's deformation induced by compressive and bending stresses (Figure 8). The developed framework was built using iSIGHT-FD as a global optimizer (www.engineous.com), where process integration is made by blocks representing its individual modules. This overseeing code system is responsible for automatically running the analysis codes, accessing the output and changing the input data, according to the pre-defined mathematical exploration schemes. The starting point of the design loop is an initial mould solution defined parametrically, which is established according to the practical guidelines followed by the mould industry (Centimfe, 2003). Then, the performance of this solution is studied through some developed analysis codes (using Excel® and MATLAB®), regarding mould phenomena behaviour. Based on that, and according to the optimization algorithm, an optimal solution is generated, in order to maximize CS.

	DP1. Conceptual	DP2. Processual	DP3. Constructive solutions	DP4. Complexity of solutions
FR1. Assures Part´s Quality	X	X	X	X
FR2. Max Process capability	X	X	X	X
FR3. Max Solution´s efficiency		X	X	X
FR4. Max Design accessibility	X		X	X

	DP1.1. Heat-exchange design	DP1.2. Partition plane location	DP1.3. Feeding design	DP1.4. Adequate temperature	DP2.1. Heat-exchange rate	DP2.2. Flow lenght	DP2.3. Structure design	DP2.4. Min volume of feed	DP3.1. Standardization/Modularity	DP3.2. Type of constructive solution	DP4.1. Complexity of design
FR1.1. Min Deflection	X	X	X	X	X		X			X	
FR1.2. Assure Tolerance	X	X	X	X	X		X			X	
FR 1.3 Min Visual marks	X	X	X		X					X	
FR 1.4. Max Properties				X	X		X			X	
FR2.1. Min Cycle time	X		X	X	X		X		X		
FR2.2. Min Pressure range		X	X			X	X				
FR2.3. Min Mold's size	X		X				X		X	X	
FR2.4. Min Volume of scrap		X	X			X		X		X	
FR3.1. Min MTTR		X							X	X	X
FR3.2. Min MTBF			X						X	X	X
FR4.1. Max Informatiom	X	X	X				X		X	X	X

Figure 7 - Design matrix for upper levels of an injection mould design

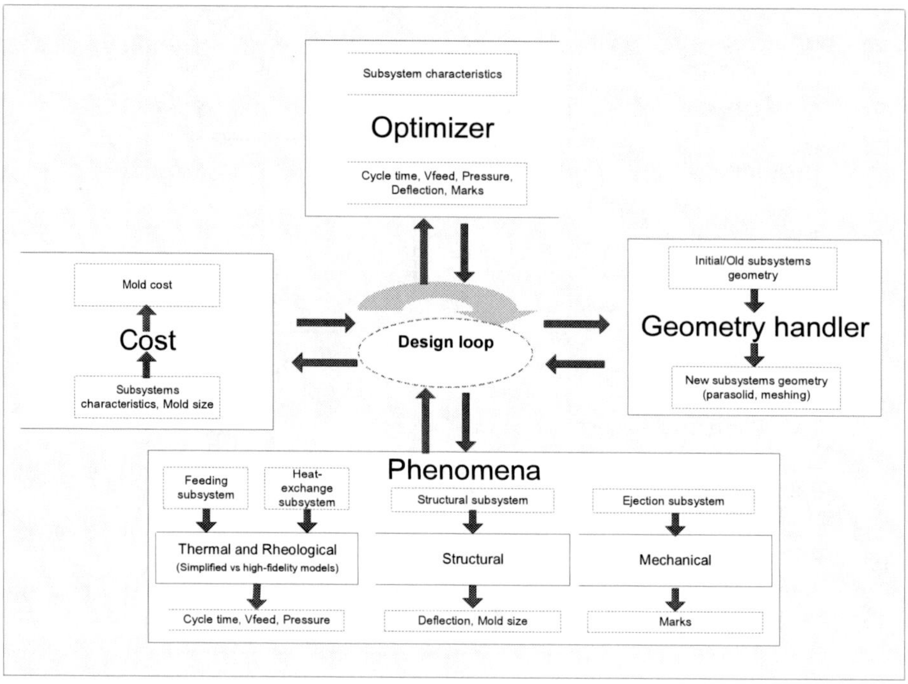

Figure 8. Framework process integration

To better illustrate how this stage is performed, the design of a particular mould, and the respective optimization procedure, will be described. In this sense, as usually happens in the mould design process, a specific geometry and material of the plastic part, as well as the injection machine parameters, are imposed. A plastic part with a simple geometry was adopted (Figure 9), because at this stage of our research the developed framework does not allow for plastic parts with undercuts. Regarding material selection, ABS Cycolac MG47 was adopted, since it is a well characterized material, for which all necessary information is included in most materials industry databases.

Figure 9. The benchmark part back

For simplicity reasons, in order to better understand the framework's procedure, it is assumed that CS increases linearly only with cycle time (i.e. the additional improvements on

CS, through other functional requirements, are neglected here). Then, the previous equations, representing the objective function, can be reduced to the following equation:

$$\Delta CS = 0.19 \times 0.422 \times \Delta cycle\ time = 0.08 \Delta cycle\ time \qquad \text{Eq. 8}$$

Theoretically, cycle time can be defined as the summation of the time for the different stages of the injection moulding process. Assuming some simplifying conditions, in order to get an analytic solution (for more details, see (Ferreira, et al., 2010)), cycle time can be mathematically expressed by Eq. 9.

$$Cycle\ time = \frac{\left[d_{Sprue} + \tan\left(\frac{D_{raftSprue}\pi}{180}\right) l_{Sprue}\right]^2}{23.1\alpha} \ln\left(0.692 \frac{(T_{melt} - T_{cool})}{(T_{demol} - T_{cool})}\right) + \frac{d_{gate}^{\ 2}}{23.1\alpha} \ln\left(0.692 \frac{(T_{melt} - T_{cool})}{(T_{demol} - T_{cool})}\right) + 2 \times \frac{d_{Release} 1 \times 10^3}{184 + 13 \log\left(\frac{P_{inj} A_{proj} 1 \times 10^{-3}}{9.8 F_{ref}}\right)} \qquad \text{Eq. 9}$$

where d_{Sprue} is the diameter of the Sprue [m], l_{Gate} is the length of the gate [m], d_{Gate} is the diameter of the gate [m], $D_{raftSprue}$ is the draft angle of the Sprue [º], l_{Sprue} is the length of the Sprue [m], P_{inj} is the Injection Pressure [Pa], $d_{Release}$ is the distance of part's release [m] and F_{ref} is a reference force of 1 ton-force. Since T_{melt} (i.e Melt temperature), T_{mold} (i.e Mould's temperature), T_{demold} (Demoulded temperature) and α (material coefficient of diffusivity) are dependent upon the material chosen, and A_{proj} (Projected area of moulding) is a function of partition plane location, which was decided at the Design stage, these items are considered as fixed parameters (constant values). Additionally, the following constraints and design variable bounds were also considered in cycle time optimization, where l_{Runner} is the runner length, l_{Gates} is the gates length and l_{part} is the part length, φ represents a constant ratio between width and thickness (which is equal to 1.5, when width is much bigger than thickness), $\bar{v}_F$ is the velocity of the flow front and η_{aeff} is the apparent effective viscosity; $MaxY$ and $MaxZ$ are the part maximum distances along the Y and Z directions, respectively; $F_{clampmax}$ is the maximum clamping force; Z_{cav}, Z_{plate_1} and Z_{mold} correspond to the distance in the Z direction for the cavity plate, plate one and for the complete mould, respectively; n is the power index of the Power Law model, $\dot{V}$ is the volumetric flow rate and $\dot{\gamma}_{max}$ is the maximum shear rate for the plastic; $MaxOpen$ is the maximum distance in the Z direction for the mould. The runner diameter is defined by d_{Runner} and $n_{downstream}$ is the number of streams for each ramification.

$-P_{inj} + \frac{32\left(l_{Sprue}+l_{Runner}+l_{Gates}+l_{part}\right)\varphi\bar{v}_F\eta_{aeff}}{\left(\frac{2MaxYM_{int}}{MaxY+M_{int}}\right)^2} \le 0$	The pressure demand to counter the resistance to flow in plate (flow length/wall thickness ratio derived from Hagen-Pouseuille's law [15]).	Eq. 10
$P_{inj} - \frac{F_{clamp\max}}{A_{proj}} \le 0$	The melt pressure acting in the projected area of mould cavities must not overcome the maximum clamp force (required to hold the mould closed during operation).	Eq. 11
$l_{Sprue} - Z_{cav} - Z_{plate\_1} = 0$	To assure geometric feasibility, the length of sprue must be equal to plate's distance starting in injection nozzle until partition plane.	Eq. 12
$-d_{Gate} + 2\left[\frac{\left(3+1/n\right)\dot{V}}{\pi\dot{\gamma}_{\max}}\right]^{1/3} \le 0$	Shear rate for flow in gates must not overcome the maximum allowable shear (power law is assumed, which is a conservative approach).	Eq. 13
$-d_{Release} + MaxOpen - Z_{mold} \le 0$	Distance of part's release must not overcome the maximum free open distance of mould.	Eq. 14
$-d_{Release} + 2.5MaxZ \le 0$	The distance of mould's open must assure part's release.	Eq. 15
$-d_{Sprue} + d_{Runner}\sqrt{n_{downstream}} \le 0$	Sprue must have enough capacity to fulfill all the downstream runners.	Eq. 16
$V_{feed} \le 0.3V_{part}$	The wasted material must be smaller than 30% of part's volume.	Eq. 17
$\Delta P \le 0.5P_{inj}$	The drop of pressure (ΔP), made by feeding system, must be smaller than 50% of Injection Pressure value.	Eq. 18

<u>Lower and upper bounds</u>:

$$d_{Sprue} \le 0.02 \quad \text{Eq. 19}$$

$$1 \le \alpha_{Sprue} \le 4 \quad \text{Eq. 20}$$

$$0.0005 \le d_{Gate} \le 0.003 \quad \text{Eq. 21}$$

$$0.0005 \le l_{Gate} \le 0.001 \quad \text{Eq. 22}$$

where V_{part} is the total volume of the moulded plastic part and V_{feed} is the wasted material, defined by the volume of the feeding system (cold runner system), which can be computed as:

$$V_{feed} = \frac{\pi}{4}\left(\bar{d}_{Sprue}^{\,2} l_{Sprue} + n_{downstream} n_{Ramif} d_{Runner}^{\,2} l_{Runner} + d_{Gate}^{\,2} l_{Gates} n_{Gates}\right) \quad \text{Eq. 23}$$

where n_{Ramif} and n_{Gates} are the number of ramifications of the runner and number of gates per each plastic part, respectively.

After that, an initial design solution, established according to best practices, was parametrically defined. This solution, which is characterized by specific DP values (Table 4), will lead to its optimal state through an optimization scheme.

Table 4. Initial mould design solution

Design Variables (DVs)	Symbol	**Units**	Initial Values
Injection Pressure	P_{inj}	Pa	1.8E+08
Distance X cavity Insert	Xins_cav	m	0.258
Distance Y cavity Insert	Yins_cav	m	0.258
Final distance X cavity and core	Xcav_core	m	0.296
Final distance Y cavity and core	Ycav_core	m	0.296
Height of core insert	Hcore_Ins	m	0.043
Height of cavity insert	Hcav_ins	m	0.044
Final distance Z for cavity	Zcav	m	0.056
Final distance Z for core	Zcore	m	0.056
Release distance	dRelease	m	0.075
Final distance Z for plate 1,9	Zplate_1,9	m	0.046
Final distance Z for plate 4	Zplate_4	m	0.046
Final distance Y for plate 5,6	Yplate_5,6	m	0.046
Final distance Z for plate 5,6	Zplate_5,6	m	0.046
Final distance Y for plate 7,8	Yplate_7,8	m	0.202
Final distance Z for plate 7	Zplate_7	m	0.016
Final distance Z for plate 8	Zplate_8	m	0.026
Length of sprue	l_{Sprue}	m	0.068
Diameter of sprue	d_{Sprue}	m	0.013
Draft angle of sprue	DraftSprue	°	1.000
Diameter of runner	d_{Runner}	m	0.009
Length of runner	l_{Runner}	m	0.120
Diameter of gates	d_{Gate}	m	0.001
Length gate	l_{Gate}	m	0.001
Diameter channel of coolant	d_{cool}	m	0.01
Distance z from cavity surface to the center of cooling line	Z_{cool}	m	0.025
Distance between turns in y	pitch_cool	m	0.05
Number of changes in position of coolant channel	n_{turns}		7
Length of coolant line	l_{Line}	m	1.196
Increase of temperature of coolant	ΔTcool	°C	0.5

Therefore, using the Generalized Reduced Gradient 2 (GRG2) algorithm, the previous mould solution was optimized through our MDO framework. It is important to note that the space of design solutions is defined by all admissible values that each design variable can assume. In this sense, the optimal solution achieved can be represented by the optimal design variables values (DPs) found. Regarding the assessment of mould solution improvement, both moulding cycle times were also computed (Table 5). Comparing the initial with the optimal solution generated, it is possible to observe a cycle time reduction of 42% over the initial value. This improvement leads to a 3.4% increase in CS (see Eq. 8). These results point out the potential of the developed framework to generate solutions leading to higher levels of CS and commitment. Based on that, it is possible to adopt the developed platform as a Decision

Support System, converting, in a consistent way, customer needs into optimal product solutions, regarding CS maximization.

Table 5. Optimal versus Initial solution

DVs	Units	Initial	Optimal DP
P_{inj}	Pa	1.80E+08	2.11E+08
$d_{Release}$	m	7.50E-02	7.50E-02
l_{Sprue}	m	9E-02	1.02E-01
d_{Sprue}	m	1.2E-03	8.49E-03
α_{Sprue}		1.0	1.0
d_{Gate}	m	0.5E-3	1.44E-03
Cycle time	s	**112.7**	**65.7**

The Fourth Stage: Validation

Finally, in the last stage, Validation, the optimal entity generated by the framework must be validated, in order to assess the real improvement achieved. To that end, both solutions, i.e. the initial and the optimal solution, were tested using the Moldflow code under the same processing conditions. Based on the data gathered (Table 6 and Figure 10), it is possible to observe that the results produced by the proposed framework are consistent with Moldflow simulation results.

Table 6. Comparison between results obtained by proposed framework and Moldflow numerical simulations

	Cycle time (s)	
Framework	112.7	65.7
Moldflow	119.8	66.72

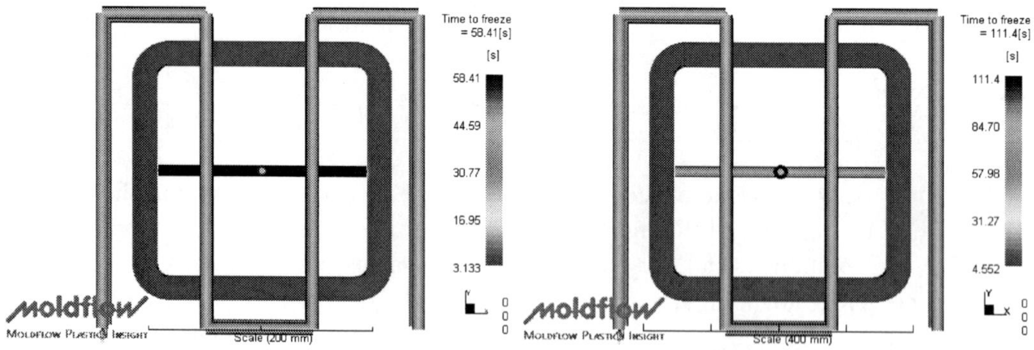

Figure 10. Time to freeze in seconds obtained by Moldflow numerical simulations: Initial (left) and Optimal (right)

Future Trends

Presently, the current challenges imposed by mould makers' customers, especially regarding lead-time and cost reduction, as well as higher levels of mould quality and reliability accomplishment, justify the effort to develop new approaches for mould design. However, due to the high complexity and mould component interactions, it is essential to develop integrative approaches for mould design in order to optimize them when seen as global systems, through their functional systems integration. For that reason, a conceptual framework, based on DFSS, AD and MDO methodologies, which tackles the design of an injection mould in a global and quantitative approach, was developed. However, so far in our work very simplified expressions to mathematically model mould phenomena were adopted. Therefore, the inclusion of more accurate and realistic models will be essential to enhance the previous framework, in order to reach even more improvements in mould design and operation. Then, to develop a more realistic model for injection moulds, in the future it will be necessary to refine the optimization model by including all important variables, especially the categorical ones, and to expand its scope, in order to cover the design of more complex elements, like sliders and lifters. The integration in our framework of some high-fidelity models, like Moldflow, in order to get more accurate results, as well as CAD tools to be able to visualize the design solutions, is also considered fundamental to achieve a fully integrated mould optimization. Finally, it is also important to build and include in our models a more reliable mould cost function. Regarding the extension of the framework to other sectors, there should be no significant barriers in adapting it to develop other products or services. In fact, considering the systematic, quantitative and rational focus of such a framework, the ultimate goal of this global methodology will be to establish a scientific basis to support product design on different fields.

Conclusion

A conceptual framework, based mainly on Design for Six Sigma (DFSS), European Customer Satisfaction Index (ECSI), Axiomatic Design (AD) and Multidisciplinary Design Optimization (MDO) methodologies, was developed, in order to guide and systematize the mould design process. This framework tackles the design of an injection mould in a global and quantitative approach, starting with a full understanding of critical customer requirements and its translation into functional requirements. Based on that mapping, an objective function, expressing customer satisfaction as a weighted function of specific functional requirements, was determined. Afterwards, AD supports the conceptual design of the mould, aiming to map the functional requirements with the corresponding design parameters. In this stage, the initial mould design decisions are established according to FR-DP mappings, developed for the upper levels of mould design. At this stage, in spite of seeking for the independence of functional requirements, some remaining coupled relations may subsist. However, they are not considered to be prohibitive. Consequently, MDO, as an appropriate methodology to design complex systems through an adequate exploitation of interacting phenomena, supports the detailed design stage. As a result, an integrated platform was developed, where all different analysis modules (e.g. structural, thermal, rheological and mechanical) were inserted and optimized through an overseeing code system, regarding the maximization of customer satisfaction levels. The results attained highlight the great potential of the proposed framework to achieve mould design improvements, with consequent reduction of re-work and time-saving for the entire mould design process. Based on that, it is possible to assume that the proposed approach can become an essential tool for future quality enhancement in the mould maker sector, acting as a decision support system, able to convert customer needs into optimal product solutions in a systematic and quantitative way. Considering its strong scientific basis, it is our conviction that the developed approach can turn into a global methodology to support more rationally different product design processes.

Key Terms

Integrated product development – An approach to product development which assumes an integration of all aspects of the design process, in such a way that the whole process becomes logical and comprehensible.

Design for Six Sigma (DFSS) – DFSS is a systematic methodology that applies tools, training and measurements to enable the organization to design products and processes that meet customer expectations and can be produced at six sigma quality levels.

Multidisciplinary Design Optimization (MDO) – MDO adopts a mathematically-based approach to find the optimal design of complex systems through the exploration of its design space.

Conceptual design - Conceptual design includes the preliminary design decisions, such as type of mould (Structure design), the types of feeding system (Feeding design), etc., which are carried out through FR-DP mappings (according to axiomatic design theory).

Detailed design – The detailed design stage involves a systematic exploration of the alternatives generated at the conceptual stage, through the exploitation of the design space, in order to determine the best solution.

Structural Equation Modelling (SEM) - SEM is a statistical methodology that takes a hypothesis-testing approach to multivariate analysis, through models which represent a series of hypothesized cause-effect relationships between variables.

References

AIAA, (1991). *White paper on current state of art: Technical Committee on Multidisciplinary Design Optimization*, (MDO)

Baake, U. F., Stratil, P. & Haussmann, D. E. (1999). Optimization and management concurrent product development processess. *Concurrent Engineering, 7(1),* 31-42.

Balakrishnan, P. V. & Jacob, V. S. (1996). Genetic Algorithms for product design. *Management Science, 42*(8), 1105-1117.

Ball, D., Vilares, M. & Coelho, P. S. (2003). A new explanation for customer loyalty: an application with the ECSI model. Paper presented at the PLS'03 *"Focus on customer" - PLS and related methods*, Lisbon.

Brown, S. L. & Eisenhardt, K. M. (1995). Product Development: Past research, present findings and future directions. *Academy of Management Review, 20*(2), 343-378.

Cassel, C., Hackl, P. & Westlund, A. H. (2000). On measurement of intangible assets: a study of robustness of Partial Least Squares. *Total Quality Management, 11*(7), 897-908. Cefamol (2009).

Centimfe, (2003). *Manual do projectista para moldes de injecção de plásticos.*

Chan, W. M., Yan, L., Xiang, W. & Cheok, B. T. (2003). A 3D CAD knowledge-based assisted injection moul design system. *International Journal Of Advanced Manufacturing Technology, 22*, 387-395.

Chin, K. S. & Wong, T. N. (1996). Knowledge- based Evaluation for the Conceptual Design Development of injection Molding Parts. *Engineering Application of Artificial Intelligence, 9*(4), 359-376.

Crawley, E., Weck, O. d., Eppinger, S., Magee, C., Moses, J. & Seering, W., et al. (2004). *The influence of architecture in engineering systems: MIT - Engineering System Divison.*

Creveling, C. M., Slutsky, J. L. & Jr., D. A. (2003). *Design for six sigma: in tecnhology &product development* (1 ed. Vol. 1). New Jersey: Prentice Hall PTR.

Dunn, G., Everitt, B. & Pickles, A. (1993). *Modelling covariances and latent variables using EQS* (1st ed. Vol. 1): Chapman&Hall.

Ferreira, I. (2002). *Caracterização da indústria de moldes da região da Marinha Grande, na óptica da Qualidade.* Unpublished MSc, Universidade do Porto.

Ferreira, I., Cabral, J. & Saraiva, P. (2008a). An integrated framework based on the ECSI approach to link mould customers' satisfaction and product design. *Total Quality Management&Business Excellence,* (Submitted).

Ferreira, I., Cabral, J. A. & Saraiva, P. (2008b). Customer´s satisfaction evaluation of Portuguese mould makers based on the ECSI approach. Paper presented at the RPD 2008 - Rapid Product Development, Oliveira de Azeméis - Portugal

Ferreira, I., Cabral, J. A. & Saraiva, P. M. (2001). Qualidade na indústria de molde. Paper presented at the 1as Jornadas de Engenharia Mecânica, Automóvel, Gestão Industrial e Ambiente, ESTG- Leiria.

Ferreira, I., Cabral, J. A. & Saraiva, P. M. (2003). *Qualidade na indústria de moldes. O Molde, 66.*

Ferreira, I., Cabral, J. A. & Saraiva, P. M. (2006, November 13th and 14th). *The Axiomatic Design applied to the injection moulds design and manufacturing for plastics parts.* Paper presented at the Rapid Product Development Event - Building the Future by Innovation, Marinha Grande, Portugal.

Ferreira, I., Cabral, J. A. & Saraiva, P. M. (2007). *A new conceptual framework based on the ECSI model to support Axiomatic Design.* Paper presented at the Virtual and Rapid Manufacturing.

Ferreira, I., Cabral, J. A. & Saraiva, P. M. (2009, March 25-27). *A new AD/MDO approach to support product design Paper presented* at the The Fifth International Conference on Axiomatic Design, Campus de Caparica, Portugal.

Ferreira, I., Weck, O. d., Saraiva, P. M. & Cabral, J. A. (2010). Multidisciplinary Optimization of Injection Molding Systems. *Structural and multidisciplinary optimization, 41*(4), 621-635.

Fornell, C. & Bookstein, F. L. (1982). Two Structural Equation Models: LISREL and PLS Applied to Consumer Exit-Voice Theory. *Journal of Marketing Research, XIX,* 440-452.

Grigoroudis, E., Nikolopoulou, G. & Zopounidis, C. (2008). Customer satisfaction barometers and economic development: an explorative ordinal regression analysis. *Total Quality Management, 19(5)*, 441-460.

Hair, J., Tatham, R., Anderson, R. & Black, W. (1998). *Multivariate Data Analysis (5th* edition ed.): Prentice Hall.

Hsu, S. H., Chen, W. H. & Hsieh, M. J. (2006). Robustness Testing of PLS, LISREL, EQS and ANN-based SEM for measuring customer satisfaction. *Total Quality Management, 17*(3), 355-371.

Korte, J. J., Weston, R. P. & Zang, T. A. (1997). *Multidisciplinary Optimization Methods for Preliminary Design.* Paper presented at the Future Aerospace in the Service of the Alliance.

Krishnan, V. & Ulrich, K. T. (2001). Product Development Decisions:A Review of the Literature. *Management Science, 47*(1), 1-21.

Lam, Y. C., Britton, G. A. & Liu, D. S. (2004). Optimisation of gate location with design constraints. *International Journal Of Advanced Manufacturing Technology, 24,* 560-566.

Lam, Y. C. & Jin, S. (2001). Optimization of gate location for plastic injection molding. *Journal of Injection Molding Technology, 5*(3), 180-192.

Lam, Y. C., Zhai, L. Y., Tai, K. & Fok, S. C. (2004). An evolutionary approach for cooling system optimization in plastic injection molding. *International Journal of Production Research, 42*(10), 2047-2061.

Lee, K. S., Li, Z., Fuh, J. Y. H., Zhang, Y. F. & Nee, A. Y. C. (1997, June). *Knowledge-based injection mold design system.* Paper presented at the International Conference and Exibition on Design and Production of Dies and Moulds, Turkey.

Lee, K. S. & Lin, J. C. (2006). Design of the runner and gating system parameters for a multi-cavity injection mould using FEM and neural network. *International Journal Of Advanced Manufacturing Technology, 27*, 1089-1096.

Lee, R. S., Chen, Y. M. & Lee, C. Z. (1997). Development of a concurrent mold design system: a knowledge-based approach. *Computer Integrated Manufacturing Systems, 10*(4), 287-307.

Lou, Z., Jiang, H. & Ruan, X. (2004). Development of an integrated knowledge-based system for mold-base design. *Journal of Materials Processing Technology, 150,* 194-199.

Loughlin, C. O. & Coenders, G. (2002). Appplication of the European Customer Satisfaction Index to Postal Services. *Structural Equation Models versus Partial Least Squares.*

Low, M. L. H. (2003). Application of standardization for initial design of plastic injection moulds. *International Journal of Production Research, 41*(10), 2301-2324.

Lu, M. H., Madu, C. N., Kuei, C. h. & Winokur, D. (1994). Integrating QFD, AHP and benchmarking in strategic marketing. *Journal of Business&Industrial Marketing, 9*(1), 41-50.

Mok, C. K., Chin, K. S. & Ho, J. K. L. (2001). An Interactive Knowledge-Based CAD System for Mould Design in Injection Moulding Processes. International Journal Of *Advanced Manufacturing Technology, 17*, 27-38.

Nebiyeloul-Kifle, Y. (2005). *Application of the Design Structure Matrix to integrated product development process. M*assachusetts Institute of Technology.

O'Loughlin, C. & Coenders, G. (2004). Estimation of the European Customer Satisfaction Index: Maximum Likelihood versus Partial Least Squares. Application to postal services. *Total Quality Management, 15*(9-10), 1231-1255.

Pandelidis, I. & Zou, Q. (1990). Optimization of injection molding design. Part I: Gate location optimization. *Polymer Engineering and Science, 30*(15), 873-882.

Qiao, H. (2006). A systematic computer-aided approach to cooling system optimal design in plastic injection molding. *International Journal of Mechanical Sciences, 48*, 430-439.

Raharjo, H., Xie, M., Goh, T. N. & Brombacher, A. (2007). A methodology to improve Higher Education quality using the Quality Function Deployment and Analytic Hierachy Process. *Total Quality Management, 18*(10), 1097-1115.

Shen, C. Y., Yu, X. R., Li, Q. & Li, H. M. (2004). Gate Location optimization in Injection Molding by using modified Hill-Climbing Algorithm. *Polymer-Plastics Technology and Engineering, 43*(3), 649-659.

Smaling, R. M. (2005). *System architecture analysis and selection under uncertainty.* Unpublished PhD Degree, Massachusetts Institute of Technology.

Sobieszczanski-Sobieski, J. & Haftka, R. T. (1997). Multidisciplinary aerospace design optimization: survey of recent developments. *Structural Optimization, 14*, 1-23.

Suh, N. P. (1990). *The principles of design* (1 ed. Vol. 1): Oxford University Press.

Tenenhaus, M. (2003). Comparison between PLS and LISREL approaches for stuctural equation modeling: application to the measure of Customer Satisfaction. Paper presented at the PLS'03 *"Focus on customer"* - PLSand related methods, Lisbon.

Tomarken, A. J. & Waller, N. G. (2005). Structural Equation Modeling: strenghts, limitations and misconceptions. *Annual Review Clinical Psychology, 1*, 31-65.

Vilares, M. & Coelho, P. (2005). Satisfação e Lealdade do Cliente: Metodologias de avaliação, gestão e análise (1st ed. Vol. 1): Escolar Editora.

Vilares, M. J., Almeida, M. H. & Coelho, P. S. (2005, September 7th-9th). *Comparison of likelihood and PLS Estimators for Structural Equation Modeling. A simulation with Customer Satisfaction Data.* Paper presented at the 4th International Symposium on PLS and related methods, Barcelona.

Weck, O. d. (2004, October 30-November 2). *Multiobjective Optimization: History and Promise. Paper presented at the The Third China-Japan-Korea Joint Symposium on Optimization of Structural and Mechanical Systems,* Kanazawa, Japan.

Yang, K. & El-Haik, B. (2003). *Design for Six Sigma: A roadmap for product development* (1 ed.): McGraw-Hill.

In: Injection Molding: Process, Design, and Applications ISBN: 978-1-61761-307-4
Editor: Phoebe H. Kauffer

Chapter 8

EXPERIMENTAL STUDY ON THE STRENGTH OF ADHESION OBTAINED BY OVER-MOLDING BETWEEN DIFFERENT MATERIALS

Miguel Sánchez-Soto*[1]*, David Arencón*[1]*,
***María Virginia Candal*[2] *and Silvia Illescas*[1]**

[1]Centre Català del Plàstic. Materials Science and Metallurgy Department. Technical University of Catalonia (UPC). Colom 114. E-08222 Terrassa (Barcelona), Spain
[2]Universidad Simón Bolívar, Departamento de Mecánica, Sección de Polímeros, Apartado 89000, Caracas 1080-A, Venezuela

ABSTRACT

In the present work we have studied the interfacial adhesion characteristics of bi-layer structures obtained by over molding. The main injection molding parameters, such as injection temperature, injection and holding pressure and injection speed have been modified to the extent permitted by both the process and materials. Several surface treatments were applied to the base substrate to evaluate the effect of the surface roughness on the adhesion level obtained with the over molding process. In addition a coupling agent (PP-g-MA) was added to the polymer to improve the adhesion level. The results showed that the adhesive fracture toughness increased as temperature and surface roughness were raised. Depending on the selected parameters, different types of failure, ranging from adhesive to cohesive, were noticed. In metal-polymer systems, maximum levels of adhesion were obtained when the coupling agent was used and over molding was performed on previously torch-heated metal plates. The creation of chemical bonds between polymer and metal leads to maximum adhesion strength and cohesive type failure. For the case of polymer-fabrics an experiment design was carried out and statistical analysis was performed. Results showed that increased adhesion was achieved at lower injection temperatures.

1. INTRODUCTION

Polymer layers on different substrate materials play an important role in many technological applications. Inert barriers such as paints or polymer coatings are commonly used to protect other materials against weathering. Epoxy coatings, for example, are widely used to inhibit corrosion in metals. Nevertheless, apart from providing protection, polymer layers can be designed to add other functionalities, like thermal or electrical insulation, toughness, scratch resistance, cushioning against vibration or even as a mean to improve the aesthetic appearance of the underlying material. Another possibility for plastics is to act as a linking element replacing joints made with mechanical fasteners, welding or adhesive bonding, reducing in this way the product cost and time to market.

There are many ways to incorporate a polymer coating into a substrate, like dispersion coating, electrostatic, flame spraying, plasma spaying, laser vapor deposition etc. Nevertheless injection over molding is one the methods that is being currently subject of a growing attention in the field of polymer processing. Over molding is a sequential process in which one plastic material in molten state is molded onto a substrate. Usually the more rigid material constitutes the substrate and the more flexible plays the role of the protective or external layer. The most common situation, also called sequential injection molding, is when both substrate and covering are of polymeric type [1-2] although the nature of the substrate can be a ceramic, metal or plastic. In the most usual case of two polymers, the first injected polymer forms the core of the component and is produced by one-stage injection [3-4]. In a subsequent step the melt of the second polymer is injected over the surface of the core forming its external outskirt.

Over molding has gained considerable importance in the field of polymer processing, because the combination of materials having very different properties is a very interesting way to obtain optimized products [1, 5]. In addition, components made by this technique can have very complex shapes making this a very versatile technique [6] However, the performance of the over-molded structures is dependent on the achievement of an adequate level of adhesion that prevents both components from detaching. For this reason, the proper material combination, the characteristics of the interface and the influence of the manufacturing process become key factors defining adhesion.

The adhesion between materials is a complex mechanism involving different physical–chemical phenomena like mechanical adhesion, chemical bonding, diffusion, absorption, and electrical attraction. The final performance of a certain union is thought to be the sum of the contributions of the above-mentioned phenomena [7-9]. For the particular case of over molded parts when both substrate and external layer are of polymeric type (e.g. fabrics), the possible reaction between molecules of the two polymers, the depth of intermolecular diffusion and the number of entanglement created at the interface are the principal factors governing adhesion.

One of the most important difficulties to overcome when bonding two different materials is the difference in their surface energies. Polymers and metals, for instance, are very dissimilar materials. The typical surface energies of polymers range from 25 to 55 mJ/m^2 [10] whereas surface free energies of metals are of about 1000-5000 mJ/m^2 and from 200 to 500 mJ/m^2 for metal oxides [11]. On the other hand, polymer surfaces are usually non-polar or have very low wet ability whereas metals are mainly polar. These are the reasons for the lack of adhesion between polymers and metals when they are directly put in contact. As a result, it is necessary to set up a methodology that allows modifying the surface characteristics of the adherent materials to make them compatible [12-13].

In general terms, the simplest method of improving the adhesion between two bodies is by increasing its common interface. When a coating is applied over a particular surface, the increase in the surface roughness leads to greater adsorption of the coating onto the substrate provided that adequate wetting exists. If the viscosity of the adhesive layer is low enough and its critical surface tension is lower than the surface free energy of the background solid, then it will cover the imperfections by capillarity creating a mechanical interlocking. Process variables can, in some extent, modify the adhesion parameters. An increase on the temperature of the injected polymer leads to a decrease on its viscosity and thus to a reduction of the contact angle favoring the adhesion. Finally an increase on the pressure between both adherents should provide and increase on the adhesion level. Nonetheless, the most powerful way of achieving adhesion is creating a strong bond between substrate and covering layer like covalent bonding.

As a result of its organic nature, the reactivity of polymers can be modified to a great extent. The creation of active functional groups in the over molded polymer can be used to promote compatibility, for instance by matching polarities or by the generation of chemical reactions between the outside layer and the substrate. The nature of the chemical modification can then be used to change the adhesion level.

The present work studies the adhesion degrees of two systems: polymer-metal and polymer fabrics. In the first case, attention is paid to promote compatibility between polymer and metals using functionalized polymers and/or the modification of the substrate by several surface metal treatments. In the second case, efforts have been put in the study of the influence of the injection molding parameters, being evaluated statistically through a design experiment matrix.

2. Materials and Methods

2.1. Polymer-Metal Study

2.1.1. Polymers and blends

A polypropylene-ethylene block copolymer (Moplen EP340 M) having a 7.8% of ethylene content was used as adherent. This polymer was kindly supplied by Basell (Spain). Polypropylene was selected because of their importance in the plastics industry and in many sectors like automotive, packaging, consumer goods, etc. The presence of ethylene basically provides toughness and higher resistance to low temperatures. In addition, polypropylene copolymer was modified with a coupling agent. Polypropylene grafted maleic anhydride, PP-g-MA, Epolene 3003 (Eastman Chemicals) was used as a coupling agent. Two blends having 5 % and 15 % wt. of PP-g-MA respectively were prepared by a extrusion in a twin-screw co rotating extruder (Collins T20) being afterwards pelletized prior to its injection molding.

2.1.2. Metal and surface treatments

Plain low carbon steel plates (96 x 96 x 0.8 mm^3) having a composition of C=0,12 %, Mn=0.5 %, P=0.04 %, S=0.04 % were used as substrate. As received the metal plates had an oil layer covering the surface to prevent oxidation, for this reason a degreasing was performed before the application of the surface treatments. Degreasing was carried out by submerging the steel plates in an acetone bath for approximately 5 min. The remaining oil layer was removed using a soft sponge and samples were submerged in a second bath of fresh acetone for 5 more min. Finally acetone was removed by drying in air.

2.1.2.1. Sanding

Some of the steel plates were polished using a 400 mesh sand paper. The metal shaves and the sand residues remaining in the steel plates were removed in the acetone bath following the procedure cited above.

2.1.2.2. Shot peening

Hard steel balls of 0.4 mm-diameter were used in the shot peening treatment of the metal plates. SAE J 444, and J827 standards were followed. The shot peened plates showed a bending deformation located in the middle area of the plates. This deformation hampers the metal plates to be fitted in the mould. As a result, to flatten the samples they were pressed in its boundary using a hydraulic press (120 MPa, 5 min). Degreasing treatment was carried out afterwards to remove impurities.

2.1.2.3. Etching

A bath containing a solution of 15% of nitric acid in ethanol (*Nital*) was prepared and degreased samples were submerged on it for 5 min. During the process an oxide layer was observed to appear in the outside layer of the metal samples. This layer was carefully removed before the over molding using a sponge wet in fresh ethanol.

2.1.2.4. Torch heating

Torch heating was applied in some of the steel samples, prior to be over molded with plastic. This process was performed as homogeneously as possible in all surfaces. Previous torch heating trials were performed at 5, 10, 15 and 20 s. Finally 15 s was chosen as optimum heating time. The average temperature of the torch steel plates after 15 s was of about 200 ± 15 °C. The temperature was measured with an infrared thermometer (*Raytek ST*).

2.2. Polymer-Fabrics Study

A commercial grade of low density polyethylene (LDPE) having a melt flow index of 7.5 dg/min (230 °C, 2.16 kg) was kindly supplied by *Polimeri Europa*. A nylon 6,6- polyurethane based fabrics supplied by *Trety S.A.* was used.

2.3. Over Molding Process

The general over molding process was to insert the over molded substrate (metal and fabrics) in the mobile side of the injection mould prior to inject the melted polymer. The dimensions were 100 x 100 x 2 mm of thickness. A fan-type gate located at the mid side was used. In order to avoid undesirable movements of the substrate specimen during the injection process, self-adhering tape was used to fix the specimen to the mould. Tape strip of 2 mm-width was used in both sides of the mould. At the top side a 30 mm tape strip was used. This latter will allow obtaining a detached polymer band necessary to perform the peeling test. For the polymer-metal case, a noozle injection temperature of 210 ºC, injection time of 2.5 s, mould temperature of 60 ºC, and holding pressure of 50 MPa applied during 5 s were employed. The set of conditions used for polymer-fabrics are detailed in Table 3.

3. Testing Procedure

3.1. Roughness Measurement of Metal Surface

The measurement of the surface metal roughness was carried out by using a computer-controlled profilometer *Mitutoyo Surftest SV512*. This profilometer has a diamond tip with a 5 μm curvature radius on its end. The procedure for the measurement was the ISO 4287 standard. Measurements were taken in different directions to compensate for possible anisotropy.

3.2. Peeling Tests

The over molded specimens were cut in 4 halves of 20 mm width and 3 mm thick (Figure 1a). With the aid of a clamp the 30 mm. surface covered with the strip band was detached to be later used to fix the specimen to the testing machine

The T-peeling test was performed on a *Galdabini Sun* universal testing machine at room temperature (23 ± 1 ºC) using a load cell of 5 kN. In accordance to the standard peeling procedure [14, 15] a speed of 10 mm/min was applied using the T-peel configuration (Figure 1b). The angle of the peeling arms was measured during the test using a goniometer. Each arm of the specimen was attached to the grips of a universal testing machine. A rigid steel strip was used to ensure the vertical alignment of the peel arm.

The general expression for peel adhesive fracture energy G in fixed peel arm configuration can be obtained from the following expression.

$$G_c = G - G_p \quad (1)$$

Where G_c is the peel strength-measured load and G_p is the plastic arm energy caused by bending the peel arm. If it is assumed that there is no tensile strain in the peel arm, G may be expressed by:

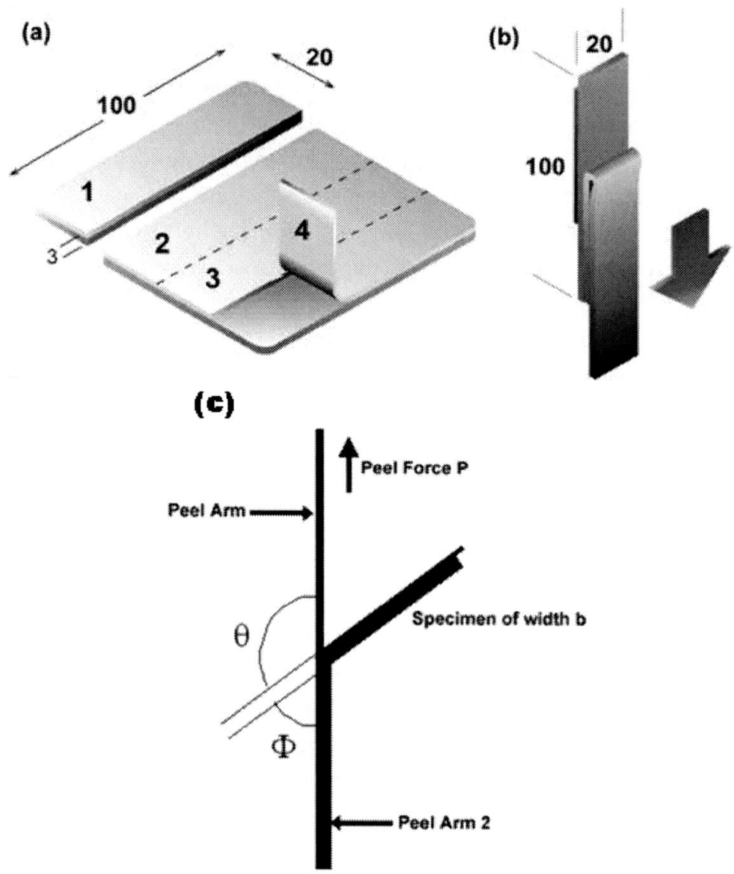

Figure 1. (a) Over molded metal plate and dimensions of the peeling specimens. (b) Scheme for the T-peel test in 180° configuration. (c) T-peel configuration and angles, if one arm is stiffer than arm 2

$$G = \frac{P}{B}(1 - \cos\theta) \tag{2}$$

This would be the case of a material with infinite stiffness and no bending stress. However, if there is an elastic-plastic deformation in the peeling arm the full expression for G_c becomes:

$$G_c = \frac{P}{B}(1 + \varepsilon - \cos\theta) - h\int_0^z \sigma \, d\varepsilon - G_{db} \tag{3}$$

Where P is the measured force, B is the specimen width, ε is the strain at a stress σ, h is the thickness of the arm, θ is the angle between arms, and G_{db} accounts for plastic or viscoelastic bending. For the particular case of T-peel configuration, if one peel arm is stiffer than the other one, two different peel angles are present (Figure 1c). Since the angles are correlated via $\Phi = \pi - \theta$, only one angle should be considered. In order to determine G_c, taken into account elastic and plastic deformation of the flexible polymer peel arm, two tests must be conducted: The peeling test and a tensile test on the polymer arm.

In accordance with the procedure detailed by Moore [15] a tensile stress–strain measurement of the peel arm up to maximum elongation at the same speed was performed in order to remove the part of the energy associated with the PP/PP-g-MA arm extension that occurs prior to the interface peel.

3.3. Microscopic Analysis

In order to investigate the morphology at the interlayer, specimen fracture surface observations were made using scanning electron microscope *JEOL 5610*. Before observation, fracture surfaces of the tested specimens were vacuum coated with a thin layer of gold of approximately 10 nm to make them electrically conductive. Optical microscopy observations were carried out in a Carton binocular lens equipped with a image capture device *Progres CT3*. The maximum zoom of the equipment was 40x.

4. Results and Discussion

4.1. Effect of Injection Molding Path

It is well-known that the effect of the gate is very important for the final performance of injection molded samples. To fill the mould cavity used in this work we selected a fan-gate type because it allows the mould cavity to be filled evenly. Nevertheless, differences in adhesion due to the particular flow created when the polymer reaches the gate may happen. In a recent work, Rossa *et al.* [1] using the same type of gate found temperature differences of about 20 ºC between adjacent points located either near the gate or at the sides of an over molded square plate. This effect was attributed to the fountain flow originated by the polymer melt entering into the mould.

Disregarding the type of surface treatment applied and the type of substrate, we observed as well differences when comparing results of peeled specimens. In Figure 2 it can be seen that symmetry from the mid-longitudinal plate axis exist. The highest adhesion strength levels were found to occur in the sections located near the centre of the plate (zones 2 and 3), whereas lower peel strength energies were found at the plate edges (zones 1 and 4). As mentioned earlier the reason for this behavior has been attributed to the polymer path developed during the injection molding. Near the gate a local heating takes place which cause a rise in the interface temperature between the steel plate and the polymer. This higher temperature means lower viscosity of the polymer and enhanced wet ability, all factors contributing to obtain greater adhesion levels with the base substrate. To avoid differences coming from the above mentioned side-effects, for the following analysis and comparisons we decided to use only the samples extracted from the central part of the molded specimens (zones 2 and 3).

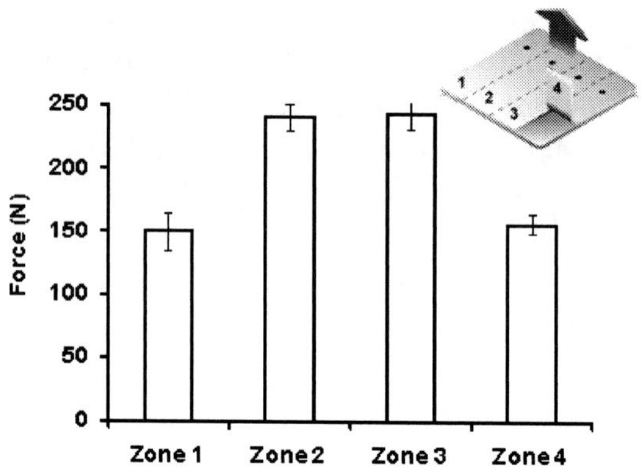

Figure 2. Peel force for the bi-later metal/polymer hybrids showing the effect of the injection molding path

4.2. Polymer to Metal Adhesion

4.2.1. Surface roughness of steel plates

Different surface treatments were applied on the steel plates with the aim of analyze its effect on adhesion. The surface appearance of steel plaques obtained after the different treatments can be observed in Figure 3. The images revealed that major variations on roughness were achieved by etching with *Nital* and shoot peening whereas sanding has the effect of diminish the initial surface roughness of the as-received metal plate due to the fine-grained mesh used (400 mesh). Shoot peening caused the maximum height of the profile of irregularities (R_z, R_y and R_a) but also the maximum value of the mean spacing between irregularities (S). This indicates that the profile irregularities caused by the peening were deep but wide as can be appreciated in figure 3d. When the steel plate was chemically etched with *Nital*, the resultant profile was slightly shallow than the peened one but in contrast it was much narrow having approximately 25 times lower S value (Table 1). This effect can be clearly seen when comparing Figure 3c and Figure 3d. As a result, the possible mechanical interlocking arisen from an increase in surface roughness is expected to be greater in the chemical treated plates than in the shoot-peened ones, decreasing its minimum for the sanded plate. Table 1 collects the measurements of the different roughness parameters. It should be noticed that for the ten-point height of irregularities (R_z) there were two cases in which it was not possible to obtain the exact measurement due to be out of scale of the profilometer. Same happened for the mean space of profile irregularities (S) of plates subjected to shoot peening.

Table 1. Roughness values of metal plates after different surface treatments; asterisk (*) represents an approximate value, out of scale

Metal treatment	R_a (μm)	R_y (μm)	R_z (μm)	S (mm)
As received	1.62	7.60	5.96	0.088
Sanded	0.47	2.87	1.95	0.013
Shoot peening	2.80	13.60	>10.00*	>0.600*
Etching	1.42	8.12	>10.00*	0.024

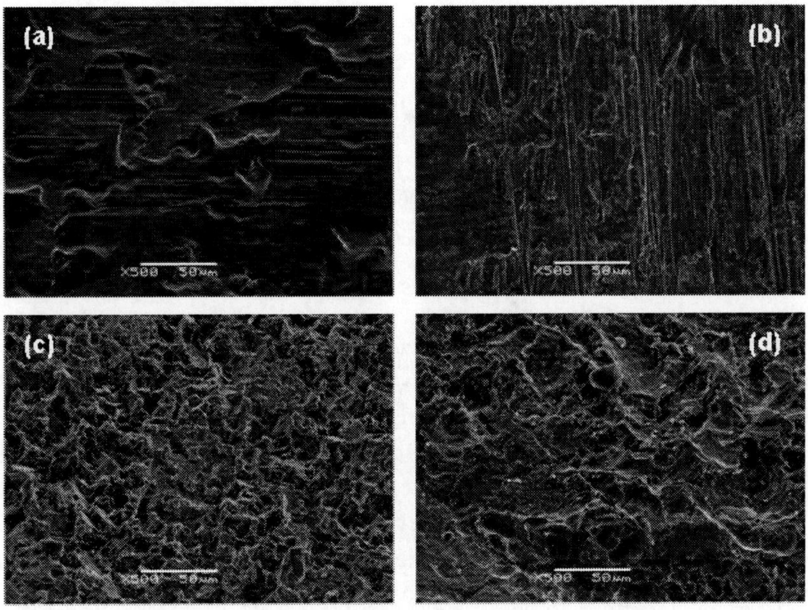

Figure 3. Surface topography of the metal plate after different surface treatments. (a) As received (b) Sanded with 400 mesh. (c) Etched with *Nital*. (d) Shoot peened

4.2.2. Effect of compatibilizing agent and torch heating

The first attempt to create union between PP copolymer and steel by direct over molding did not provide enough force of adhesion. The coating layer of PP detached from the steel surface after applying a very low force whatever the surface treatment applied on the steel. The detached surface do not show any trace of polymer adhered on it. The reason for such behavior was believed to be mainly the lack of adhesion between polymer and metal due to the inexistence of chemical links. In addition the rapid cooling of the injected polymer was also a factor contributing to the low metal-polymer interlocking. Polypropylene is known to be a very stable, non reactive polymer as a result no interaction between PP and steel was expected. On the other hand, in these tests the surface temperature of the metal substrate was the same of the mould, 60 °C, as a result when the melt touched the steel plate it cooled down very rapidly forming a viscous skin that can hardly be mechanically anchored on the steel plate.

Due to the lack of interaction, the addition of a coupling agent was considered. PP-g-MA is a kind of adhesion promoter that is formed by two parts. One part is compatible with the polymer (PP) and the other contains the maleic anhydride group that is able to react with several chemical groups like amine, epoxy or hydroxyl among others. Tan *et al.* [16] have used maleic anhydride to promote adhesion between stainless steel fibers and polypropylene prepared by melt mixed at 190 °C. They found that when PP-g-MA was compounded with the steel fibers strong interfacial adhesion was found due to the formation of chemical bonding. In comparison, when neat PP resin was used the interfacial adhesion was poor because of the low energy of the secondary van der Waals forces.

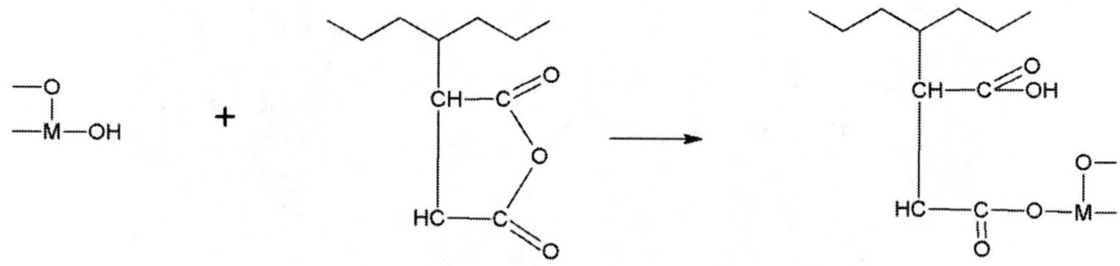

Figure 4. Scheme of PP/PP-g-MA chemical reactions with active groups created on the steel surface

The generation of chemical bonds between polymer and steel requires the presence of reactive groups in both adherent sides. Carbonyl groups appear naturally in the PP/PP-g-MA system as a result of maleic anhydride opening with temperature. On the other hand, it is possible to create active groups on the surface of the steel plate using an adequate treatment of the surface. The easiest variation in the steel surface is caused by the oxidation of Fe, usually in the form of Fe_2O_3. Iron oxidizes very quickly at ambient temperature or slightly higher temperatures to give a thin layer of oxide that grows with time. However, this layer of rust has natural tendency to grow and detach from the surface being thus a weak layer with no adhesion to the base metal. To avoid the presence of oxide layers not strongly adhered to the steel plate, is necessary to perform the oxidation in a short period of time. Metals possessing more than one valence state like iron can form complex layers of oxide, containing oxides of different compositions. If the iron is heated at temperatures lower than 570 ºC the layer of oxide formed is composed by Fe_3O_4/Fe_2O_3. At lower temperatures only Fe_2O_3 is found. An adequate contact between steel and a humid atmosphere yields the formation of a complex net of oxides in the inner layer (Fe_2O_3, Fe_3O_4) while the outermost layers is rich in hydroxides, FeOOH and $Fe(OH)_3$ [17]. These hydroxides are capable of react with the maleic groups introduced in the polymer through the reaction shown in Figure 4.

In spite of the addition of maleic anhydride the adhesion between oxidized steel and over molded PP/PP-g-MA did not yield the expected results giving in all cases a weak interface. The reason for such behavior was attributed to the absence of enough energy to promote the reaction between the hydroxyl group linked to the metal and the carboxylic group generated at the polymer side. One easy way to overcome the energy barrier necessary for the linking of both groups was the application of an external heating. Torch heating was applied to the surface of the steel plate during 15 s just before closing the mould. The external heating caused different effects. First, heating reduced the viscosity of the entering melt and gives additional mobility to the polymer chains, contributing to the increase of the reactivity of PP/PP-g-MA. In second term premature cooling at the metal interface was avoided improving polymer wet ability. Finally, heating was the energy source required to make possible the reactions shown in Figure 4.

4.2.3. Peeling results

The force–displacement traces obtained from peeling test have a similar shape than those ideal reported in the ESIS protocol as can be appreciated in figure 5. The initial part of the curve corresponds to the tensile stretching of the peeling arm without detaching between the steel substrate and the plastic layer. The maximum force (zone A) of the graph indicates the onset of the detaching. Usually this is follow by a drop on the force curve and a rapid increase of the detached surface (zone B). In section C of same figure oscillations on the peeling force appear. These oscillations are common when large surfaces are put in contact. They are caused by differences in the degree of adhesion mainly to the presence of either punctual defects of adhesion or local areas having weak layers. Finally the break of the adhesive union occurs being characterized by the sharp drop on the adhesion force that happen in point D and subsequently plastic strip separates from metal base plate.

The photographs represented in Figure 5 shows that the advance of the peeling front was uniform and perpendicular to the applied force. It can be appreciated that detached layers of polymer remain stuck in the metal side indicating the development of an adhesive-cohesive fracture (zone A) that changes to fully cohesive in zones C and D. The results of the adhesion energies obtained from the peeling tests are displayed in Table 2. Within the range of concentrations of compatibilizing agent studied, the greater adhesion energies were always found on the 15 % wt. PP-g-MA, independently on the metal surface treatment applied implying that a higher number of chemical bonds have been created between polymer and steel. Overall the best combination was achieved with Nital etching of the steel surface and 15 % wt. PP-g-MA in the presence of external heating. Nital etching showed the highest values of adhesion energy, which matches well with the surface roughness measurements presented before. The high adhesion levels achieved are the result of both chemical bonding and mechanical interlocking.

Cohesive adhesion has been detected both in the cases of sanding and Nital etching. The fracture of the polymer layer indicates that the force of adhesion at the polymer-metal interface is higher than the force necessary to break the bulk polymer, indicating that maximum possible adhesion level was reached.

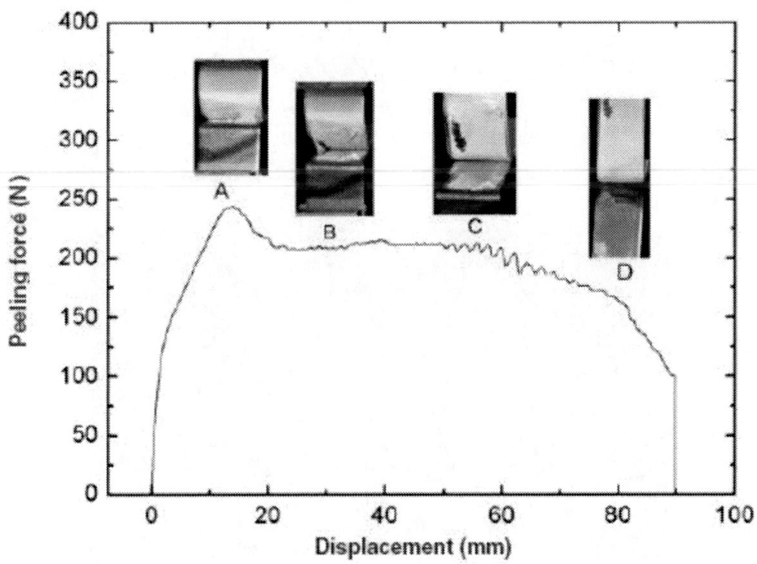

Figure 5. Typical peeling force-displacement graph for the case of a PP/PP-g-Ma 15% over molded onto the *Nital* etching metal plate

Table 2. Peel adhesion energies as a function of the surface treatment applied to the steel plate. *AR*= As received, *SN*= Sanding *NE*= *Nital* etching, *SP*= Shoot peening, n.a.= no adhesion

Material	Adhesion energy (kJ/m^2)							
	Non heated				Heated			
	AR	*SN*	*NE*	*SP*	*AR*	*SN*	*NE*	*SP*
PP	n.a.	n.a.	n.a.	n.a.	n.a.	n.a.	n.a.	n.a.
PP-g-MA (5 %)	n.a.	n.a.	n.a.	n.a.	2.3 ± 0.2	3.1 ± 0.3	3.3 ± 0.1	0.9 ± 0.3
PP-g-MA (15 %)	n.a.	n.a.	n.a.	n.a.	3.2 ± 0.2	4.2 ± 0.2	4.5 ± 0.2	3.7 ± 0.1

4.2.4. Microscopical analysis

The microscopical observations of the peeled steel surfaces revealed the coexistence of easily-peeled zones with others in which the polymer remained well attached to the steel plate. This fact occurred preferment in the samples with no treatment (*AR*) and those shoot peened coated with PP/PP-g-MA 5 % wt. The remaining polymer layer was in all cases elongated in the stretching direction applied by the peeling test. As received and shoot-peened steel plates over moulded with PP/PP-g-MA at 15 % wt. showed mainly cohesive fracture adhesion although in some particular regions both types of adhesion, adhesive and cohesive, coexists as can be seen in Figure 6.

When the steel surface was sanded of etched with *Nital* the entire surface was covered with a thin layer of polymer as can be seen on the optical micrographs of figure 5. A detailed SEM analysis confirmed that even on the apparently not well adhered areas this thin layer of dettached polymer is found on the surface (Figure 7a). The comparison between Figure 3c and Figure 7a shows clearly that the initial roughness has been notably diminished because steel surface is, in the latter case, coated with PP/PP-g-MA. Moreover in the detail shown in Figure 7b it can be appreciated that apart of the surface covering, polymer is anchored on the depressions caused by the etching, clearly indicating that the Nital etching treatment generated an oxide interlayer strong adhered to both steel substrate and PP/PP-g-MA layers.

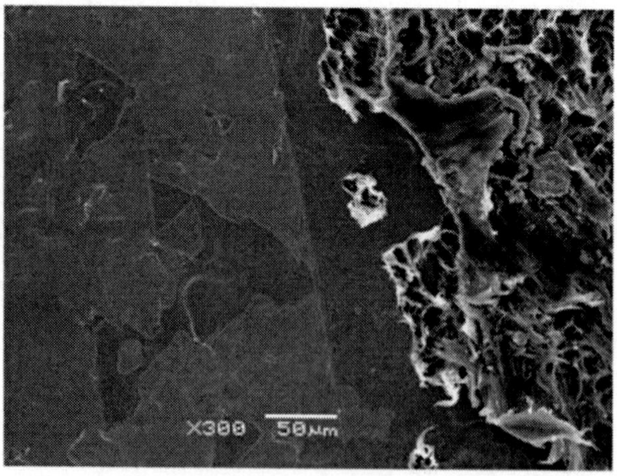

Figure 6. Fracture surface of as received plate over molded with PP/PP-g-MA 5 % wt. Half detached – half adherent zone is shown

Table 3. Values of the peeling force (F_p) obtained from the experiment design matrix for the polymer-fabrics system, with several levels of injection speed (V_i), holding pressure (P_h) and injection temperature (T_i)

Condition	V_i (cm^3/s)	P_h (MPa)	T_i (ºC)	F_p (N)	Condition	V_i (cm^3/s)	P_h (MPa)	T_i (ºC)	F_p (N)	Condition	V_i (cm^3/s)	P_h (MPa)	T_i (ºC)	F_p (N)
L1	3	2,2	200	52,20	M1	3	2,2	215	57,90	H1	3	2,2	230	45,00
L2	4	2,2	200	59,80	M2	4	2,2	215	54,30	H2	4	2,2	230	54,00
L3	5	2,2	200	60,85	M3	5	2,2	215	52,45	H3	5	2,2	230	48,10
L4	3	3,3	200	58,15	M4	3	3,3	215	55,10	H4	3	3,3	230	47,95
L5	4	3,3	200	57,50	M5	4	3,3	215	50,30	H5	4	3,3	230	46,90
L6	5	3,3	200	60,85	M6	5	3,3	215	60,30	H6	5	3,3	230	50,60
L7	3	4,4	200	61,05	M7	3	4,4	215	58,55	H7	3	4,4	230	51,70
L8	4	4,4	200	57,85	M8	4	4,4	215	62,10	H8	4	4,4	230	45,55
L9	5	4,4	200	48,30	M9	5	4,4	215	53,65	H9	5	4,4	230	46,00

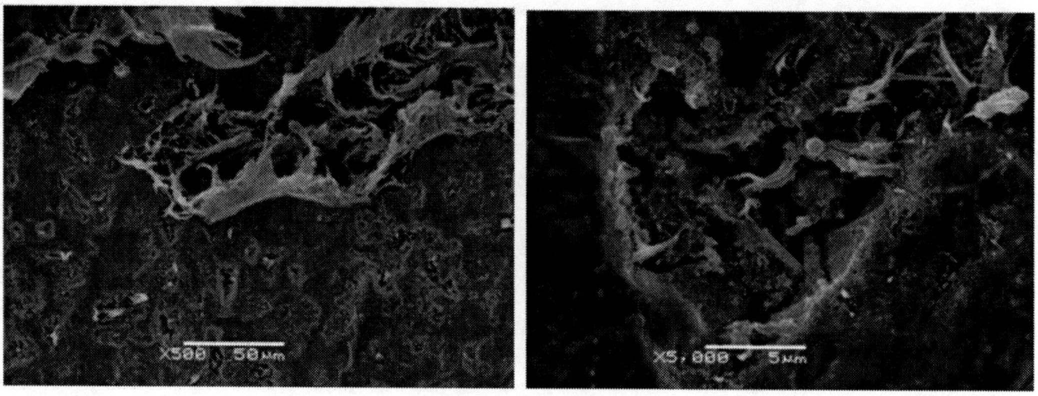

Figure 7. (a) Fracture surface of *Nital* etched steel plate over molded with PP/PP-g-MA 15 % wt. and (b) detail of the surface fracture showing PP traces adhered to the holes created by the *Nital* etching

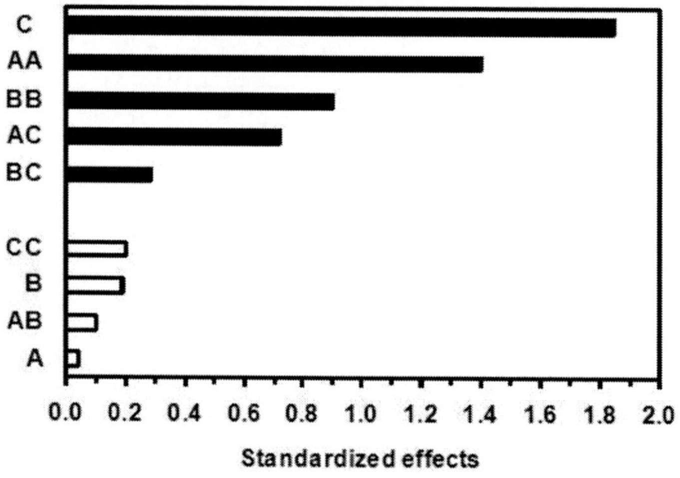

Figure 8. *Pareto* chart obtained for the studied experiment design; A (injection speed), B (holding pressure), C (injection temperature)

Table 4. P values obtained from the experiment design matrix: A (injection speed), B (holding pressure), C (injection temperature); the double terms reflect an interactive effect

	P values
A	0.9592
B	0.8142
C	0.0790
AA	0.1729
AB	0.9301
AC	0.4846
BB	0.4243
BC	0.7583
CC	0.8061

4.3. Polymer-Fabrics

In order to study the influence of the injection temperature, injection speed and holding pressure on the over molding process, a design experiment matrix (Table 3) was prepared. Previous studies of Candal *et al.* [9] showed that these variables are the most affecting the over molding process. The statistical analysis of the design experiment matrix is shown in Figure 8.

Although the differences in the peeling force values are not very remarkable, some trends can be extracted from the experiment design. Hollow bars denotes an increase in the rupture force with the increase of the studied variable; by other hand, bold bars represent an inverse effect, that is a reduction of the rupture force with the increase of the studied parameter. The statistical constants (Table 4) give an idea of the homogeneity degree of the individual differences between the samples with regard to one or several parameters. Values of P close to 1 are related to the processing parameters having lower influence, whereas P values close to zero correspond with the more influencing parameters. This inverse effect of the temperature could de explained on a basis of the incompatibility between fabrics substrate and the polymer.

CONCLUSIONS

Over molding process has been successfully used to create bi-layered hybrid structures consisting of metal/polymer and fabrics/polymer. The interface adhesion of the formed hybrids was modified by changing either the surface characteristics of the specimens or the process variables. To obtain high adhesion in metal-plastic systems a necessary condition was the achievement of chemical bonding between both materials. The addition of PP-g-MA to the polymeric matrix along with the presence of well adhered oxides onto the metal surface leaded to an improved interfacial adhesion strength. Heating of the metal surface was necessary to overcome the energy barrier needed for the activation of chemical reaction between oxides and maleic anhydride. The higher level of adhesion was displayed by the combination of Nital etching and a 15 % wt. of PP-g-MA. In this case fully cohesive peeling fractures were obtained indicating that maximum adhesion levels were achieved.

Regarding to fabrics, the experimental design analysis did not reveal remarkable differences in the absolute values of the peeling force. However, the injection temperature seemed to have the most influencing effect on adhesion. In this sense, an increase in the injection temperature leaded to a decrease in the peeling force, mainly due to the incompatibility between the fabrics substrate and the over molded polymer.

ACKNOWLEDGMENTS

Authors thank to the Ministerio de Ciencia e Innovación for the funding received through the project "Integrauto" PSS-370000-2009-36.

REFERENCES

[1] Rossa-Sierra, A; Sánchez-Soto, MA; Illescas, S; Maspoch, Ml. Study of the interface behavior between MABS/TPU bi-layer structures obtained through over molding. *Materials and Design*, 2009, 30, 3979-3988.
[2] Brown, HR. Adhesion between polymers. *IBM J Res Dev*, 1994, 38 (4), 379-389. 1994.
[3] Gordillo, A; Ariza, D; Sánchez-Soto M; Maspoch, Ml. Shrinkage predictions of injection molded parts in semi-crystalline polymers: experimental verifications. *In IEEE Symposium on emerging technologies and factory Automation, EFTA*, vol. 2, 1999, 1281-7.
[4] Sánchez-Soto, M; Gordillo, A; Arasanz, B; Aurrekoetxea, J; Aretxabaleta, L. Optimising the gas-injection molding of an automobile plastic cover using an experimental design procedure. *J Mater Process Technol*, 2006, 178, 369-78.
[5] Islam, A; Hansen, HN; Bondo, M. Experimental investigation of the factors influencing the polymer-polymer bond strength during two-component injection molding. *Int J Adv Manuf Tech.*, doi: 10.1007/s00170-009-2507-8.
[6] Jiang, J; Wu, H; Yan, B; Guo, S; Huang, J. Reinforcement of solid-melt interfaces for semicrystalline polymers in a sequential two-staged injection molding process. *J Polym Sci Pol Phys.*, 2009, 47, 1112-1124.
[7] Kinloch, AJ. In Adhesion and adhesives, 1st Ed.London, *Chapman and Hall*, 1987.
[8] Weng, D; Andries, J; Morin, P; Saunders, K; Politis, J. Fundamentals and material development for thermoplastic elastomer (TPE) overmolding. *J Inj Mold Tech.*, 2001, 4(1), 22-28.
[9] Candal, MV; Gordillo, A; Santana, O; Sánchez, J. Study of adhesion strength on overmolded plastic materials using the essential work of interfacial fracture (EWIF) concept. *J Mater Sci.*, 2008, 43, 5052-5060.
[10] Lewin, M; Mey-Marom, A; Reuven, F. Surface free energies of polymeric materials, additives and minerals. *Polym Advan Technol.*, 2005, 16, 429-441.
[11] Castle, JE. The composition of metal surfaces after atmospheric exposure: An historical perspective. *J Adhesion*, 2008, 84, 368-388.
[12] Amancio-Filho, ST; dos Santos, JF. Joining of polymers and polymer-metal hybrid structures: Recent developments and trends. *Polym Engng Sci.*, 2009, 49, 1461-1479.
[13] Ghosh, A; Schiraldi, DA. Improving interfacial adhesion between thermoplastic polyurethane and copper foil using amino carboxylic acids. *J Appl Polym Sci.*, 2009, 112, 1738-1744.
[14] Kawashita, L; Moore, DR; Williams, JG. Protocols for the measurement of adhesive fracture toughness by peel test. *J Adhesion*, 2006, 82, 973-95.
[15] Moore, DR; Williams, JG. Peel testing of flexible laminates. *In Fracture mechanics testing methods for polymers adhesives and composites*, vol. 28. DR; Moore, A; Pavan JG. Williams, Eds. ESIS Publication, 2001.
[16] Tan, ST; Zhang, MQ; Rong, MZ; Zeng, HM; Zhao FM. Interfacial Interaction in stainless steel fiber-filled polypropylene composites. *J Appl Polym Sci.*, 2000, 78, 2174-2179.
[17] Chilton, JP. Principles of metallic corrosion. *The Chemical Society*, 1973.

INDEX

A

B

D

E

I

J

K

L

M

N

O

P

Q

R

S

T